Advances in Hydro-Meteorological Monitoring

Special Issue Editors

Tommaso Moramarco
Roberto Ranzi

MDPI • Basel • Beijing • Wuhan • Barcelona • Belgrade

MDPI

Special Issue Editors

Tommaso Moramarco
Research Institute for Geo-Hydrological Protection
Italy

Roberto Ranzi
Università degli Studi di Brescia
Italy

Editorial Office
MDPI
St. Alban-Anlage 66
Basel, Switzerland

This edition is a reprint of the Special Issue published online in the open access journal *Water* (ISSN 2073-4441) from 2016–2017 (available at: http://www.mdpi.com/journal/water/special_issues/ Hydro-Meteorological).

For citation purposes, cite each article independently as indicated on the article page online and as indicated below:

Lastname, F.M.; Lastname, F.M. Article title. *Journal Name* **Year**, *Article number*, page range.

First Edition 2018

ISBN 978-3-03842-977-7 (Pbk)
ISBN 978-3-03842-978-4 (PDF)

Table of Contents

About the Special Issue Editors

Tommaso Moramarco is a Senior Researcher with the CNR Research Institute for Geo-Hydrological Protection (CNR IRPI), in Perugia (Italy). Since 1989, he has been conducting research in the field of hydrological processes, addressing flood forecasting and hydraulic risk mitigation. He has developed original works in these areas, producing more than 135 original contributions in the leading hydrologic and hydraulic journals. He was visiting scientist at Massachusetts Institute of Technology (MIT) in Cambridge (Boston) and at the Department of Civil and Environmental Engineering of Louisiana State University in Baton Rouge. Recently, the American Society of Civil Engineers (ASCE) awarded him the Normal Medal for two of his works on flood routing in natural channels. In June 2012 he was appointed by CNR as the National Representative of the International Association of Hydrogeological Sciences (IAHS/IUGG). In 2013 he was appointed as a Full Professor in Hydrology and Hydraulics. In January 2017 he was elected President of the Italian Hydrological Society.

Roberto Ranzi, Prof. Ph.D., graduated cum laude in Civil and Environmental Engineering at the Politecnico di Milano, Italy, where he obtained a PhD in Hydraulic Engineering in 1994. He is the author of over 60 scientific papers in refereed international journals and over 200 articles about hydrology, environmental and water engineering. His preferred research topics are monitoring and modelling of the space–time variability of hydrological processes, mainly precipitation, radiation, heat and moisture fluxes, snowmelt and ice-melt runoff, flood formation and routing, especially in mountain watersheds and under anthropic changes and climate scenarios. He studied methods and criteria for the design of flood retention basins, for the determination of the design storm in urban and agricultural basins, and hydrological models coupled with mesoscale meteorological models for flood forecasting. He chairs the Climate Change Working Group of IAHR and is the Rector's Delegate for International Affairs at the University of Brescia.

Preface to "Advances in Hydro-Meteorological Monitoring"

This book presents a reprint of the Special Issue of the journal Water dedicated to "Advances in Hydro-Meteorological Monitoring". As we planned, as editors, this Special Issue aimed to shed light on the more recent advances in ground observations and remote sensing products, as well as on the benefit that comes from the integration of technological innovation and the development of new ideas in hydrology science. This original objective was achieved, and in the eleven papers collected here the readers can appreciate theoretical and applied contributions dealing with the development and comparison of different monitoring procedures based on ground and satellite data. The papers are arranged with a process-oriented criterion that follows the hydrologic cycle. Starting from soil moisture monitoring from satellites and in-situ, we move to pan evaporimeter measurements. Runoff measurement using advanced methods from both satellites and in the field are then addressed. Three papers are dedicated to rainfall measurement and spatial representation and the two final papers address the problem of systematic errors in snow precipitation measurement and the design of snow measurement networks. The geographic distribution of the case studies is wide enough to attract the interest of an international audience of readers. This Special Issue will hopefully provide different, useful insights into advancements in the emerging technologies for the monitoring of key hydrological variables and will be of support for the design of a scalable system of operational tools leading to suitable flood mitigation measures and reliable real-time warning systems.

Tommaso Moramarco and Roberto Ranzi

Special Issue Editors

Article

Soil Moisture for Hydrological Applications: Open Questions and New Opportunities

Luca Brocca *, Luca Ciabatta, Christian Massari, Stefania Camici and Angelica Tarpanelli

Research Institute for Geo-Hydrological Protection, National Research Council, 06128 Perugia, Italy;
luca.ciabatta@irpi.cnr.it (L.C.); christian.massari@irpi.cnr.it (C.M.); stefania.camici@irpi.cnr.it (S.C.);
angelica.tarpanelli@irpi.cnr.it (A.T.)
* Correspondence: luca.brocca@irpi.cnr.it; Tel.: +39-075-501-4418

Academic Editors: Roberto Ranzi and Tommaso Moramarco
Received: 10 January 2017; Accepted: 15 February 2017; Published: 20 February 2017

Abstract: Soil moisture is widely recognized as a key parameter in the mass and energy balance between the land surface and the atmosphere and, hence, the potential societal benefits of an accurate estimation of soil moisture are immense. Recently, scientific community is making great effort for addressing the estimation of soil moisture over large areas through in situ sensors, remote sensing and modelling approaches. The different techniques used for addressing the monitoring of soil moisture for hydrological applications are briefly reviewed here. Moreover, some examples in which in situ and satellite soil moisture data are successfully employed for improving hydrological monitoring and predictions (e.g., floods, landslides, precipitation and irrigation) are presented. Finally, the emerging applications, the open issues and the future opportunities given by the increased availability of soil moisture measurements are outlined.

Keywords: soil moisture; hydrology; in situ measurements; remote sensing; floods; landslides; precipitation; irrigation

---

1. Introduction

The importance of soil moisture in the hydrological cycle has been stressed in a number of papers [1–4] and scientific projects (European Space Agency Climate Change Initiative Soil Moisture, ESACCISM [5], Soil Moisture Active and Passive mission, SMAP [6], International Soil Moisture Network, ISMN [7], and Cosmic-ray Soil Moisture Observing System, COSMOS [8]). Soil moisture governs the partitioning of the mass and energy fluxes between the land and the atmosphere, thus playing a key role in the assessment of the different components of the water and energy balance. Soil moisture is an important variable for flood and landslide modelling and prediction [9–12], for drought assessment and forecasting [13–15], and for numerical weather prediction [16–19], to cite a few. Soil moisture is also the water source for plants and, hence, its knowledge is required for irrigation management and agricultural studies [20]. The scientific community has well recognized the very important role of soil moisture in Earth Science applications and in the last 40 years new approaches and techniques for monitoring, modelling and using soil moisture data have been developed.

In this review, an overview of the potential of soil moisture observations for improving hydrological applications is presented. Firstly, we briefly describe the main techniques employed for soil moisture estimation at different spatial scales. Secondly, the understanding that we gained in the assessment of soil moisture spatial-temporal variability is illustrated. Thirdly, we provide a list of the hydrological applications that have benefited (and will benefit) from the use of soil moisture observations. Finally, the open questions and the novel opportunities that would be important to address in future investigations are described.

We underline that this paper reflects the opinion and the experience of the authors on this research topic and we mainly focus on the applications and the research questions that we investigated in the past. This paper should not be considered as a review paper. Indeed, several recent reviews already investigated different hydrological aspects related to soil moisture, e.g., the estimation of soil moisture through remote sensing [3,21,22], the different techniques for in situ soil moisture monitoring at different scales [23,24], the use of satellite soil moisture data in hydrological and climatic applications [25,26], and the assessment and modelling of soil moisture spatial-temporal variability [2,27,28]. Therefore, we believe there is no need of an additional review on soil moisture. Differently, the manuscript is conceived to be a comprehensive summary of our research activity on soil moisture and our view for future opportunities to be analysed. We wanted to identify the main knowledge gaps that need to be filled in future investigations for improving the monitoring, and the use, of soil moisture observations in hydrological (and others) applications.

2. How Do We Estimate Soil Moisture?

Notwithstanding soil moisture is largely recognized as a fundamental physical variable, e.g., soil moisture is included at the 2nd place among the Essential Climate Variable by the Global Climate Observing System (GCOS), its monitoring over large areas is still an open research activity [23,29]. Indeed, differently from precipitation, in situ monitoring networks for soil moisture are much less developed and only in few countries (e.g., United States) a good coverage of in situ stations is present (see Figure 1). This should be attributed to the very recent availability, with respect to precipitation, of ground sensors measuring soil moisture, and partly to the lower knowledge of the large benefits that we can obtain from soil moisture observations. Currently, three different approaches are used for providing soil moisture estimates: (1) in situ observations; (2) remote sensing; and (3) modelled data. In the following, we describe the main features of each approach in order to underline their pros and cons. Overall, it is widely accepted that the integration of in situ, satellite and modelled data is the best approach to exploit the potential of soil moisture in hydrological applications [25,30].

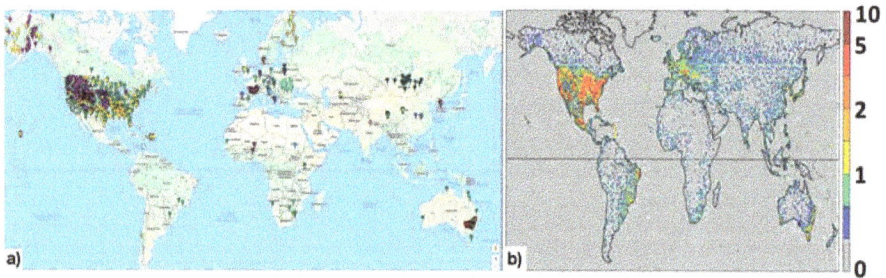

Figure 1. (**a**) Global distribution of soil moisture stations available at the International Soil Moisture Network from January 2000 ([7]); different colours refer to different networks; (**b**) Density of rain gauges underlying the Climate Prediction Center Unified Gauge-Based Analysis of Global Daily Precipitation (number of gauges/0.5°). Even though the two maps are not directly comparable, it is evident that the number of soil moisture stations is much lower than precipitation gauges, and, except in USA and Europe, soil moisture data are nearly absent.

2.1. In Situ Measurements

The first approaches for monitoring soil moisture through in situ observations are based on gravimetric, tensiometric and nuclear techniques [24]. Although invasive and time consuming, the gravimetric technique is still the reference method to which the other techniques and new methods are calibrated and tested. In the 1980s, Topp et al. [31] proposed the Time Domain Reflectometry

(TDR) technique. TDR was found to provide accurate measurements for a wide range of soils and settings thus becoming the new standard approach for measuring soil moisture and replacing the gravimetric approach (TDR is less time consuming and invasive). More recently, the emergence of low-cost capacitance sensors, based on Frequency Domain Reflectometry (FDR) technique, as well as impedance, time domain transmission sensors, etc., strongly promoted the usage of soil moisture sensors in environmental research [32]. Currently, FDR is likely the most used approach for in situ monitoring of soil moisture thanks to its lower cost with respect to TDR, even though at the expense of lower accuracy. Moreover, recent advances for the calibration and testing of electromagnetic sensors were proposed by Bogena et al. [33], thus further improving the accuracy and reliability of soil moisture sensors.

However, all the approaches listed above provide only point measurements that are representative of a small volume of soil [34]. The emergence of wireless sensor networks made it possible to cover larger areas with low-budget soil moisture sensors [35], but still this technology is only used for scientific studies. In the last decade, novel technologies were developed for providing soil moisture measurements over larger areas [29]. Currently, cosmic ray neutron sensors, Global Positioning System (GPS), and geophysical measurements (e.g., electrical resistivity and electromagnetic induction) are the most promising techniques [33,36–38]. Indeed, they are able to provide measurements with a support scale much larger than few cubic centimetres as point scale measurements. Cosmic ray and GPS techniques are well developed and some networks are already established in various countries (COSMOS [8], COSMOS-UK [39], and Plate Boundary Observatory to study the water cycle: PBO H2O [40]). However, these techniques are very recent, e.g., COSMOS network in UK was set up starting from 2016, and their use as established technique still requires further research and technical investigations.

Point in situ measurements of soil moisture (e.g., gravimetric, TDR, and FDR) are surely the most accurate techniques; however, they suffer from low spatial representativeness. The new techniques (e.g., cosmic ray and GPS) partly address this issue. Specifically, several papers demonstrated that the accuracy of the cosmic ray is similar to that point sensors, even under unfavourable conditions (e.g., [41–43]). Moreover, the maintenance of in situ soil moisture networks requires significant economic and human resources (even though lower than climatic stations) and, frequently, it is a non-trivial task to make a network operational for several (e.g., >5) years [34]. On this basis, it is evident that the development of established techniques providing measurements at 0.1–1 km spatial scale is vital [29].

2.2. Remote Sensing

Remote sensing is definitely the most appropriate technique to obtain large scale soil moisture measurements. Different methods were developed in the last 40 years for the retrieval of soil moisture from microwave, optical and thermal satellite sensors. Several review papers on this topic were published recently [21,22,44] that well describe the algorithms developed for the retrieval of soil moisture from the different bands of the electromagnetic spectrum. In this paper, we focus only on active and passive microwave-based products as they are the most used methods, also providing operational soil moisture products [45]; the reader is referred to Rahimzadeh-Bajgiran and Berg [44] for a review on recent advances in thermal and optical remote sensing for soil moisture estimation. Specifically, it is important to distinguish among active microwave sensors, i.e., between Synthetic Aperture Radars (SARs) and scatterometers. The former (SAR) can provide high spatial resolution (<1 km) but coarse temporal resolution (i.e., >10 days). The latter (scatterometer) provides coarse spatial resolution (~20 km) and high temporal coverage (~daily). Passive microwave radiometers are characterized by nearly the same spatial-temporal resolution of scatterometers. We note that the temporal coverage of satellite data can be significantly improved by using multiple sensors or even a constellation of sensors [46,47]. Another important aspect is the specific band in the microwave range of the electromagnetic spectrum. Usually, X-, C- and L-band are used for estimating soil moisture with the latter theoretically more suitable [48]. However, other aspects such as the radiometric accuracy and the radio frequency interferences can significantly affect the final quality of soil moisture retrievals, even more than the microwave band [45].

At the time of writing (January 2017), four quasi-operational, i.e., available either in near real time (NRT) or few days after sensing, coarse resolution satellite surface soil moisture products are available: (1) the Soil Moisture Active and Passive (SMAP) mission (L-band radiometer) starting from April 2015 with ~36 km/1-day spatial/temporal resolution [49]; (2) the Advanced Microwave Scanning Radiometer 2 (AMSR2) onboard the Global Change Observation Mission for Water, GCOM-W, satellite (C- and X-band radiometers) starting from July 2012 with ~25 km/1-day spatial/temporal resolution [50]; (3) the Soil Moisture and Ocean Salinity (SMOS) mission product (L-band radiometer) starting from January 2010 with ~50 km/2-day spatial/temporal resolution [51]; and (4) the Advanced SCATterometer (ASCAT) onboard Metop-A and Metop-B satellites (C-band scatterometer) starting from January 2007 with ~25 km/1-day spatial/temporal resolution [45]. Recently, higher resolution (~1 km) soil moisture products are becoming available based on the disaggregation of coarse resolution products [52–54] and, in the near future, from Sentinel-1 satellites (e.g., [55]). However, thus far a high-resolution soil moisture product from SAR sensors has never been made available.

The most striking benefit of remote sensing observations is related to their spatial-temporal coverage, and the relatively lower costs for large scale applications. The accuracy of satellite soil moisture products is surely lower than in situ observations, even though the recent products listed above have achieved a high level of reliability and maturity, as demonstrated in several validation studies comparing satellite data with in situ observations and land surface/hydrological modelling ([56–59] to cite a few). Moreover, the number of applications already using these products is an indirect assessment of their good quality level, as it was demonstrated their utility for improving and supporting hydrological and climatic predictions (see below and [26,60,61]). The three major limitations of satellite soil moisture products are: (1) the very shallow soil layer (2–7 cm) that is sensed from satellites; (2) the coarse spatial resolution (~20 km) of the currently available products; and (3) the very low quality under certain surface conditions (dense vegetation, frozen soils, snow, mountainous terrain). While these limitations may prevent their use in several hydrological applications [25], some approaches were already developed to overcome these issues, i.e., spatial downscaling techniques (e.g., [53]) and simplified methods for root-zone soil moisture estimation from surface measurements (e.g., [62,63]).

2.3. Hydrological and Land Surface Modelling

Hydrological and land surface models are largely used for estimating soil moisture at different spatial and temporal scales (e.g., [64]). Basically, both hydrological and land surface models use the same set of equations for simulating the water and energy balance, and hence soil moisture [65]. Indeed, different models use different structures for simulating each of the components (e.g., Richard's equation or variable infiltration curve for infiltration), but the main governing balance equations are the same. The models differ with respect to the spatial (horizontal and vertical) and temporal discretization, to the simulated physical processes and to the corresponding parameterization. More frequently, hydrological models are discretized at basin level and land surface model considers regular grids, but it is not always the case. Potentially, through modelling we are able to obtain soil moisture estimates at the desired temporal and spatial resolution (e.g., sub-hourly and 100 m, [66]).

The accuracy of modelled soil moisture data is strongly dependent on the employed model. However, besides to the model accuracy, the quality of meteorological observations used as input data also plays a very important role. Even a perfect model will fail if the quality, or the density, of meteorological inputs is low. Similarly, also the spatial and temporal discretization depends on the resolution of meteorological forcings, as well as of static information such as land use and soil texture maps [67]. Sometimes, high resolution modelled data (e.g., 500 m) are obtained by using as inputs coarse scale meteorological forcings (e.g., precipitation data at 10–100 km resolution). In these cases, the effective resolution of modelled data should be the one of the meteorological inputs (i.e., 10–100 km and not 500 m). A common issue faced by modeller is related to the parameterization of complex modelling structure and of the model parameter values. For instance, the parameterization of soil hydraulic parameters is a very difficult task, even in well gauged or experimental sites [68], and mainly over large areas [69]. Additionally, many key hydrologic processes are extremely difficult to parameterize (e.g., irrigation, dam operation, snow melting, and interception), especially in challenging regions (deserts, pluvial forests, and high altitudes). Therefore, modelled soil moisture data surely represents an important dataset that, however, needs to be used with caution [70,71].

3. How Does the Soil Moisture Vary in Space and Time?

In the last 40 years, a large number of studies were dedicated to the understanding of the spatial and temporal variability of soil moisture from the local to the regional and global scale (e.g., [2,72–75]). Indeed, after the development of the first measurement techniques, several researchers carried out detailed field campaigns in different geomorphological and climatic settings to assess soil moisture variability [24]. Different methods were applied including statistical [2,75] and geostatistical approaches [76], temporal stability analysis [73,77], and regression and wavelet techniques [78]. In this section, we provide a short overview of the major results obtained in the different studies trying to summarize the status of the knowledge we gained about the soil moisture spatial-temporal variability.

A large body of scientific literature analysed the statistical properties of soil moisture datasets obtained by field campaigns. The values of the first two statistical moments of absolute soil moisture data (in volumetric units, m^3/m^3), i.e., mean and variance, as well as their mutual relationship, were largely investigated (e.g., [79]). In most cases, a convex upward relationship is obtained between mean and variance thus concluding that soil moisture variability peaks for intermediates wetness conditions (e.g., [2,75,80,81]). Recently, Mittelbach and Seneviratne [82] proposed a new framework to decompose the soil moisture variability of the temporal mean and temporal anomalies. Interestingly, they obtained that the variability of soil moisture anomalies is larger for wet and dry conditions, with the minimum occurring at intermediate wetness conditions (see also [83,84]). Therefore, the behaviour of absolute and anomaly soil moisture values is not consistent and additional investigations are required.

A fundamental concept was introduced by [73], i.e., the soil moisture temporal stability. Vachaud et al. [73] found that the spatial patterns of soil moisture are stable over time and they named it as "temporal stability" (see [77] for a recent review). In other words, while the mean soil moisture for a given area might be changing over time, the spatial location of wetter and drier areas does not change. Figure 2 shows an example of soil moisture spatial patterns obtained by portable soil moisture measurements in the Colorso experimental plot in central Italy [2]. As can be seen, the different maps show the same pattern with wetter areas in the centre and southern part of the plot, and drier values in the northern part. The patterns are also quite consistent with the topography [2]. The simple concept of temporal stability has tremendous implications in hydrological and climate applications. It means that the monitoring of soil moisture with in situ observations over large areas (>100–200 km^2) can be carried out with a limited number of stations [75,77,83,85], with evident economic and time savings. Similarly, a limited number of stations can be used for the validation of coarse scale satellite soil moisture products, as indirectly demonstrated by the high correlations that are frequently obtained between coarse scale (~20 km) and point scale (~5 cm) measurements (e.g., [57–59]). Vice versa, coarse scale measurements, as obtained by remote sensing, can be used for regional or local studies (e.g., [11,86]). In summary, the temporal stability concept allows to mitigate the spatial discrepancies between point measurements and the scale needed in the applications.

A series of publications investigated the factors influencing soil moisture variability, both in time [87] and in space [88]. In time, soil moisture is mainly driven by precipitation and evapotranspiration, and its temporal variability is also a function of soil characteristics, vegetation, topography and groundwater [87,89]. Several modelling studies obtained very good performances in simulating point- (grid-) scale soil moisture temporal evolution [65,67,90], thus showing that a good knowledge has been gained about the factors influencing point scale soil moisture temporal variability. Figure 3 shows a comparison between observed and modelled data at two sites in central Italy using the Soil Water Balance Model [67]. The agreement with observations highlights the good model performances, and similar results were obtained at several sites in Europe (e.g., [57,91]). In space, the same meteorological factors, i.e., precipitation and evapotranspiration, have a clear impact on soil moisture patterns at large scales (>500–1000 km^2). At smaller scales, static factors such as land cover, topography and soil texture/structure affects soil moisture spatial variability (see Figure 1 in [75]). Several authors compared soil moisture spatial patterns with these static factors (e.g., [2,92]) obtaining moderate to low predictive capability, largely varying across sites and climates. Therefore, our knowledge about the factors influencing spatial variability of soil moisture is still limited. This important point is evident if we compare the capability of hydrological models in reproducing soil moisture spatial variability with respect to temporal variability. Frequently, the magnitude of variability is severely underestimated [90,93,94], and at regional scale soil moisture spatial patterns from modelling and satellite observations are found to be not consistent [95]. In this context, some recent modelling studies are demonstrating the value of soil moisture data for a more in-depth validation of distributed hydrological models [96,97].

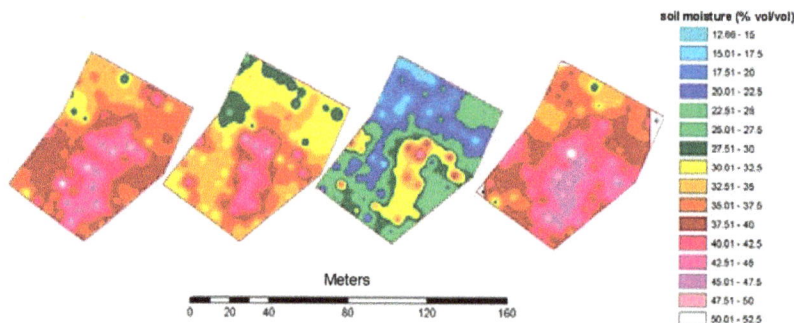

Figure 2. Spatial soil moisture maps (% vol/vol) obtained for the Colorso experimental plot in Italy at fuor different dates (from left to right): 21 April 2005, 28 April 2005, 5 May 2005, and 2 December 2005 (data from [2]). Even though the mean soil moisture conditions are changing over time, the spatial pattern of wetter and drier areas remains the same.

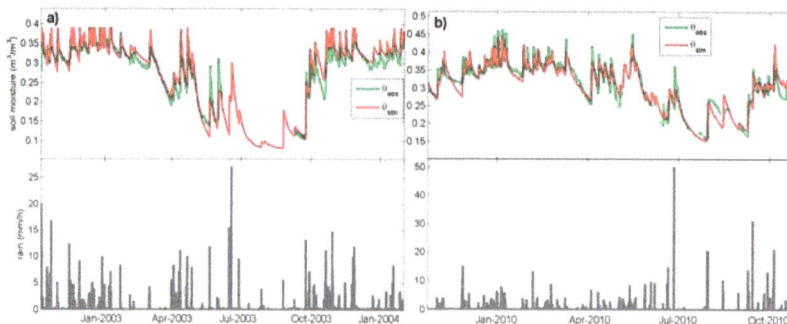

Figure 3. Modelled, θ_{sim}, through the Soil Water Balance Model [67], versus observed, θ_{obs}, soil moisture for: Colorso (**a**); and Ingegneria (**b**) sites in central Italy. The agreement between observed and modelled data reveals the good capability we achieved in simulating point scale soil moisture temporal evolution.

4. Which Hydrological Applications Are Benefiting (and Will Benefit) from Soil Moisture Data?

As evidenced in Section 2, several in situ, satellite and modelled soil moisture datasets are currently available, with large spatial and temporal coverage, and with appropriate spatial and temporal resolutions for hydrological, climatic and agricultural applications. Drought monitoring, runoff modelling and flood forecasting, numerical weather prediction, land surface and climate models assessment, wildfire risk assessment, agricultural monitoring and crop yield forecast [25,26,60,61] are among the most important applications benefiting from in situ and satellite soil moisture observations.

In this section, we mainly focus on a single hydrological application that was the object of our previous and on-going studies, i.e., runoff modelling and flood forecasting. Moreover, we list a number of "emerging" applications in which soil moisture data are being used in the last couple of years: (1) landslides and erosion prediction; (2) epidemic risk monitoring; (3) rainfall estimation; and (4) irrigation detection and assessment.

The purpose of this section is to briefly describe the main outcomes of previous studies on the selected applications in order to highlight the open issues to be addressed.

4.1. Runoff Modelling

Soil moisture is the key variable for the partitioning of rainfall into infiltration and runoff, thus playing a fundamental role in runoff modelling and flood forecasting [9,10]. On the one hand, several studies employed in situ soil moisture observations for enhancing: (1) the estimation of initial conditions in flood modelling [98,99]; (2) the calibration of hydrological models [100,101]; and (3) flood simulation through data assimilation approaches [86,102,103]. Overall, these studies demonstrated the value of in situ soil moisture observations for improving flood simulation even though a comprehensive assessment was not possible due to the very limited spatial and temporal coverage of in situ observations. On the other hand, due the recent availability of satellite soil moisture products, the number of studies considering satellite soil moisture observations in this topic is significantly increasing in the last five years. For instance, under the Climate Change Initiative (CCI)—Soil Moisture of the European Space Agency (ESA), a global-scale and long-term soil moisture product is currently available from 1979 to 2015 (i.e., 36 years [46]), and next year it is foreseen that it will be available in near real-time [104]. The global-scale ASCAT soil moisture product is already available in near real-time (130 min after satellite pass) through the Satellite Application Facility on support to Operational Hydrology and Water Management (H SAF) project of EUMETSAT (European Organization for the Exploitation of Meteorological Satellites) since 2007. These new products have large potential to be employed for long-term and real-time flood forecasting. Similarly to in situ observations, satellite data have been used for model initialization [105–109], for hydrological model calibration [110,111], and in a data assimilation framework [9,86,112–115]. By way of example, Figure 4 shows the results of the assimilation of the ESA CCI soil moisture product [46] in MISDc (Modello Idrologico Semi-Distribuito in continuo) rainfall-runoff model [116] for the Niccone basin in central Italy in the period 1995–2010. For the latest period 2007–2010, in which the temporal density of satellite observations is higher, the assimilation provides a significant increase in the model performance with a Nash-Sutcliffe efficiency value from 0.79 to 0.88 when the ESA CCI soil moisture product is assimilated. The improvement in discharge simulation is more evident for larger flood events.

However, the analysis of the scientific literature on this topic reveals that the actual added-value in using satellite soil moisture for runoff modelling is still unclear [86,113]. Some authors obtained moderate to significant improvement through the assimilation of satellite data soil moisture in hydrological modelling (e.g., [9,112,117]) while other studies obtained a deterioration of the performances (e.g., [118]). These contrasting results have to be attributed to the inherent uncertainties and issues involved in the use of satellite (and in situ) soil moisture data in hydrological modelling. By considering the example of data assimilation, Massari et al. [113] clarified well that several choices should be made (i.e., the "cooking techniques") in a data assimilation study and they might have a significant impact on final results, with even the same relevance of the considered hydrological model, observation to be assimilated, and assimilation approach (i.e., the "ingredients"). These choices include the assessment of the magnitude and the structure of the errors in the hydrological model and in the observations, and the selection of the rescaling (e.g., linear and non-linear bias correction, triple collocation, [119]) and/or filtering (e.g., the Soil Water Index method [62]) techniques. Moreover, several inherent issues are related to the spatial mismatch between observations, i.e., point scale for in situ data and coarse scale (plus shallow soil depth) for satellite data, and modelled data, to the structure of hydrological models [103,112], and to the error characterization of (satellite) observations. For that, a strong effort is required from the hydrologic and remote sensing communities to overcome these issues. On the one hand, hydrologists should modify hydrological models that usually consider a single soil layer and are spatially lumped (basin scale). Both in situ and satellite data requires a model tailored to their use. It would mean including a shallow soil layer close to the surface and a distributed spatial discretization for fully exploiting the potential of satellite soil moisture observations [112,120]. On the other hand, remote sensing scientists should improve the characterization of the errors associated to the soil moisture products, increase their spatial and temporal resolution, their temporal coverage and temporal consistency. Indeed, long-term soil moisture datasets (e.g., >10–15 years) are needed not only in the climate community but also from hydrologists to assess the impact of using such datasets through robust analyses.

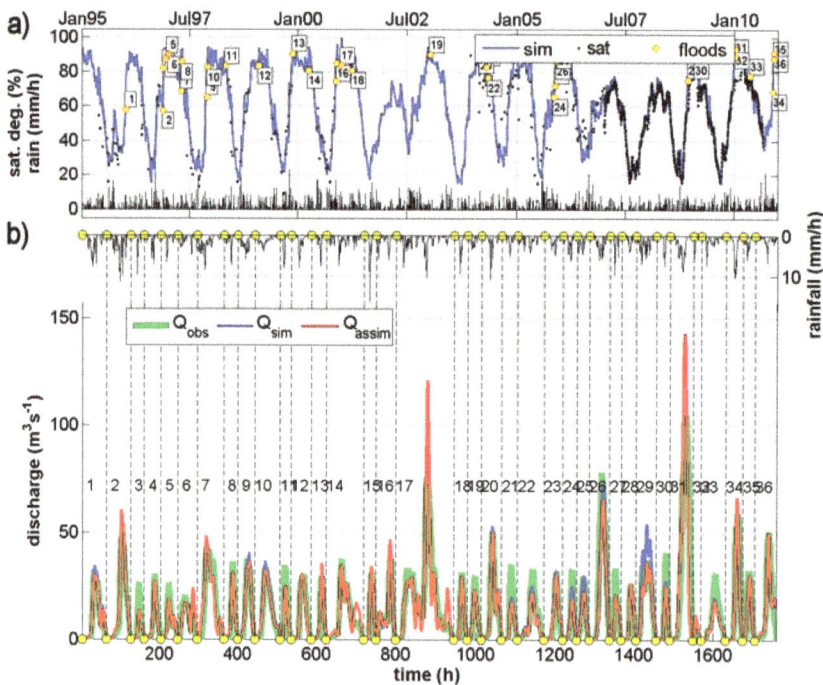

Figure 4. MISDc model results without and with the assimilation of ESA CCI soil moisture product in the Niccone river basin (central Italy). (**a**) Root zone soil moisture simulated by the model, sim, and obtained by ESA CCI soil moisture product, sat, for the period 1995–2010. The filled circles, floods, represent the initial conditions of the selected flood events simulated by MISDc model and identified by the numbers in the lower panel; (**b**) Observed, Q_{obs}, versus simulated, Q_{sim}, and simulated with assimilation, Q_{assim}, discharge for the sequence of the flood events occurred in the study period. The corresponding hourly rainfall is also shown in both panels (note that time axis is different in the upper and lower panel).

4.2. Emerging Applications

Soil moisture observations play an important role for the mitigation of many natural hazards other than floods, i.e., for landslide and erosion prediction [121,122]. Several studies used in situ observations for studying the linkage between soil moisture conditions and landslide occurrence [12,123]. For instance, Hawke and McConchie [124] analysed this relationship in a small area of New Zealand affected by landslide and obtained that the slope failure occurred when maximum soil moisture conditions were observed. Based on these studies, the use of soil moisture stations for monitoring slope stability conditions is becoming a standard (e.g., [125], United States Geological Survey website [126]). Differently from in situ observations, satellite soil moisture data were rarely employed in this context and only three studies were published [11,127,128]. Brocca et al. [11] demonstrated that the ASCAT soil moisture product is able to improve the performances of a statistical regression model aimed at predicting the temporal evolution of landslide movement for a small landslide in central Italy. Similarly, in the context of erosion prediction, Todisco et al. [122] recently suggested that ASCAT soil moisture data might be used for predicting event soil losses at plot scale in central Italy. The very small number of studies employing satellite soil moisture data for these applications is related firstly to the spatial mismatch between the targeted areas, typically a hillslopes or a small catchment, and the spatial resolution of satellite products, and secondly, to the difficulties in the exploitation of these technologies from the hydrological community [129,130]. Based on the studies mentioned above, we think that satellite soil moisture products have a high potential for improving the prediction of landslide and erosion processes, therefore these aspects deserve further investigations [26].

Soil moisture information might be helpful for the assessment and the monitoring of the epidemic risk. In the scientific literature, meteorological observations have been widely employed for determining spatial and temporal occurrence of epidemic risk (e.g., [131]), while only recently soil moisture data from modelling and observations have been considered in this respect [45,132,133]. For instance, Montosi et al. [132] developed an ecohydrological model for identifying the factors influencing malaria dynamics and highlighted the important role played by soil moisture. By using published data of malaria incidence rates in Mpumalanga and Botswana regions (Africa), we performed a simple correlation analysis between European Remote Sensing Scatterometer (ESCAT) and ASCAT satellite soil moisture products and malaria incidence rates (Figure 5). Although mainly driven by the seasonal cycle, a moderate agreement is observed between the monthly time series in the two regions, thus highlighting the potential of using satellite soil moisture data in this application that is extremely important in developing countries.

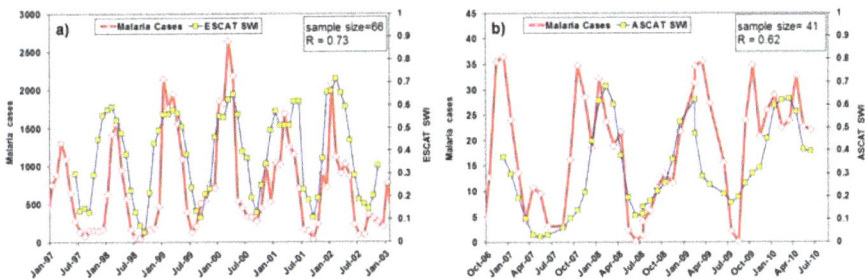

Figure 5. Comparison between malaria cases and soil moisture time series obtained by ESCAT Soil Water Index: SWI, in Mpumalanga (**a**); and ASCAT SWI in Botswana (**b**). The data on malaria cases are obtained by: Montosi et al. [132] (**a**); and by Chirebvu et al. [131] (**b**). R is the Pearson's correlation coefficient.

Pellarin et al. [134] and Crow et al. [135] made the first studies using satellite soil moisture observations for correcting satellite precipitation estimates. Thanks to the strong relationship between the temporal evolution of rainfall and soil moisture, the latter can be employed for improving the quality of satellite precipitation products that are affected by several issues (e.g., [136]). More recently, the number of publications on this topic is remarkably increasing likely due to the work by Brocca et al. [137,138] who developed a "bottom-up" approach, called SM2RAIN, for directly estimating precipitation rates from soil moisture observations only. The method has been applied on a local scale with in situ observations [137,139] and on a regional/global scale with satellite data [138,140,141]. Moreover, the "bottom-up" approach was integrated with state-of-the-art rainfall products (i.e., "top-down" approach) for obtaining a superior rainfall product by Brocca et al. [141] and Ciabatta et al. [142] in Australia and Italy. Finally, the precipitation product corrected through soil moisture data were used in several recent studies for improving flood prediction [98,143,144]. This new application is receiving more and more attention in the recent time, as also confirmed by a number of research projects funded by ESA, EUMETSAT, and NASA on this topic. Therefore, we believe that its further development and widespread application is highly beneficial for contributing to efforts in global precipitation estimation [140].

Soil moisture observations can be highly important for the detection and the quantification of irrigation, and specifically satellite datasets thanks to their large scale coverage [145,146]. For instance, Singh et al. [147] used Advanced Microwave Scanning Radiometer-Earth Observing System (AMSR-E) soil moisture data for discerning the shifting over time in the irrigation practices in north western India while Qiu et al. [146] highlighted that the ESA CCI SM product can be used to detect irrigated areas in eastern China by comparing trends in satellite precipitation and soil moisture. For instance, by using the SM2RAIN method described above, the quantification of irrigation from soil moisture data can be also performed. Figure 6 shows an example in Nebraska in which soil moisture data from ASCAT satellite and European Re Analysis -Interim (ERA-Interim) reanalysis were considered. The higher values of satellite soil moisture data compared to ERA-Interim reanalysis during summer are clearly visible in the Figure 6a (highlighted by grey areas). As ERA-Interim reanalysis does not incorporate irrigation data, the higher values obtained through satellite observations identify the occurrence and the amount of irrigation (Figure 6b,c). As the amount of water used for irrigation is largely unknown on a global scale, the capability of satellite soil moisture data to estimate irrigation might have a tremendous impact in future applications. We note, however, that dedicated studies still need to be carried out to demonstrate the feasibility of this application.

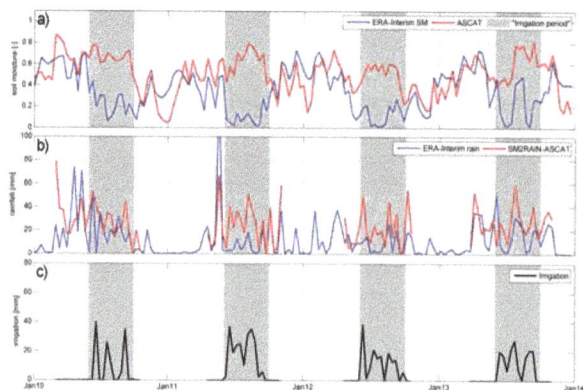

Figure 6. Soil moisture and rainfall time series for a pixel in Nebraska (Longitude, Latitude = −98 E, 40 N). (**a**) Temporal evolution of ASCAT and ERA-Interim soil moisture, SM, data (10-day moving average). Grey areas represent the months during which irrigation takes place; (**b**) Temporal evolution of ERA-Interim and SM2RAIN-ASCAT rainfall (10-day accumulations); (**c**) Temporal evolution of irrigation obtained as the difference between SM2RAIN-ASCAT and ERA-Interim rainfall (10-day accumulations).

5. Soil Moisture for Hydrological Applications: A Scientific Roadmap

In the previous three sections, we provided a short overview on the techniques used for soil moisture monitoring, on the spatial-temporal variability of soil moisture and on (some of) the hydrological applications in which soil moisture data are beneficial. Specifically, we highlighted the main limitations and research issues to be addressed together with the new hydrological applications that, in our opinion, should be investigated in the near future. We want to underline here that some of the open issues we raised might be considered as not doable, at least from the theoretical viewpoint. Theoretically, coarse resolution satellite soil moisture observations, or point scale in situ observations, cannot be used for catchment-scale applications; but several studies demonstrated the opposite in contrast with expectations (e.g., [9,11,86,113,122]). Similarly, it was not expected that soil moisture data could be used for rainfall estimation (as shown in [138]) or that C-band scatterometer data (i.e., ASCAT) could provide an accurate and usable satellite soil moisture product as it was shown in [26] and [45]. Research activities in the recent years demonstrated that our a priori, or theoretical, knowledge might fail when we started using actual observations. We believe that, as scientists, we have to experimentally demonstrate our hypothesis, and it was the rationale behind our past and on-going research activities. Bearing that in mind, we have defined three open issues to be addressed in the near future and the corresponding research opportunities that come from the attempt to answer to these scientific questions.

5.1. High Spatial-Temporal Resolution Soil Moisture Measurements

Currently, we are able to estimate accurately soil moisture at the point scale through in situ sensors. Moreover, less accurate measurements can be obtained at coarse scale (~20 km) using satellite sensors. Daily and sub-daily temporal resolutions can be obtained through these techniques. However, soil moisture measurements at high spatial (i.e., <5 km) and temporal (sub-daily) resolution are not available at the time of writing (January 2017). To get this target, we must improve, test and integrate the new ground-based monitoring techniques such as GPS, cosmic ray and geophysical methods with more accurate point measurements techniques (e.g., TDR). At the same time, satellite sensors are rapidly increasing their capabilities in terms of spatial and temporal resolution, and accuracy. By way of example, in the near future (2020), the launch of next generation scatterometer sensors by EUMETSAT [148] will provide 5 km/daily soil moisture data by using a well-established technology and a soil moisture retrieval algorithm largely tested with ASCAT data. Moreover, Sentinel-1 is expected to deliver soon high spatial resolution (1 km) soil moisture measurements every 6 days [55] and new technologies such as cubesat and nano-satellites, if well designed, have the potential to provide high spatial-temporal resolution at low costs [130]. Finally, spatial downscaling/upscaling approaches can be used to integrate the different techniques, as well as observations with modelling. Data assimilation and merging methods can be also considered to optimally integrate in situ, satellite and modelled data.

On this basis, an exciting future is foreseen in a couple of years from now in which high resolution soil moisture datasets are expected to be developed. The availability of these datasets to the hydrological (and climate) communities and users, also in an operational context (e.g., data availability in real-time) will open a number of new and unexplored research opportunities.

5.2. Soil Moisture Modelling in Space

As highlighted in Section 3, modelling soil moisture in space is an important issue that is largely unexplored. Currently, this issue is magnified by: (1) the unavailability (only for small regions and limited time periods) of spatially detailed soil moisture datasets; (2) the difficulties in obtaining static (e.g., soil type) and dynamic (e.g., precipitation) inputs at high spatial resolution for modelling; and (3) the larger interest of hydrological and climate communities to simulate soil moisture in time with respect to in space. In our opinion, we should improve our capability in modelling soil moisture both in time and in space.

Several research questions, and corresponding opportunities, should be addressed to achieve this target. Firstly, the spatial comparison between satellite and modelled data would reveal the similarities and the differences between the derived patterns. Mainly looking at the differences, we would gain knowledge on where (and when) satellite and modelled datasets need improvements. Secondly, a stronger link between studies that analysed soil moisture spatial variability and studies performing satellite datasets validation and/or using these datasets in hydrological and climatic applications should be carried out. For instance, lower (higher) performances in the comparison between in situ and satellite datasets can be expected when the spatial variance is higher (lower). Similarly, the error characterization of satellite-based soil moisture measurements (with coarse resolution) should be related to soil moisture spatial variability. We believe that detailed investigations on these two points would provide new important insights and, consequently, new research chance.

5.3. Comprehensive Assessment of the Value of Soil Moisture Data

The robust assessment of the effective value of in situ and satellite soil moisture data in hydrological applications is still missing. Theoretically, we are fully aware of the fundamental role of soil moisture but, with actual observations, contrasting results have been obtained. Open questions are: (1) What is the required accuracy of soil moisture data to be of benefit in hydrological applications? (2) What is the needed spatial and temporal resolution? Do we need higher spatial or temporal resolution? (3) What is the added-value of soil moisture data with respect to other physical quantities (e.g., river discharge, precipitation)? The answers to these questions are not trivial as, they are application-, model- and scale-dependent, and require dedicated studies encompassing a wide range of climates, models, etc. In this context, we strongly support the investigation of soil moisture data assimilation into flood modelling for different basins, climates, hydrological models and datasets (e.g., from different satellite sensors). For instance, we still need to assess if it is preferable to assimilate in situ or satellite soil moisture data (or alternatively, soil moisture or river discharge data). Moreover, another interesting research topic is related to the estimation of hydraulic soil properties using data assimilation of soil moisture (e.g., [149]). As mentioned above, the benefit in assimilating soil moisture data should also be related to soil moisture spatial-temporal variability studies.

Besides a comprehensive assessment in flood applications, we foster the use of soil moisture data in emerging applications. Indeed, we obtained that in situ and satellite soil moisture data can be extremely useful for landslides and erosion prediction, for malaria risk monitoring and for rainfall and irrigation assessment. Each of these new applications deserves further investigations in order to test the effective usability of soil moisture data on these topics. If successful, these new research activities can potentially open a number of new scientific and practical opportunities. For instance, the quantitative irrigation assessment from soil moisture data would be highly important for food security and management on a global scale, which is becoming more and more challenging under climate changing scenarios.

Acknowledgments: The study was partly supported by the EUMETSAT Satellite Application Facility in Support of Operational Hydrology and Water Management (H-SAF) project, the European Space Agency SMOS+rainfall and WACMOS-MED projects, the Italian Civil Protection Department. This work was partly funded by The Project of Interest NextData (MIUR-CNR). The authors wish to thank Umbria Region, Department of Environment, Planning and Infrastructure, for providing Tiber River basin data.

Author Contributions: All authors contributed extensively to the work presented in this paper. Luca Brocca conceived and designed the paper. Luca Ciabatta, Christian Massari, Stefania Camici and Angelica Tarpanelli contributed to the acquisition and processing of ground and satellite-based datasets, to the analysis and the elaboration of the data, and to the preparation and discussion of results. All co-authors contributed to the editing of the manuscript and to the discussion and interpretation of the results.

Conflicts of Interest: The authors declare no conflict of interest.

References

1. Seneviratne, S.I.; Corti, T.; Davin, E.L.; Hirschi, M.; Jaeger, E.B.; Lehner, I.; Orlowsky, B.; Teuling, A.J. Investigating soil moisture-climate interactions in a changing climate: A review. *Earth-Sci. Rev.* **2010**, *99*, 125–161. [CrossRef]

2. Brocca, L.; Morbidelli, R.; Melone, F.; Moramarco, T. Soil moisture spatial variability in experimental areas of central Italy. *J. Hydrol.* **2007**, *333*, 356–373. [CrossRef]
3. Wagner, W.; Blöschl, G.; Pampaloni, P.; Calvet, J.C.; Bizzarri, B.; Wigneron, J.P.; Kerr, Y. Operational readiness of microwave remote sensing of soil moisture for hydrologic applications. *Hydrol. Res.* **2007**, *38*, 1–20. [CrossRef]
4. Corradini, C. Soil moisture in the development of hydrological processes and its determination at different spatial scales. *J. Hydrol.* **2014**, *516*, 1–5. [CrossRef]
5. ESA Climate Change Initiative. Available online: http://www.esa-soilmoisture-cci.org/ (accessed on 13 February 2017).
6. Soil Moisture Active and Passive Mission. Available online: http://smap.jpl.nasa.gov/ (accessed on 13 February 2017).
7. International Soil Moisture Network. Available online: http://ismn.geo.tuwien.ac.at/ (accessed on 13 February 2017).
8. Cosmic-ray Soil Moisture Observing System. Available online: http://cosmos.hwr.arizona.edu/ (accessed on 13 February 2017).
9. Brocca, L.; Melone, F.; Moramarco, T.; Wagner, W.; Naeimi, V.; Bartalis, Z.; Hasenauer, S. Improving runoff prediction through the assimilation of the ASCAT soil moisture product. *Hydrol. Earth Syst. Sci.* **2010**, *14*, 1881–1893. [CrossRef]
10. Koster, R.D.; Mahanama, S.P.P.; Livneh, B.; Lettenmaier, D.; Reichle, R.H. Skill in Streamflow Forecasts Derived from Large-Scale Estimates of Soil Moisture and Snow. *Nat. Geosci.* **2010**, *3*, 613–616. [CrossRef]
11. Brocca, L.; Ponziani, F.; Moramarco, T.; Melone, F.; Berni, N.; Wagner, W. Improving landslide forecasting using ASCAT-derived soil moisture data: A case study of the Torgiovannetto landslide in central Italy. *Remote Sens.* **2012**, *4*, 1232–1244. [CrossRef]
12. Bittelli, M.; Valentino, R.; Salvatorelli, F.; Rossi Pisa, P. Monitoring soil-water and displacement conditions leading to landslide occurrence in partially saturated clays. *Geomorphology* **2012**, *173–174*, 161–173. [CrossRef]
13. Rahmani, A.; Golian, S.; Brocca, L. Multiyear monitoring of soil moisture over Iran through satellite and reanalysis soil moisture products. *Int. J. Appl. Earth Obs. Geoinf.* **2016**, *48*, 85–95. [CrossRef]
14. Enenkel, M.; Steiner, C.; Mistelbauer, T.; Dorigo, W.; Wagner, W.; See, L.; Atzberger, C.; Schneider, S.; Rogenhofer, E. A combined satellite-derived drought indicator to support humanitarian aid organizations. *Remote Sens.* **2016**, *8*, 340. [CrossRef]
15. Sánchez, N.; González-Zamora, Á.; Piles, M.; Martínez-Fernández, J. A New Soil Moisture Agricultural Drought Index (SMADI) Integrating MODIS and SMOS products: A case of study over the Iberian Peninsula. *Remote Sens.* **2016**, *8*, 287. [CrossRef]
16. De Rosnay, P.; Drusch, M.; Vasiljevic, D.; Balsamo, G.; Albergel, C.; Isaksen, L. A simplified Extended Kalman Filter for the global operational soil moisture analysis at ECMWF. *Q. J. R. Meteorol. Soc.* **2013**, *139*, 1199–1213. [CrossRef]
17. De Rosnay, P.; Balsamo, G.; Albergel, C.; Muñoz-Sabater, J.; Isaksen, L. Initialisation of land surface variables for Numerical Weather Prediction. *Surv. Geophys.* **2014**, *35*, 607–621. [CrossRef]
18. Dharssi, I.; Bovis, K.; Macpherson, B.; Jones, C. Operational assimilation of ASCAT surface soil wetness at the Met Office. *Hydrol. Earth Syst. Sci.* **2011**, *15*, 2729–2746. [CrossRef]
19. Capecchi, V.; Brocca, L. A simple assimilation method to ingest satellite soil moisture into a limited-area NWP model. *Meteorol. Z.* **2014**, *23*, 105–121.
20. Rodríguez-Iturbe, I.; Porporato, A. *Ecohydrology of Water-Controlled Ecosystems: Soil Moisture and Plant Dynamics*; Cambridge University Press: Cambridge, UK, 2007; p. 460.
21. Fang, B.; Lakshmi, V. Soil moisture at watershed scale: Remote sensing techniques. *J. Hydrol.* **2014**, *516*, 258–272. [CrossRef]
22. Muñoz-Sabater, J.; Al Bitar, A.; Brocca, L. Soil moisture retrievals based on active and passive microwave data: State-of-the-art and operational applications. In *Satellite Soil Moisture Retrievals: Techniques and Applications*; Petropoulos, G.P., Srivastava, P., Kerr, Y., Eds.; Elsevier: Amsterdam, The Netherlands, 2016; Volume 18, pp. 351–378.
23. Ochsner, T.E.; Cosh, M.H.; Cuenca, R.H.; Dorigo, W.A.; Draper, C.S.; Hagimoto, Y.; Kerr, Y.H.; Njoku, E.G.; Small, E.E.; Zreda, M. State of the art in large-scale soil moisture monitoring. *Soil Sci. Soc. Am. J.* **2013**, *77*, 1888–1919. [CrossRef]

24. Romano, N. Soil moisture at local scale: Measurements and simulations. *J. Hydrol.* **2014**, *516*, 6–20. [CrossRef]

25. Tebbs, E.; Gerard, F.; Petrie, A.; De Witte, E. Emerging and Potential Future Applications of Satellite-Based Soil Moisture Products. In *Satellite Soil Moisture Retrievals: Techniques and Applications*; Petropoulos, G.P., Srivastava, P., Kerr, Y., Eds.; Elsevier: Amsterdam, The Netherlands, 2016; Volume 19, pp. 379–400.

26. Brocca, L.; Crow, W.T.; Ciabatta, L.; Massari, C.; de Rosnay, P.; Enenkel, M.; Hahn, S.; Amarnath, G.; Camici, S.; Tarpanelli, A.; et al. A review of the applications of ASCAT soil moisture products. *IEEE J. Sel. Top. Appl. Earth Obs. Remote Sens.* **2017**. [CrossRef]

27. Vereecken, H.; Huisman, J.A.; Pachepsky, Y.; Montzka, C.; van der Kruk, J.; Bogena, H.; Weihermüller, L.; Herbst, M.; Martinez, G.; Vanderborght, J. On the spatiotemporal dynamics of soil moisture at the field scale. *J. Hydrol.* **2014**, *516*, 76–96. [CrossRef]

28. Vereecken, H.; Schnepf, A.; Hopmans, J.; Javaux, M.; Or, D.; Roose, T.; Vanderborght, J.; Young, M.; Amelung, W.; Aitkenhead, M.; et al. Modelling Soil Processes: Key challenges and new perspectives. *Vadose Zone J.* **2016**, *15*. [CrossRef]

29. Bogena, H.R.; Huisman, J.A.; Güntner, A.; Hübner, C.; Kusche, J.; Jonard, F.; Vey, S.; Vereecken, H. Emerging methods for noninvasive sensing of soil moisture dynamics from field to catchment scale: A review. *WIREs Water* **2015**, *2*, 635–647. [CrossRef]

30. Crow, W.T.; Yilmaz, M.T. The Auto-Tuned Land Assimilation System (ATLAS). *Water Resour. Res.* **2014**, *50*, 371–384. [CrossRef]

31. Topp, G.C.; David, J.L.; Annan, A.P. Electromagnetic Determination of Soil Water Content: Measurement in Coaxial Transmission Lines. *Water Resour. Res.* **1980**, *16*, 574–582. [CrossRef]

32. Robinson, D.A.; Campbell, C.S.; Hopmans, J.W.; Hornbuckle, B.K.; Jones, S.B.; Knight, R.; Ogden, F.; Selker, J.; Wendroth, O. Soil moisture measurement for ecological and hydrological watershed-scale observatories: A review. *Vadose Zone J.* **2008**, *7*, 358–389. [CrossRef]

33. Bogena, H.; Huisman, J.A.; Schilling, B.; Weuthen, A.; Vereecken, H. Effective calibration of low-cost soil water content sensors. *Sensors* **2017**, *17*, 208. [CrossRef] [PubMed]

34. Dorigo, W.A.; Xaver, A.; Vreugdenhil, M.; Gruber, A.; Hegyiová, A.; Sanchis-Dufau, A.D.; Zamojski, D.; Cordes, C.; Wagner, W.; Drusch, M. Global automated quality control of in situ soil moisture data from the International Soil Moisture Network. *Vadose Zone J.* **2013**, *12*. [CrossRef]

35. Bogena, H.R.; Herbst, M.; Huisman, J.A.; Rosenbaum, U.; Weuthen, A.; Vereecken, H. Potential of wireless sensor networks for measuring soil water content variability. *Vadose Zone J.* **2010**, *9*, 1002–1013. [CrossRef]

36. Calamita, G.; Perrone, A.; Brocca, L.; Onorati, B.; Manfreda, S. Field test of a multi-frequency electromagnetic induction sensor for soil moisture monitoring in southern Italy test sites. *J. Hydrol.* **2015**, *529*, 316–329. [CrossRef]

37. Franz, T.E.; Wang, T.; Avery, W.; Finkenbiner, C.; Brocca, L. Combined analysis of soil moisture measurements from roving and fixed cosmic-ray neutron probes for multi-scale real-time monitoring. *Geophys. Res. Lett.* **2015**, *42*, 3389–3396. [CrossRef]

38. Larson, K.M.; Small, E.E.; Gutmann, E.D.; Bilich, A.L.; Braun, J.J.; Zavorotny, V.U. Use of GPS receivers as a soil moisture network for water cycle studies. *Geophys. Res. Lett.* **2008**, *35*, L24405. [CrossRef]

39. Cosmic-ray Soil Moisture Monitoring Network-UK. Available online: http://cosmos.ceh.ac.uk/ (accessed on 13 February 2017).

40. PBO H$_2$O Data Portal. Available online: http://xenon.colorado.edu/portal/ (accessed on 13 February 2017).

41. Bogena, H.R.; Huisman, J.A.; Baatz, R.; Hendricks Franssen, H.-J.; Vereecken, H. Accuracy of the cosmic-ray soil water content probe in humid forest ecosystems: The worst case scenario. *Water Resour. Res.* **2013**, *49*, 5778–5791. [CrossRef]

42. Baatz, R.; Bogena, H.; Hendricks Franssen, H.-J.; Huisman, J.A.; Wei, Q.; Montzka, C.; Vereecken, H. Calibration of a catchment scale cosmic-ray soil moisture network: A comparison of three different methods. *J. Hydrol.* **2014**, *516*, 231–244. [CrossRef]

43. Heidbüchel, I.; Güntner, A.; Blume, T. Use of Cosmic-Ray Neutron Sensors for Soil Moisture Monitoring in Forests. *Hydrol. Earth Syst. Sci.* **2016**, *20*, 1269–1288. [CrossRef]

44. Rahimzadeh-Bajgiran, P.; Berg, A. Soil Moisture Retrievals Using Optical/TIR Methods. In *Satellite Soil Moisture Retrievals: Techniques & Applications*; Petropoulos, G.P., Srivastava, P., Kerr, Y., Eds.; Elsevier: Amsterdam, The Netherlands, 2016; Volume 3, pp. 47–72.

45. Wagner, W.; Hahn, S.; Kidd, R.; Melzer, T.; Bartalis, Z.; Hasenauer, S.; Figa, J.; de Rosnay, P.; Jann, A.; Schneider, S.; et al. The ASCAT soil moisture product: A review of its specifications, validation results, and emerging applications. *Meteorol. Z.* **2013**, *22*, 5–33.

46. Liu, Y.Y.; Parinussa, R.M.; Dorigo, W.A.; De Jeu, R.A.M.; Wagner, W.; van Dijk, A.I.J.M.; McCabe, M.F.; Evans, J.P. Developing an improved soil moisture dataset by blending passive and active microwave satellite-based retrievals. *Hydrol. Earth Syst. Sci.* **2011**, *15*, 425–436. [CrossRef]

47. Brocca, L.; Massari, C.; Ciabatta, L.; Wagner, W.; Stoffelen, A. Remote sensing of terrestrial rainfall from Ku-band scatterometers. *IEEE J. Sel. Top. Appl. Earth Obs. Remote Sens.* **2016**, *9*, 533–539. [CrossRef]

48. Kerr, Y.H.; Waldteufel, P.; Wigneron, J.-P.; Martinuzzi, J.; Font, J.; Berger, M. Soil moisture retrieval from space: The Soil Moisture and Ocean Salinity (SMOS) mission. *IEEE Trans. Geosci. Remote Sens.* **2001**, *39*, 1729–1735. [CrossRef]

49. Entekhabi, D.; Njoku, E.G.; Neill, P.E.; Kellogg, K.H.; Crow, W.T.; Edelstein, W.N.; Entin, J.K.; Goodman, S.D.; Jackson, T.J.; Johnson, J.; et al. The soil moisture active passive (SMAP) mission. *Proc. IEEE* **2010**, *98*, 704–716. [CrossRef]

50. Kim, S.; Liu, Y.Y.; Johnson, F.M.; Parinussa, R.M.; Sharma, A. A global comparison of alternate AMSR2 soil moisture products: Why do they differ? *Remote Sens. Environ.* **2015**, *161*, 43–62. [CrossRef]

51. Kerr, Y.H.; Waldteufel, P.; Richaume, P.; Wigneron, J.P.; Ferrazzoli, P.; Mahmoodi, A.; Juglea, S.E.; Leroux, D.; Mialon, A.; Delwart, S. The SMOS soil moisture retrieval algorithm. *IEEE Trans. Geosci. Remote Sens.* **2012**, *50*, 1384–1403. [CrossRef]

52. Malbéteau, Y.; Merlin, O.; Molero, B.; Rüdiger, C.; Bacon, S. DisPATCh as a tool to evaluate coarse-scale remotely sensed soil moisture using localized in situ measurements: Application to SMOS and AMSR-E data in Southeastern Australia. *Int. J. Appl. Earth Obs. Geoinf.* **2016**, 221–234. [CrossRef]

53. Piles, M.; Sánchez, N.; Vall-llossera, M.; Camps, A.; Martinez-Fernandez, J.; Martinez, J.; Gonzalez-Gambau, V. A downscaling approach for SMOS land observations: Evaluation of high-resolution soil moisture maps over the Iberian Peninsula. *IEEE J. Sel. Top. Appl. Earth Obs. Remote Sens.* **2014**, *7*, 3845–3857. [CrossRef]

54. Santi, E.; Paloscia, S.; Pettinato, S.; Brocca, L.; Ciabatta, L. Robust assessment of an operational algorithm for the retrieval of soil moisture from AMSR-E data in central Italy. *IEEE J. Sel. Top. Appl. Earth Obs. Remote Sens.* **2016**, *9*, 2478–2492. [CrossRef]

55. Paloscia, S.; Pettinato, S.; Santi, E.; Notarnicola, C.; Pasolli, L.; Reppucci, A. Soil moisture mapping using Sentinel-1 images: Algorithm and preliminary validation. *Remote Sens. Environ.* **2013**, *134*, 234–248. [CrossRef]

56. Al-Yaari, A.; Wigneron, J.-P.; Ducharne, A.; Kerr, Y.H.; Wagner, W.; De Lannoy, G.; Reichle, R.H.; Al-Bitar, A.; Dorigo, W.; Richaume, P.; et al. Global-scale comparison of passive (SMOS) and active (ASCAT) satellite based microwave soil moisture retrievals with soil moisture simulations (MERRA-Land). *Remote Sens. Environ.* **2014**, *152*, 614–626. [CrossRef]

57. Brocca, L.; Hasenauer, S.; Lacava, T.; Melone, F.; Moramarco, T.; Wagner, W.; Dorigo, W.; Matgen, P.; Martínez-Fernández, J.; Llorens, P.; et al. Soil moisture estimation through ASCAT and AMSR-E sensors: An intercomparison and validation study across Europe. *Remote Sens. Environ.* **2011**, *115*, 3390–3408. [CrossRef]

58. Dorigo, W.A.; Gruber, A.; De Jeu, R.A.M.; Wagner, W.; Stacke, T.; Loew, A.; Albergel, C.; Brocca, L.; Chung, D.; Parinussa, R.M.; et al. Evaluation of the ESA CCI soil moisture product using ground-based observations. *Remote Sens. Environ.* **2015**, 380–395. [CrossRef]

59. Paulik, C.; Dorigo, W.; Wagner, W.; Kidd, R. Validation of the ASCAT Soil Water Index using in situ data from the International Soil Moisture Network. *Int. J. Appl. Earth Obs. Geoinf.* **2014**, *30*, 1–8. [CrossRef]

60. Escobar, V.M.; Srinivasan, M.; Arias, S.D. Improving NASA's Earth Observation Systems and Data Programs through the Engagement of Mission Early Adopters. In *Earth Science Satellite Applications*; Hossain, F., Ed.; Springer International Publishing: Argovia, Switzerland, 2016; Volume 9, pp. 223–267.

61. Moran, M.S.; Doorn, B.; Escobar, V.; Brown, M.E. Connecting NASA science and engineering with earth science applications. *J. Hydrometeorol.* **2015**, *16*, 473–483. [CrossRef]

62. Wagner, W.; Lemoine, G.; Rott, H. A method for estimating soil moisture from ERS scatterometer and soil data. *Remote Sens. Environ.* **1999**, *70*, 191–207. [CrossRef]

63. Manfreda, S.; Brocca, L.; Moramarco, T.; Melone, F.; Sheffield, J. A physically based approach for the estimation of root-zone soil moisture from surface measurements. *Hydrol. Earth Syst. Sci.* **2014**, *18*, 1199–1212. [CrossRef]

64. Balsamo, G.; Albergel, C.; Beljaars, A.; Boussetta, S.; Brun, E.; Cloke, H.; Dee, D.; Dutra, E.; Muñoz-Sabater, J.; Pappenberger, F.; et al. ERA-Interim/Land: A global land surface reanalysis data set. *Hydrol. Earth Syst. Sci.* **2015**, *19*, 389–407. [CrossRef]

65. Famiglietti, J.S.; Wood, E.F. Multiscale modeling of spatially variable water and energy balance processes. *Water Resour. Res.* **1994**, *11*, 3061–3078. [CrossRef]

66. Bierkens, M.F.; Bell, V.A.; Burek, P.; Chaney, N.; Condon, L.E.; David, C.H.; de Roo, A.; Döll, P.; Drost, N.; Famiglietti, J.S.; et al. Hyper-resolution global hydrological modelling: What is next? Everywhere and locally relevant. *Hydrol. Process.* **2015**, *29*, 310–320. [CrossRef]

67. Brocca, L.; Camici, S.; Melone, F.; Moramarco, T.; Martinez-Fernandez, J.; Didon-Lescot, J.-F.; Morbidelli, R. Improving the representation of soil moisture by using a semi-analytical infiltration model. *Hydrol. Process.* **2014**, *28*, 2103–2115. [CrossRef]

68. Morbidelli, R.; Corradini, C.; Govindaraju, R.S. A field-scale infiltration model accounting for spatial heterogeneity of rainfall and soil saturated hydraulic conductivity. *Hydrol. Process.* **2006**, *20*, 1465–1481. [CrossRef]

69. Hopmans, J.W.; Nielsen, D.R.; Bristow, K.L. *How Useful are Small-Scale Soil Hydraulic Property Measurements for Large-Scale Vadose Zone Modeling?* In *Environmental Mechanics: Water, Mass and Energy Transfer in the Biosphere: The Philip Volume*; Raats, P.A.C., Smiles, D., Warrick, A.W., Eds.; American Geophysical Union: Washington, DC, USA, 2002; pp. 247–258.

70. Juglea, S.; Kerr, Y.; Mialon, A.; Lopez-Baeza, E.; Braithwaite, D.; Hsu, K. Soil moisture modelling of a SMOS pixel: Interest of using the persiann database over the Valencia anchor station. *Hydrol. Earth Syst. Sci.* **2010**, *14*, 1509–1525. [CrossRef]

71. Montzka, C.; Bogena, H.R.; Zreda, M.; Monerris, A.; Morrison, R.; Muddu, S.; Vereecken, H. Validation of Spaceborne and Modelled Surface Soil Moisture Products with Cosmic-Ray Neutron Probes. *Remote Sens.* **2017**, *9*, 103. [CrossRef]

72. Bell, K.R.; Blanchard, B.J.; Schmugge, T.J.; Witczak, M.W. Analysis of surface moisture variations within large field sites. *Water Resour. Res.* **1980**, *16*, 796–810. [CrossRef]

73. Vachaud, G.; Passerat de Silans, A.; Balabanis, P.; Vauclin, M. Temporal stability of spatially measured soil water probability density function. *Soil Sci. Soc. Am. J.* **1985**, *49*, 822–828. [CrossRef]

74. Vinnikov, K.Y.; Robock, A.; Speranskaya, N.A.; Schlosser, C.A. Scales of temporal and spatial variability of midlatitude soil moisture. *J. Geophys. Res.* **1996**, *101*, 7163–7174. [CrossRef]

75. Crow, W.T.; Berg, A.A.; Cosh, M.H.; Loew, A.; Mohanty, B.P.; Panciera, R.; Rosnay, P.D.; Ryu, D.; Walker, J.P. Upscaling sparse ground-based soil moisture observations for the validation of coarse-resolution satellite soil moisture products. *Rev. Geophys.* **2012**, *50*, RG2002. [CrossRef]

76. Western, A.W.; Zhou, S.; Grayson, R.B.; McMahon, T.A.; Blöschl, G.; Wilson, D.J. Spatial correlation of soil moisture in small catchments and its relationship to dominant spatial hydrological processes. *J. Hydrol.* **2004**, *286*, 113–134. [CrossRef]

77. Vanderlinden, K.; Vereecken, H.; Hardelauf, H.; Herbst, M.; Martínez, G.; Cosh, M.H.; Pachepsky, Y.A. Temporal stability of soil water contents: A review of data and analyses. *Vadose Zone J.* **2012**, *11*. [CrossRef]

78. Biswas, A. Scaling analysis of soil water storage with missing measurements using the second-generation continuous wavelet transform. *Eur. J. Soil Sci.* **2014**, *65*, 594–604. [CrossRef]

79. Pan, F.; Peters-Lidard, C.D. On the Relationship between Mean and Variance of Soil Moisture Fields. *J. Am. Water Resour. Assoc.* **2008**, *44*, 235–242. [CrossRef]

80. Teuling, A.J.; Troch, P.A. Improved understanding of soil moisture variability dynamics. *Geophys. Res. Lett.* **2005**, *32*, L05404. [CrossRef]

81. Lawrence, J.E.; Hornberger, G.M. Soil moisture variability across climate zones. *Geophys. Res. Lett.* **2007**, *34*, L20402. [CrossRef]

82. Mittelbach, H.; Seneviratne, S.I. A new perspective on the spatio-temporal variability of soil moisture: Temporal dynamics versus time invariant contributions. *Hydrol. Earth Syst. Sci.* **2012**, *16*, 2169–2179. [CrossRef]

83. Brocca, L.; Zucco, G.; Mittelbach, H.; Moramarco, T.; Seneviratne, S.I. Absolute versus temporal anomaly and percent of saturation soil moisture spatial variability for six networks worldwide. *Water Resour. Res.* **2014**, *50*, 5560–5576. [CrossRef]

84. Hu, W.; Si, B.C. Estimating spatially distributed soil water content at small watershed scales based on decomposition of temporal anomaly and time stability analysis. *Hydrol. Earth Syst. Sci.* **2016**, *20*, 571–587. [CrossRef]

85. Cosh, M.H.; Jackson, T.J.; Starks, P.; Heathman, G. Temporal stability of surface soil moisture in the Little Washita River watershed and its applications in satellite soil moisture product validation. *J. Hydrol.* **2006**, *323*, 168–177. [CrossRef]

86. Matgen, P.; Fenicia, F.; Heitz, S.; Plaza, D.; de Keyser, R.; Pauwels, V.R.N.; Wagner, W.; Savenije, H. Can ASCAT-derived soil wetness indices reduce predictive uncertainty in well-gauged areas? A comparison with in situ observed soil moisture in an assimilation application. *Adv. Water Resour.* **2012**, *44*, 49–65. [CrossRef]

87. Porporato, A.; Daly, E.; Rodriguez-Iturbe, I. Soil water balance and ecosystem response to climate change. *Amer. Nat.* **2004**, *164*, 625–632. [CrossRef] [PubMed]

88. Riley, W.J.; Shen, C. Characterizing coarse-resolution watershed soil moisture heterogeneity using fine-scale simulations. *Hydrol. Earth Syst. Sci.* **2014**, *18*, 2463–2483. [CrossRef]

89. Rosenbaum, U.; Bogena, H.R.; Herbst, M.; Huisman, J.A.; Peterson, T.J.; Weuthen, A.; Western, A.; Vereecken, H. Seasonal and event dynamics of spatial soil moisture patterns at the small catchment scale. *Water Resour. Res.* **2002**, *48*, W10544. [CrossRef]

90. Chaney, N.W.; Roundy, J.K.; Herrera-Estrada, J.E.; Wood, E.F. High-resolution modeling of the spatial heterogeneity of soil moisture: Applications in network design. *Water Resour. Res.* **2015**, *51*, 619–638. [CrossRef]

91. Lacava, T.; Matgen, P.; Brocca, L.; Bittelli, M.; Moramarco, T. A first assessment of the SMOS soil moisture product with in-situ and modelled data in Italy and Luxembourg. *IEEE Trans. Geosci. Remote Sens.* **2012**, *50*, 1612–1622. [CrossRef]

92. Western, A.W.; Grayson, R.B.; Blöschl, G.; Willgoose, G.R.; McMahon, T.A. Observed spatial organization of soil moisture and its relation to terrain indices. *Water Resour. Res.* **1999**, *35*, 797–810. [CrossRef]

93. Li, B.; Rodell, M. Spatial variability and its scale dependency of observed and modeled soil moisture over different climate regions. *Hydrol. Earth Syst. Sci.* **2013**, *17*, 1177–1188. [CrossRef]

94. Cornelissen, T.; Diekkrüger, B.; Bogena, H.R. Significance of scale and lower boundary condition in the 3D simulation of hydrological processes and soil moisture variability in a forested headwater catchment. *J. Hydrol.* **2014**, *516*, 140–153. [CrossRef]

95. Polcher, J.; Piles, M.; Gelati, E.; Barella-Ortiz, A.; Tello, M. Comparing surface-soil moisture from the SMOS mission and the ORCHIDEE land-surface model over the Iberian Peninsula. *Remote Sens. Environ.* **2016**, *174*, 69–81. [CrossRef]

96. Koch, J.; Stisen, S.; Fang, Z.; Bogena, H.R.; Cornelissen, T.; Diekkrüger, B.; Kollet, S. Inter-comparison of three distributed hydrological models with respect to the seasonal variability of soil moisture patterns at a small forested catchment. *J. Hydrol.* **2016**, *533*, 234–249. [CrossRef]

97. Fang, Z.; Bogena, H.R.; Kollet, S.; Koch, J.; Vereecken, H. Spatio-temporal validation of long-term 3D hydrological simulations of a forested catchment using empirical orthogonal functions and wavelet coherence analysis. *J. Hydrol.* **2015**, *529*, 1754–1767. [CrossRef]

98. Massari, C.; Brocca, L.; Moramarco, T.; Tramblay, Y.; Didon Lescot, J.-F. Potential of soil moisture observations in flood modelling: Estimating initial conditions and correcting rainfall. *Adv. Water Resour.* **2014**, *74*, 44–53. [CrossRef]

99. Tramblay, Y.; Bouvier, C.; Martin, C.; Didon-Lescot, J.F.; Todorovik, D.; Domergue, J.M. Assessment of initial soil moisture conditions for event-based rainfall-runoff modelling. *J. Hydrol.* **2010**, *387*, 176–187. [CrossRef]

100. Wooldridge, S.A.; Kalma, J.D.; Walker, J.P. Importance of soil moisture measurements for inferring parameters in hydrologic models of low-yielding ephemeral catchments. *Environ. Model. Softw.* **2003**, *18*, 35–48. [CrossRef]

101. Koren, V.; Moreda, F.; Smith, M. Use of soil moisture observations to improve parameter consistency in watershed calibration. *Phys. Chem. Earth* **2008**, *33*, 1068–1080. [CrossRef]

102. Aubert, D.; Loumagne, C.; Oudin, L. Sequential assimilation of soil moisture and streamflow data in a conceptual rainfall-runoff model. *J. Hydrol.* **2003**, *280*, 145–161. [CrossRef]

103. Chen, F.; Crow, W.T.; Starks, P.J.; Moriasi, D.N. Improving hydrologic predictions of a catchment model via assimilation of surface soil moisture. *Adv. Water Resour.* **2011**, *34*, 526–536. [CrossRef]

104. Enenkel, M.; Reimer, C.; Dorigo, W.; Wagner, W.; Pfeil, I.; Parinussa, R.; De Jeu, R. Combining satellite observations to develop a global soil moisture product for near-real-time applications. *Hydrol. Earth Syst. Sci.* **2016**, *20*, 4191–4208. [CrossRef]

105. Brocca, L.; Melone, F.; Moramarco, T.; Morbidelli, R. Antecedent wetness conditions based on ERS scatterometer data. *J. Hydrol.* **2009**, *364*, 73–87. [CrossRef]

106. Massari, C.; Brocca, L.; Ciabatta, L.; Moramarco, T.; Gabellani, S.; Albergel, C.; de Rosnay, P.; Puca, S.; Wagner, W. The use of H-SAF soil moisture products for operational hydrology: Flood modelling over Italy. *Hydrology* **2015**, *2*, 2–22. [CrossRef]

107. Beck, H.E.; Jeu, R.A.M.D.; Schellekens, J.; Dijk, A.I.J.M.V.; Bruijnzeel, L.A. Improving curve number based storm runoff estimates using soil moisture proxies. *IEEE J. Sel. Top. Appl. Earth Obs. Remote Sens.* **2009**, *2*, 250–259. [CrossRef]

108. Tramblay, Y.; Bouaicha, R.; Brocca, L.; Dorigo, W.; Bouvier, C.; Camici, S.; Servat, E. Estimation of antecedent wetness conditions for flood modelling in northern Morocco. *Hydrol. Earth Syst. Sci.* **2012**, *16*, 4375–4386. [CrossRef]

109. Zhuo, L.; Dai, Q.; Han, D. Evaluation of SMOS soil moisture retrievals over the central United States for hydro-meteorological application. *Phys. Chem. Earth* **2015**, *83*, 146–155. [CrossRef]

110. Parajka, J.; Naemi, V.; Bloschl, G.; Wagner, W.; Merz, R.; Scipal, K. Assimilating scatterometer soil moisture data into conceptual hydrologic models at coarse scales. *Hydrol. Earth Syst. Sci.* **2006**, *10*, 353–368. [CrossRef]

111. Silvestro, F.; Gabellani, S.; Rudari, R.; Delogu, F.; Laiolo, P.; Boni, G. Uncertainty reduction and parameter estimation of a distributed hydrological model with ground and remote-sensing data. *Hydrol. Earth Syst. Sci.* **2015**, *19*, 1727–1751. [CrossRef]

112. Brocca, L.; Moramarco, T.; Melone, F.; Wagner, W.; Hasenauer, S.; Hahn, S. Assimilation of surface and root-zone ASCAT soil moisture products into rainfall-runoff modelling. *IEEE Trans. Geosci. Remote Sens.* **2012**, *50*, 2542–2555. [CrossRef]

113. Massari, C.; Brocca, L.; Tarpanelli, A.; Moramarco, T. Data assimilation of satellite soil moisture into rainfall-runoff modelling: A complex recipe? *Remote Sens.* **2015**, *7*, 11403–11433. [CrossRef]

114. Lievens, H.; Tomer, S.K.; Al Bitar, A.; De Lannoy, G.J.M.; Drusch, M.; Dumedah, G.; Hendricks Franssen, H.-J.; Kerr, Y.H.; Martens, B.; Pan, M.; et al. SMOS soil moisture assimilation for improved hydrologic simulation in the Murray Darling Basin, Australia. *Remote Sens. Environ.* **2015**, *168*, 146–162. [CrossRef]

115. Alvarez-Garreton, C.; Ryu, D.; Western, A.W.; Crow, W.T.; Robertson, D.E. The impacts of assimilating satellite soil moisture into a rainfall–runoff model in a semi-arid catchment. *J. Hydrol.* **2014**, *519*, 2763–2774. [CrossRef]

116. Masseroni, D.; Cislaghi, A.; Camici, S.; Massari, C.; Brocca, L. A reliable rainfall-runoff model for flood forecasting: Review and application to a semiurbanized watershed at high flood risk in Italy. *Hydrol. Res.* **2017**. [CrossRef]

117. Cenci, L.; Laiolo, P.; Gabellani, S.; Campo, L.; Silvestro, F.; Delogu, F.; Boni, G.; Rudari, R. Assimilation of H-SAF soil moisture products for flash flood early warning systems. Case study: Mediterranean catchments. *IEEE J. Sel. Top. Appl. Earth Obs. Remote Sens.* **2016**, *9*, 5634–5646. [CrossRef]

118. Alvarez-Garreton, C.; Ryu, D.; Western, A.W.; Su, C.-H.; Crow, W.T.; Robertson, D.E.; Leahy, C. Improving operational flood ensemble prediction by the assimilation of satellite soil moisture: Comparison between lumped and semi-distributed schemes. *Hydrol. Earth Syst. Sci.* **2015**, *19*, 1659–1676. [CrossRef]

119. Gruber, A.; Su, C.-H.; Zwieback, S.; Crow, W.; Dorigo, W.; Wagner, W. Recent advances in (soil moisture) triple collocation analysis. *Int. J. Appl. Earth Obs. Geoinf.* **2016**, *45*, 200–211. [CrossRef]

120. Zhuo, L.; Han, D. Could operational hydrological models be made compatible with satellite soil moisture observations? *Hydrol. Process.* **2016**, *30*, 1637–1648. [CrossRef]

121. Godt, J.W.; Baum, R.L.; Chleborad, A.F. Rainfall characteristics for shallow landsliding in Seattle, Washington, USA. *Earth Surf. Process. Landf.* **2006**, *31*, 97–110. [CrossRef]

122. Todisco, F.; Brocca, L.; Termite, L.; Wagner, W. Use of satellite and modelled soil moisture data for predicting event soil loss at plot scale. *Hydrol. Earth Syst. Sci.* **2015**, *19*, 3845–3856. [CrossRef]

123. Ponziani, F.; Pandolfo, C.; Stelluti, M.; Berni, N.; Brocca, L.; Moramarco, T. Assessment of rainfall thresholds and soil moisture modeling for operational hydrogeological risk prevention in the Umbria region (central Italy). *Landslides* **2012**, *9*, 229–237. [CrossRef]

124. Hawke, R.; McConchie, J. In situ measurement of soil moisture and pore-water pressures in an 'incipient' landslide: Lake Tutira, New Zealand. *J. Environ. Manag.* **2011**, *92*, 266–274. [CrossRef] [PubMed]

125. Baum, R.L.; Godt, J.W. Early warning of rainfall-induced shallow landslides and debris flows in the USA. *Landslides* **2010**, *7*, 259–272. [CrossRef]

126. USGS Real-Time Monitoring of Landslides. Available online: http://landslides.usgs.gov/research/rtmonitoring/ (accessed on 13 February 2017).

127. Ray, R.L.; Jacobs, J.M.; Cosh, M.H. Landslide susceptibility mapping using downscaled AMSR-E soil moisture: A case study from Cleveland Corral, California, US. *Remote Sens. Environ.* **2010**, *114*, 2624–2632. [CrossRef]

128. Ray, R.L.; Jacobs, J.M.; Ballestero, T.P. Regional landslide susceptibility: Spatiotemporal variations under dynamic soil moisture conditions. *Nat. Hazards* **2011**, *59*, 1317–1337. [CrossRef]

129. AghaKouchak, A.; Farahmand, A.; Melton, F.S.; Teixeira, J.; Anderson, M.C.; Wardlow, B.D.; Hain, C.R. Remote sensing of drought: Progress, challenges and opportunities. *Rev. Geophys.* **2015**, *53*, 452–480. [CrossRef]

130. McCabe, M.F.; Rodell, M.; Alsdorf, D.E.; Miralles, D.G.; Uijlenhoet, R.; Wagner, W.; Lucieer, A.; Houborg, R.; Verhoest, N.E.C.; Franz, T.E.; et al. The future of earth observation in hydrology. *Hydrol. Earth Syst. Sci.* **2017**. [CrossRef]

131. Chirebvu, E.; Chimbari, M.J.; Ngwenya, B.N.; Sartorius, B. Clinical malaria transmission trends and its association with climatic variables in tubu village, Botswana: A retrospective analysis. *PLoS ONE* **2016**, *11*, e0139843. [CrossRef] [PubMed]

132. Montosi, E.; Manzoni, S.; Porporato, A.; Montanari, A. An ecohydrological model of malaria outbreaks. *Hydrol. Earth Syst. Sci.* **2012**, *16*, 2759–2769. [CrossRef]

133. Peters, J.; Conte, A.; Verhoest, N.E.; De Clercq, E.; Goffredo, M.; De Baets, B.; Hendrickx, G.; Ducheyne, E. On the relation between soil moisture dynamics and the geographical distribution of Culicoides imicola. *Ecohydrology* **2014**, *7*, 622–632. [CrossRef]

134. Pellarin, T.; Ali, A.; Chopin, F.; Jobard, I.; Bergs, J.-C. Using spaceborne surface soil moisture to constrain satellite precipitation estimates over West Africa. *Geophys. Res. Lett.* **2008**, *35*, L02813. [CrossRef]

135. Crow, W.T.; Huffman, G.F.; Bindlish, R.; Jackson, T.J. Improving satellite rainfall accumulation estimates using spaceborne soil moisture retrievals. *J. Hydrometeorol.* **2009**, *10*, 199–212. [CrossRef]

136. Trenberth, K.E.; Asrar, G.R. Challenges and opportunities in water cycle research: WCRP contributions. *Surv. Geophys.* **2014**, *35*, 515–532. [CrossRef]

137. Brocca, L.; Melone, F.; Moramarco, T.; Wagner, W. A new method for rainfall estimation through soil moisture observations. *Geophys. Res. Lett.* **2013**, *40*, 853–858. [CrossRef]

138. Brocca, L.; Ciabatta, L.; Massari, C.; Moramarco, T.; Hahn, S.; Hasenauer, S.; Kidd, R.; Dorigo, W.; Wagner, W.; Levizzani, V. Soil as a natural rain gauge: Estimating global rainfall from satellite soil moisture data. *J. Geophys. Res.* **2014**, *119*, 5128–5141. [CrossRef]

139. Brocca, L.; Massari, C.; Ciabatta, L.; Moramarco, T.; Penna, D.; Zuecco, G.; Pianezzola, L.; Borga, M.; Matgen, P.; Martínez-Fernández, J. Rainfall estimation from in situ soil moisture observations at several sites in Europe: An evaluation of SM2RAIN algorithm. *J. Hydrol. Hydromech.* **2015**, *63*, 201–209. [CrossRef]

140. Koster, R.D.; Brocca, L.; Crow, W.T.; Burgin, M.S.; De Lannoy, G.J.M. Precipitation Estimation Using L-Band and C-Band Soil Moisture Retrievals. *Water Resour. Res.* **2016**, *52*, 7213–7225. [CrossRef]

141. Brocca, L.; Pellarin, T.; Crow, W.T.; Ciabatta, L.; Massari, C.; Ryu, D.; Su, C.-H.; Rudiger, C.; Kerr, Y. Rainfall estimation by inverting SMOS soil moisture estimates: A comparison of different methods over Australia. *J. Geophys. Res.* **2016**, *121*, 12062–12079. [CrossRef]

142. Ciabatta, L.; Brocca, L.; Massari, C.; Moramarco, T.; Puca, S.; Rinollo, A.; Gabellani, S.; Wagner, W. Integration of satellite soil moisture and rainfall observations over the Italian territory. *J. Hydrometeorol.* **2015**, *16*, 1341–1355. [CrossRef]

143. Ciabatta, L.; Brocca, L.; Massari, C.; Moramarco, T.; Gabellani, S.; Puca, S.; Wagner, W. Rainfall-runoff modelling by using SM2RAIN-derived and state-of-the-art satellite rainfall products over Italy. *Int. J. Appl. Earth Obs. Geoinf.* **2016**, *48*, 163–173. [CrossRef]

144. Alvarez-Garreton, C.; Ryu, D.; Western, A.W.; Crow, W.T.; Su, C.-H.; Robertson, D.R. Dual assimilation of satellite soil moisture to improve streamflow prediction in data-scarce catchments. *Water Resour. Res.* **2016**, *52*, 5357–5375. [CrossRef]

145. Kumar, S.V.; Peters-Lidard, C.D.; Santanello, J.A.; Reichle, R.H.; Draper, C.S.; Koster, R.D.; Nearing, G.; Jasinski, M.F. Evaluating the utility of satellite soil moisture retrievals over irrigated areas and the ability of land data assimilation methods to correct for unmodeled processes. *Hydrol. Earth Syst. Sci.* **2015**, *19*, 4463–4478. [CrossRef]
146. Qiu, J.; Gao, Q.; Wang, S.; Su, Z. Comparison of temporal trends from multiple soil moisture data sets and precipitation: The implication of irrigation on regional soil moisture trend. *Int. J. Appl. Earth Obs. Geoinf.* **2016**, *48*, 17–27. [CrossRef]
147. Singh, D.; Gupta, P.K.; Pradhan, R.; Dubey, A.K.; Singh, R.P. Discerning shifting irrigation practices from passive microwave radiometry over Punjab and Haryana. *J. Water Clim. Chang.* **2016**. [CrossRef]
148. Fois, F.; Lin, C.C.; Loiselet, M.; Scipal, K.; Stoffelen, A.; Wilson, J.J.W. The Metop second generation scatterometer. In Proceeding of the International Geoscience and Remote Sensing Symposium, Quebec City, QC, Canada, 13–18 July 2014; pp. 1–4.
149. Montzka, C.; Moradkhani, H.; Weihermuller, L.; Franssen, H.-J.H.; Canty, M.; Vereecken, H. Hydraulic parameter estimation by remotely-sensed top soil moisture observations with the particle filter. *J. Hydrol.* **2011**, *399*, 410–421. [CrossRef]

Article

Local- and Plot-Scale Measurements of Soil Moisture: Time and Spatially Resolved Field Techniques in Plain, Hill and Mountain Sites

Giulia Raffelli [1], Maurizio Previati [1,*], Davide Canone [1], Davide Gisolo [1], Ivan Bevilacqua [1], Giorgio Capello [2], Marcella Biddoccu [2], Eugenio Cavallo [2], Rita Deiana [3], Giorgio Cassiani [4] and Stefano Ferraris [1]

[1] Interuniversity Department of Regional and Urban Studies and Planning (DIST), Polytechnic and University of Torino, viale Mattioli, 39, 10125 Torino, Italy; giulia.raffelli@gmail.com (G.R.); davide.canone@unito.it (D.C.); davide.gisolo92@gmail.com (D.G.); Ivan.bevilacqua@gmail.com (I.B.); stefano.ferraris@unito.it (S.F.)

[2] Institute for Agricultural and Earthmoving Machines, Italian National Research Council, Strada delle Cacce 73, 10135 Torino, Italy; g.capello@ima.to.cnr.it (G.C.); m.biddoccu@ima.to.cnr.it (M.B.); e.cavallo@imamoter.cnr.it (E.C.)

[3] Department of Cultural Heritage (dBC), University of Padova, 35127 Padova, Italy; rita.deiana@unipd.it

[4] Department of Geosciences, University of Padova, 35127 Padova, Italy; giorgio.cassiani@unipd.it

* Correspondence: maurizio.previati@unito.it; Tel.: +39-011-0907428

Received: 2 June 2017; Accepted: 11 September 2017; Published: 15 September 2017

Abstract: Soil moisture measurement is essential to validate hydrological models and satellite data. In this work we provide an overview of different local and plot scale soil moisture measurement techniques applied in three different conditions in terms of altitude, land use, and soil type, namely a plain, a mountain meadow and a hilly vineyard. The main goal is to provide a synoptic view of techniques supported by practical case studies to show that in such different conditions it is possible to estimate a time and spatially resolved soil moisture by the same combination of instruments: contact-based methods (i.e., Time Domain Reflectometry—TDR, and two low frequency probes) for the time resolved, and hydro-geophysical minimally-invasive methods (i.e., Electromagnetic Induction—EMI, Ground Penetrating Radar—GPR, and the Electrical Resistivity Tomography—ERT) for the spatially resolved. Both long-term soil moisture measurements and spatially resolved measurement campaigns are discussed. Technical and operational measures are detailed to allow critical factors to be identified.

Keywords: soil moisture measurements; TDR; FDR Sensors; EMI; ERT; GPR; case studies

1. Introduction

1.1. Soil Moisture and Soil–Water Relations

Soil moisture can be defined as the water in the uppermost part of a field soil, and it is strictly affected by soil physical properties, such as texture, organic matter content and stone presence, but also by land cover (vegetation), land use, topography and rainfall [1–3]. Soil moisture represents a key state variable to understand surface hydrological processes (such as drainage, evaporation and plant uptake) and it controls water and energy exchanges between the land surface and the atmosphere [4,5] contributing also, as key factor, in soil–snow interactions and snow gliding [6]. Through direct evaporation and plant transpiration, soil moisture regulates the partitioning of the incoming solar energy at the land surface into the outgoing sensible, latent, and ground heat fluxes. Furthermore, soil moisture plays a significant role in the organization of natural ecosystems and biodiversity [4], also with consequences for human disease (e.g., malaria transmission [7]). This is especially important in water-scarce environments where more frequent drought events occur [8].

Considering agricultural and irrigation management practices, since soil moisture is extremely variable in space and time, it is very important to accurately evaluate its spatiotemporal dynamics in the root zone at the field scale. In fact, soil water content strongly influences crop production, crop health, and soil salinization. In addition, understanding the factors controlling soil moisture variability allows for an improvement in irrigation management strategies with respect to crop production and optimal use of water resources [5,9,10].

During the last decades new approaches and techniques for monitoring, modelling and using soil moisture data have been developed e.g., [11–13]. With reference to soil moisture monitoring, several studies have been focused on surveying test sites having very different morphological conditions. Lin et al. [1] did year-round monitoring at different sites in the Shale Hills Catchment in central Pennsylvania and found a temporal stability of the soil moisture spatial pattern, which was governed by soil types and landforms. Tromp-van Meerveld and McDonnell [2] measured soil moisture content at different depths between the soil surface and the soil-bedrock and in different locations across the trenched hillslope in the Panola Mountain Research Watershed, GA, USA. They observed that spatial differences in soil depth and total water available at the end of the wet season and during the summer period appeared to be responsible for the spatial differences in basal area and species distribution between upslope and mid-slope sections of the hillslope. Teuling et al. [3] investigated the role of interannual climate variability on soil moisture spatial dynamics for a field site in Louvain-la-Neuve, Belgium, considering three different years (intermediate conditions—1999; wet conditions—2000; extremely dry conditions—2003). They observed that climate variability induces non-uniqueness and two distinct hysteresis modes in the yearly relation between the spatial mean soil moisture and its variability.

In this work we show the soil moisture estimation at local and plot scale in three different field conditions: a plain permanent meadow, a mountain permanent meadow and a hilly vineyard (all located in Northwestern Italy). The main goal is to provide a synoptic view of techniques supported by practical case studies to show that in very different conditions it is possible to estimate a time and spatially resolved soil moisture by the same combination of instruments: contact-based methods for the time resolved, and minimally invasive hydro-geophysical methods for the spatially resolved. Both long-time measurement series and single-measurement campaigns performed simultaneously with different techniques are presented.

The paper is organized as follows. In the following part a general overview of the field measurement technique is presented. In Section 2 we describe some of the available measurement techniques (also employed for the purpose of this paper) to evaluate soil water content both at the local and at the plot scale, focusing on how they work from a theoretical point of view. In Section 3 we present three different case studies (together with the obtained results), showing how we applied the methods. In Section 4 we discuss the results. In Section 5 we draw conclusions and provide an outlook on current opportunities and challenges.

1.2. Soil Moisture and Field Measurement Techniques: A General Overview

According to Robinson et al. [11], soil moisture in situ measurement methods can be divided into contact-based and contact-free methods. Contact-based methods imply direct contact with the soil. They are mostly applied at the local scale, and include both the destructive sampling (e.g., gravimetric methods—also defined as properly "direct" method—[14]), and indirect methods that account for the effects exerted by water content in the soil on its dielectric properties such as: Time Domain Reflectometry—TDR e.g., [15–17]; Frequency Domain Reflectometry—FDR e.g., [11]; capacitance sensors e.g., [18,19]; and Time Domain Transmission (TDT) sensors, e.g., [20]. The sensor networks (wireless in particular) are generally constituted by capacitance, FDR, and TDT sensors, e.g., [5,21,22]; however, more complex but reliable solutions are represented by TDR multiplexing systems, e.g., [23,24], similar to the one presented in the plain permanent meadow case study. These techniques give spatially and resolved temporally highly measurements at the local scale (e.g., vertical soil moisture profile) and the spatiotemporal dynamics of soil water content at the field scale [4,25]. Contact-free methods include hydrogeophysical methods and remote sensing. Hydrogeophysical methods provide multi-point measurements using electromagnetic soil moisture sensors, with a spatial resolution down to several meters [5]. The most used hydrogeophysical methods are: (1) Ground Penetrating Radar (GPR) e.g., [26–32], in which the propagation velocity of high frequency (1 MHz to 1 GHz) electromagnetic waves is used to determine the soil dielectric permittivity and thus soil moisture content. It can be used with different setups, such as crosshole, air-launched and surface-based [5,26]; (2) Electromagnetic Induction (EMI) e.g., [33–35], and Electrical Resistivity Tomography (ERT) e.g., [32,35–40], which measure the apparent electrical conductivity (EC$_a$) of the soil, and will be treated more in detail in the following chapters. Note that strictly speaking ERT is rather a non-invasive method, as galvanic contact with the ground must be ensured. Remote sensing methods include: (a) passive microwave remote sensing e.g., [5,41–43], in which radiometers measure the thermal radiance emitted from the earth surface of a bare or cropped soil, using frequencies between 1 and 12 GHz (L- to X-band), obtaining the dielectric permittivity [5]; (b) airborne and spaceborne active microwave remote sensing (i.e., radar), such as TerraSAR-X, European Remote Sensing 2 (ERS-2), Advanced Synthetic Aperture Radar (ENVISAT/ASAR), Radarsat 2, and Phased Array type L-band Synthetic Aperture Radar (ALOS/PALSAR), which estimate surface soil moisture by the interpretation of the backscattering coefficient (expressed in decibels and depending on the dielectric permittivity of soil and surface roughness) [44,45]; (c) a cosmic-ray probe [46,47], which counts secondary fast neutrons near the soil surface created by primary cosmic-ray particles in the atmosphere and in the soil. The cosmic-ray probe was applied in different environmental and agricultural settings, e.g., [48–50]. Despite effortz, e.g., [51–54], to downscale the remote sensing soil moisture products (from several km to the local scale), the promising results obtained in scale root-zone soil moisture, e.g., [55–58], and the challenging opportunities offered by the future missions under launch in the coming decades, e.g., [59], remote sensing methods have typically been used to detect surface and near-surface soil moisture (topsoil—0–0.05 m) with coarse spatial and temporal resolution. Extensive reviews of these topics are offered by Peng et al. [60] and Mohanty et al. [61]. Within this context, a cosmic-ray probe is actually the only one used at the field scale, because it has the ability to measure integral soil moisture content with an acceptable temporal resolution [5]. However, it is essential to highlight that (i) the cosmic ray footprint radius is some hundreds of meters depending on the humidity, soil moisture and vegetation; (ii) the detector is extraordinarily sensitive for the first few meters close to the detector; and (iii) the penetration depth (on the order of a few decimetres) decreases exponentially with the distance to the sensor, e.g., [46,62–64].

In the following sections we will describe in more detail the soil measurement methods we adopted in our experimental sites. Namely: TDR and two different low-frequency types of soil moisture sensors (ECH$_2$O—5TM (Decagon Devices Inc., Pullman, WA, USA) and CS616 (Campbell Scientific Inc., Logan, UT, USA)) directly connected to data loggers, which give local measurements of soil water content in time and space; and EMI, GPR, and ERT, which represent plot-scale measurement techniques.

2. Soil Moisture Field Measurement Methods: A Focus on the Techniques Used

2.1. Time Domain Reflectometry (TDR)

As it is non-destructive and less time-consuming with respect to gravimetric measurements, Time Domain Reflectometry (TDR) is a widely used method for soil moisture measurement. Thanks to the accurate results achievable, together with its applicability to a large range of soils and settings, TDR progressively strengthened its position, replacing in the end the gravimetric technique [13]. In this work, a Tektronix 1502 C TDR cable tester coupled with two wire rods probes has been used.

TDR estimates the apparent dielectric permittivity of the soil by measuring the travel time a step voltage pulse takes to propagate along the probe and back. The probe must be placed in the soil either vertically or horizontally at the selected measuring depth. The dielectric permittivity of the system that surrounds the probe depends, in turn, on the soil moisture [65–67]. Calibration requirements are minimal (in many cases soil-specific calibration is not needed), but soil-specific calibration is possible for applications that demand high accuracy.

Once a soil permittivity value has been calculated starting from the travel time measurements, the soil moisture can be obtained by employing either empirical relationships [15] or quasi-physically based equations [68,69]. Considering that soils can have rather different electrical properties, the choice of the proper relationship/equation is connected with the physico-chemical characteristics of the surveyed material.

Different types of TDR field probes were studied in the literature, e.g., [70–77]. Robinson et al. [17] indicated the probes with two parallel rods as those to be preferred in field conditions, since they investigate a bigger volume in a more homogeneous way. Probes with two parallel rods can also be inserted more easily into the soil. Nissen et al. [78] suggested that a balun transformer could be omitted due to the results they obtained in their study.

2.2. Soil Moisture Sensors Directly Connected to Data Loggers

An increasing number of soil moisture sensors directly connected to data loggers are deployed to measure volumetric soil water content (θ) for agricultural, ecological, and geotechnical applications. While time-domain reflectometry (TDR) and transmissometry (TDT) operate at GHz range frequencies, the above mentioned sensors generally operate between 20 and 300 MHz [79]. With respect to time domain techniques, lower-frequency sensors are less expensive and, despite TDR still being more accurate, modern electronics coupled with a better understanding of the theory progressively improve low-frequency sensors' performance [80]. In this work, we present two contact-based sensor types: ECH_2O—5TM capacitance/frequency domain sensors (Decagon) and CS 616 water content reflectometers (Campbell Scientific). The frequency of 70 MHz at which both sensors are operating minimizes textural effects (and salinity) [81–83].

ECH_2O—5TM sensors employ a capacitance technique to determine soil moisture. In particular, these sensors determine the dielectric permittivity of the medium by measuring the charge time of a capacitor, using the soil as a dielectric medium. CS616 measures a period value. The period in air is approximately 14.7 microseconds, and the period in saturated soil with porosity equal to 0.4 is approximately 31 microseconds. The output is a ± 0.7 volt square wave with the frequency dependent on water content. This frequency is scaled down in the water content reflectometer circuit output to a frequency easily measured by a data logger. The probe output frequency or period is empirically related to water content using a calibration equation. Concerning the geometry, 5TM probes have three flat 52 mm long prongs spaced 5 mm apart, while CS616s have two 300 mm long rods with a diameter of 3.2 mm, spaced 32 mm apart. Measurement rods of CS616s are therefore approximately six times longer and six times wider apart than those of 5TMs. With reference to the porous medium measurement volume, considering that it is restricted to the direct surroundings material of the prongs/rods, the 5TMs investigate a soil volume much smaller than the CS616 [82,84]. Particular attention must be paid to the probe–soil contact.

2.3. Electromagnetic Induction (EMI)

As already mentioned, soil is a three-phase (solid–water–air) system, in which the main conducting phase is the aqueous solution. Due to this assumption, apparent Electrical Conductivity (EC_a) measurements can be used to evaluate soil water content [85]. Friedman [85] also states that there are three main categories of factors that can affect soil EC_a: (i) factors describing the bulk soil and the respective volumetric fractions occupied by the three phases and possible secondary structural configurations (aggregation): porosity, water content and structure; (ii) factors quantifying solid particle, which are relatively time-invariable: particle shape and orientation, particle size distribution, cation exchange capacity (CEC) and wettability; (iii) factors representing soil solution attributes, changing quickly in response to alterations in management and environmental conditions (e.g., environmental factors): ionic strength, cation composition and temperature. Electrical resistivity thus can change in space and time as an effect of (a) lithology (time-independent); (b) moisture content; and (c) pore water salinity. This is accounted for in all constitutive models describing the dependence of bulk electrical resistivity on soil parameters and state variables, starting with Archie's law [86]. The possible misleading effects must be accounted for, as shown in a number of recent literature reports, e.g., [87]. The same problem applies of course to EMI and ERT (see next section) as both measure electrical resistivity/conductivity.

Since it is a fast, non-invasive technique easy to use in field conditions, Electromagnetic Induction (EMI) is often used to map soil properties [88], but also soil moisture changes [35]. EMI follows the principle that within a time-varying electromagnetic field, any conductive object carries a current. Each instrument has two coils, a transmitter and a receiver, placed at either a fixed or variable spacing. Vereecken et al. [5] report typical fixed coil separations between 0.3 and 4.0 m, and a frequency operational range of between 5 kHz and 50 kHz.

EMI induces an electrical current into the soil; the depth of penetration is influenced by the separation of the coils and by the frequency of the current. Since EC_a is affected by soil's properties, the signal reaches a specific depth, also related to the uniformity of the soil. If the soil is very conductive close to the surface, then the signal will be locally dissipated without going deeper. Therefore, the depth of investigation of EMI equipment depends on four principal factors: (1) the resistivity of the investigated soil; (2) the coil separation of the instrument; (3) the coil orientation (vertical or horizontal); and (4) the waveform frequency. By varying the conditions of (2), (3) and (4), multiple depths of investigation can be achieved. The depth of investigation can be obtained as a function of the waveform frequency, the magnetic permeability and the electrical conductivity.

Advanced inversion techniques hold the potential of being applied effectively on small-scale EMI data, such as described above, e.g., [89,90]. While this is not yet common practice for soil moisture monitoring, some notable applications already exist, e.g., [91]. A recent review of EMI methods for hydrological and environmental applications is given by Boaga [92].

A number of EMI instruments are available in the market. For this study, we used a commercial GF Instruments CMD1 system (GF Instruments, Brno, Czech Republic) as well as a commercial GSSI Profiler EMP-400 (Geophysical Survey Systems Inc., Nashua, NH, USA). No inversion was carried out in this study, and maps show apparent electrical conductivity as traditionally estimated using a small induction number approach [93].

2.4. Electrical Resistivity Tomography (ERT)

The ERT is a non-destructive, cost-effective, indirect method that can also acquire data concerning soil moisture data within the subsurface, e.g., [5,40,94,95]. ERT uses a set of electrodes: two that inject electrical current into the soil, and two others that monitor the resulting voltage difference. Given a certain number of deployed electrodes, a very large number of measurement combinations is possible, giving rise to the measured dataset. Different configurations are possible, with pros and cons [94]. The measured voltage differences versus injected currents are summarized in terms of the relative ratio, i.e., the measured resistances (one for each measurement quadripole). A good practice consists of collecting both direct and reciprocal resistances (i.e., swapping current and voltage electrode pairs). Theoretically the two configurations shall give the same resistance value, so differences can be taken as an estimate of measurement error, to be used later for a more educated inversion procedure, e.g., [96]. Data are then inverted to give an estimate of the electrical resistivity spatial distribution that causes the observed resistances. Inversion requires the minimization of the overall difference between measured and simulated resistances, the latter being modelled using a physico-mathematical simulator that reproduces a DC current in a heterogeneous medium. This least-squares approach must generally be supplemented by some form of regularization [94], i.e., some a priori information about the spatial variability of electrical resistivity. In the most common case, the smoothest spatial distribution is sought, according to the so-called Occam's inversion approach that makes use of error estimates derived, e.g., from reciprocal analysis [94–97]. The resolution of the resulting resistivity distribution estimates decreases with depth, but is also dependent on the resistivity values themselves. The resulting distribution is generally reconstructed in 2D, but 3D imaging is possible if dense electrode distributions are available in 2D at the ground surface or in boreholes [5].

In order to reconstruct soil moisture patterns from ERT data, petro-physical relations such as Archie's law are used [40,98]. ERT measurements are widely employed to monitor vadose zone water dynamics, e.g., [32,94,99,100]. Since ERT allows providing images of spatial scale soil water content variability along either 2D transects or 3D soil volumes (considering high spatial resolution and daily temporal resolution) [5], it has been widely used in the studies of root water uptake by plants and temporal and spatial variations of soil–moisture interactions, e.g., [5,35,101–103]. However, there is evidence that this inversion procedure can produce mass balance errors due to a rapid decrease of ERT resolution with depth [40]. This problem may be solved through a coupled hydro-geophysical approach [40,101]. However, if only qualitative assessment of spatial distribution and time variations of soil moisture content are sufficient, an uncoupled approach can be used [32] based on the following steps: (1) inversion of geophysical field data gives the spatial distribution of electrical resistivity; (2) application of a petro-physical relationship to obtain an estimation of moisture content distribution from the electrical resistivity.

Many ERT instruments are commercially available. In this study we used a commercial IRIS Instruments SYSCAL Pro 10 resistivimeter with 72 channels (IRIS Instruments, Orleans, France). Inversion was carried out using the freely accessible inversion codes developed by Andrew Binley [32,104–106].

2.5. Ground Penetrating Radar (GPR)

GPR is a non-invasive and non-destructive geophysical method that uses radar pulses to image the subsurface. Given that the velocity of an electromagnetic wave in a soil is connected to its bulk dielectric constant (which is, in turn, related to the soil moisture), over the last three decades GPR has also been widely applied to estimate soil moisture in the unsaturated zone. A transmitter emits pulses (high-frequency radio waves typically in the range between 10 MHz and 2.6 GHz) that travel in the medium; when pulses encounter materials having different permittivities they are reflected, refracted or scattered back to the surface where a receiving antenna records the signal. Due to the frequency-dependent attenuation, lower frequencies lead to higher soil depth penetrations. However, higher frequencies provide higher resolutions. Two classes of GPR methods to estimate soil moisture can be distinguished: single-antenna separation methods (from scattering objects and traditional GPR sections), and different antenna separation methods (which require multiple measurements). For a comprehensive presentation of each single method, Huisman et al. [26] provide a detailed review.

Considering that the use of GPR systems is difficult in uneven irregular areas, an application is here presented in the plain permanent meadow case study. A commercial PulseEkko Pro radar system (Sensors and Software Inc., Mississauga, ON, Canada) with 100 MHz antennas has been used. For more details please refer to Rossi et al. [32].

3. Results: Testing Sites Description

In this section we describe the main characteristics of the three experimental sites together with the field campaigns layouts and the measurement methods applied in each site. All sites are located in Northwestern Italy. However, they are representative of very different conditions: the first is a permanent meadow located in a plain, while the second is a meadow located in a sloping mountain abandoned pasture; finally, the third is a vineyard in a hilly zone. A synoptic scheme of the different equipment used in the different testing sites is given in Table 1.

Table 1. Synoptic table summarizing the different equipment used in the different testing sites.

	Plain [Permanent Meadow]		Mountain [Permanent Meadow]		Hill [Vineyard]	
	Long Time Surveys	Single Campaigns	Long Time Surveys	Single Campaigns	Long Time Surveys	Single Campaigns
Time Domain Reflectometry (TDR)	√	√		√		√
Soil moisture sensors directly connected to data loggers			√		√	
Electromagnetic Induction (EMI)				√		√
Electrical Resistivity Tomography (ERT)		√		√		
Ground Penetrating Radar (GPR)		√				

Authors make the datasets collected within this work available. Readers or researchers interested in receiving and/or analysing one or more dataset shown in this work must address their specific request to the corresponding author.

3.1. Plain Permanent Meadow

Site and instrumental description. The experimental site is located in Grugliasco (Torino), in the northwestern part of the Po Plain, Italy (45°03′ N, 7°35′ E) at 290 m a.s.l. (Figure 1). In this area, rainfall climatology is characterized by two maxima, respectively in spring (April–May) and fall (October–November), and by a relatively dry winter and summer [107,108]. The vegetation growing season lasts, approximately, from late March to mid-October. The monitored area is 1500 m^2, constituted by a permanent meadow. The soil is loamy–sand to sandy ($\approx$80% of sand [32], which progressively rises to $\approx$95% below the 1st meter), without any gravel, and with a slope of about 1%. These last conditions are particularly suitable for TDR and GPR applications for both practical operation simplification (e.g., TDR probes insertion; GPR surveying process) and porous medium characteristics (i.e., low signal attenuation) e.g., [109,110].

Figure 1. Experimental site located in Grugliasco (Torino–Italy) and a picture shot in April 2005.

Long-term soil moisture measurements. The soil water content monitoring station is based on a TDR Tektronix 1502C and a personal computer controlling 11 multiplexers connected to 160 probes vertically inserted into the soil. The great number of installed probes was conceived to consent the calibration of the 3D numerical codes described by Paniconi et al. [111]. The sandy soil and the absence of stones allowed the vertical insertion without disturbance and/or induced lack of contact. Probes are made out of two parallel stainless steel rods (diameter 6 mm) with the following lengths: 150, 300, 600,

1000 and 2000 mm. To prevent geometrical probes = deformations and keep rods parallel during the insertion process, a steel guide was used. The TDR probes' layout in the experimental field is depicted in Figure 2.

The measurements started in 1997 and, since summer 2005, the TDR measurements have been automatic. The TDR signals are sampled and acquired using WinTDR software [112]. Volumetric soil water content is computed using the composite dielectric approach described by Roth et al. [68]. This relation has been validated by comparison with gravimetric sample measurements in the oven, obtaining an average error of 2%. The sampling time interval between TDR measurements can vary from one hour (during and after rainfall events, or during intense exfiltration periods) up to one day. In the same field, a meteorological station collects meteorological data at hourly intervals.

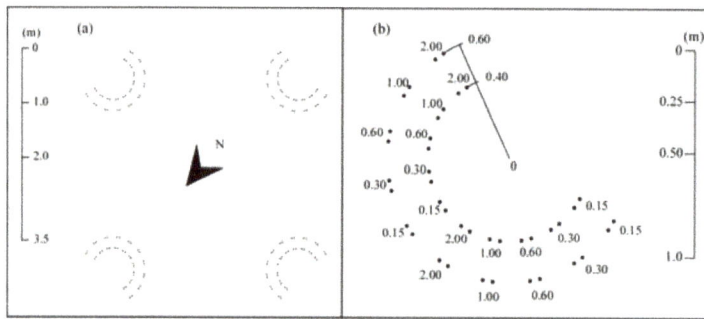

Figure 2. Grugliasco site. (**a**) Layout of the TDR probes in the experimental field; and (**b**) detail of the disposition of the probes of different lengths within one of the four groups.

ERT, GPR and TDR single measurement campaign. To evaluate the soil moisture evolution along time and space, on 28 August 2009 an infiltration experiment (over a rectangular area of 18 m by 2.6 m) was performed at the Grugliasco site. The scheme and a picture of the experimental setup is shown in Figure 3 (for details, see Rossi et al. [32]).

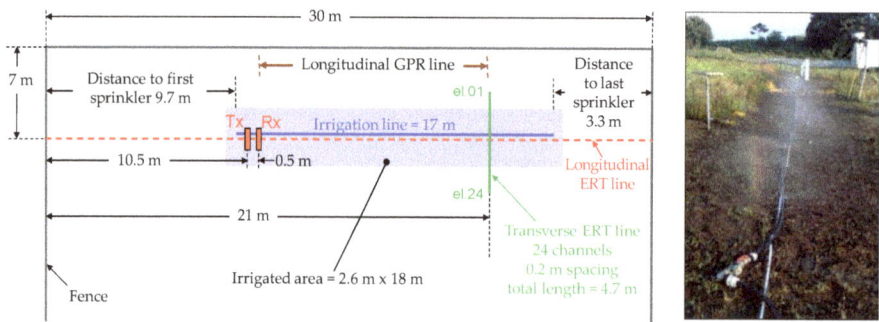

Figure 3. Grugliasco site, August 2009. Scheme of the irrigation experiment setup and a picture taken during the experiment.

The irrigation was performed in three steps: about 2.5 m^3 of water were distributed during the first irrigation step and about 1.5 m^3 during the others. ERT acquisitions occurred during irrigation breaks and with a precise scheduled time. Eight TDR probes with a length of 0.30 and 2.00 m were also vertically inserted into the soil, along the sprinkler line centred on the ERT profile. The absence of ponding was guaranteed by ensuring that the irrigation intensity was always lower than the soil infiltration capacity.

A longitudinal ERT line was acquired along the transect—see Figure 3 for the location—and measurements were taken only before and after irrigation. The line is composed of 48 electrodes with 0.6 m spacing, for a total length of 28.2 m. A dipole–dipole skip zero scheme was acquired, e.g., [113]

3.2. Mountain Permanent Meadow

Site and instrumental description. The Cogne Valley is a side valley of Valle d'Aosta (Northwestern Italy). The experimental site (45°36′ N, 7°21′ E) is located at 1730 m a.s.l. on a 26° slope facing south–southeast (169°)—Figure 4. The vegetation is characterized by herbaceous and shrub components typical of degraded pastures at high altitudes (e.g., *Hippophaë rhamnoides* L.), therefore it is representative of wide areas in the Alps (here, the pastures were abandoned due to the slope and low accessibility). This region is characterized by rainfall occurring mainly in the spring and autumn, with an average of 650 mm per year. The average annual temperature is about 4 °C. The soil depth ranges from 0.4 to 1.5 m, with a sandy loam texture and some gravel ($\approx$73% of sand).

The experimental site is characterized by intense direct solar radiation due to its aspect, resulting in large daily temperature variations. The incident radiation accelerates snow melt and the drying of soil (compared to the plain).

Figure 4. Experimental site located in Cogne (Aosta–Italy) and a picture shot in October 2016.

Long-term soil moisture measurements. Soil water content is measured with three CS616 sensors (Campbell) connected to a CR1000 Data logger. In 2010, data were collected hourly by one probe horizontally inserted at 0.1 m depth. In 2015 two more CS616 probes horizontally inserted into the soil at 0.2 and 0.4 m depths were installed. The gravimetric calibration method and temperature correction were adopted.

EMI, ERT and TDR measurement campaigns. EMI surveys aiming at mapping the spatial and temporal variability of soil resistivity along the steep mountain meadow testing site were performed in 2010, on the 26 September and the 23 October.

Data were collected using a GF Instruments CMD1 in high penetration mode (hence the estimated investigation depth ranges from the surface down to −1.5 m). During the same investigation dates, an ERT survey was performed along a transect located on the slope. A dipole–dipole skip zero acquisition scheme, with full reciprocal acquisition, was adopted. Data were inverted using an Occam inversion approach, as implemented in the Profile R/R2/R3 software package, accounting for the error level estimated from the data themselves [114]; here 5% was chosen as a relative error. During the 26 September 2010 EMI and ERT monitoring campaigns, soil moisture values were also measured by vertically inserted TDR probes. Two transects were investigated: the first was realized with measurements at the beginning, at the centre and at the end of the ERT transect (0.75 m long TDR probes vertically inserted into the soil); the second was more superficial (0.15 and 0.30 m long TDR probes vertically inserted into the soil), conducted perpendicularly to the ERT profile.

3.3. Hilly Vineyard

Site and instrumental description. The "Tenuta Cannona" (44°40′ N, 8°37′ E, 296 m a.s.l.) is located in Carpeneto (AL), 85 km southeast of Torino, Northwest Italy (Figure 5). The area is in the Alto Monferrato hilly region, which is a valuable vine-growing and DOC (Controlled Designation of Origin) wine production area. The soil has a clay to clay–loam texture (≈35% of clay and ≈40% of silt). The climate presents an average annual precipitation of 965 mm (based on the Ovada meteorological station data), mainly concentrated from October to March. Mean annual air temperature measured at the experimental site in the period 2000–2013 was 13 °C and the average annual precipitation was 849 mm [115,116].

Figure 5. Experimental site located in Carpeneto (Alessandria–Italy) and a picture shot in March 2016.

Long-term soil moisture measurements. A continuous monitoring was performed starting from August 2011 to provide information about the space–time variability of hydrological processes with different cultivation techniques. The test sites, which include conventional tillage ("TI") and grass-covered soil ("GC"), are adjacent to each other on a parcel located on a hill with a SE aspect and average slope of 15%. Each site includes six vine rows aligned along the slope, where the vines are spaced 1.0 m along the row and there is 2.75 m between the rows (Figure 6).

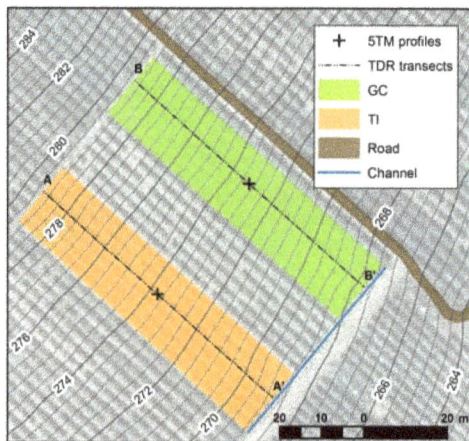

Figure 6. Carpeneto site. Plot of the experimental field. Grass Cover (GC) and Conventional Tillage (TI) represent the two investigated cultivation techniques. The two dotted lines A–A′ and B–B′ represent the investigated TDR transects.

Soil moisture data were collected by measuring the soil dielectric constant using four series of 5TM capacitance probes. Sensors have been horizontally inserted both in the central part of the inter-row (no soil compaction due to the passage of tractors—"NT") and in the track position ("T"), affected by the passage of tractor wheels. During the monitored period (winter 2015–2016), the soil was never frozen.

As regards to the sensors, we followed the calibration suggested by Decagon Devices which apply the Starr and Paltineanu approach [117], and resulting in an accuracy of ± 0.03 m^3/m^3 ($\pm 3\%$ VWC). Measurements are recorded and stored by a Decagon EM50 Data logger every 60 min.

EMI and TDR combined campaign. In this field campaign, we used the EMP400 multi-frequency EMI sensor, which can collect data up to three frequencies from 1 to 16 kHz, in horizontal dipole orientation (HDO) and vertical dipole orientation (VDO), with a coil spacing equal to 1.22 m. In this study, the EMP400 was used in VDO operating mode, with selected frequencies of 5, 10, and 15 kHz. The EMP400 is very sensitive at greater depths because of the larger coil distance. The EMP400, equipped with an integrated WAAS GPS, was carried by hand.

The TDR soil moisture was detected by a TDR Tektronix 1502C with three probes (made out of two parallel stainless steel rods with diameter 6 mm) vertically inserted into the soil from the surface down to the following depths: 0–0.15 m, 0–0.30 m, and 0–0.75 m. Measurements were performed every 10 m along two transects both in conventional tillage (TI), and in grass cover (GC) rows.

4. Results: Measurement Campaigns and Surveys Outputs

4.1. Plain Permanent Meadow

Long-term soil moisture measurements. The long-time survey depicted in Figure 7 shows the soil moisture data from 1 January 2004 to 31 December 2006 at the following vertical depths: 0–0.3, 0–0.6, and 0–1.0 m. Values are obtained through the arithmetic average of all the probes with the same rod lengths (namely, 16 probes per each considered length).

During the whole of 2004, until summer 2005, measurements were collected on a weekly base. Despite the general agreement with the precipitation data, the weekly sampling time is not exhaustive in terms of hydrological processes description (especially with coarse texture soils characterized by fast dynamics). Under such condition, minor meteorological events characterized by low magnitudes, are often not detected by the soil moisture profiles. At the same time soil moisture peaks corresponding to high precipitation events appear often damped (i.e., rarely exceeding the 0.20 m^3/m^3). Starting from August 2005, data have been automatically collected at hourly steps. With this sampling interval, minor events are also detected, and soil moisture peaks are better described, arriving to reach at the 0.30 m^3/m^3 threshold (which can be considered close to saturation for the investigated sandy soil).

With reference to the adopted experimental set-up, characterized by TDR probes vertically inserted into the soil, all the different lengths of probes react to meteorological forcing. Obviously this reaction is inversely proportional to the investigated soil volume/layer in relation to the magnitude of the events (both in terms of wetting and drying). Accordingly, the most intense reaction is observable in the data collected by the shortest probes (in Figure 7 represented by the 0.30 m lengths), which investigate the shallowest layer. Moreover, during the winter months, the presence of frozen topsoil can lead to TDR measurement complications due to the TDR's inability to detect simultaneously liquid and ice contents in frozen soils within the bandwidth included between a few megahertz and 1.5 GHz [118,119]. These complications, if not properly considered, can cause measurement errors proportional to the soil depth reached by the frost in relation to the depth investigated by the probes.

ERT, GPR and TDR single measurement campaign. Figure 8 shows how visible is the patch of soil where irrigation had taken place with respect to the non-irrigated zone. However, at this measurement scale, it is hard to identify the depth of infiltration as the resolution cannot be any better (and is actually worse) than the electrode spacing (0.6 m in this case).

In order to capture in more detail the changes in moisture content with depth, ERT measurements were collected also using 24 electrodes (spaced 20 cm), set on a transect perpendicular to the sprinkler's line over a total length of 4.6 m. Surface data were obtained with a Syscal-Pro resistivimeter (IRIS Instruments, Orleans, France) and a dipole–dipole skip zero (dipoles with minimal distance equal to one electrode spacing) acquisition scheme. Time-lapse measurements were taken periodically, using a dipole–dipole skip 0 scheme and full acquisition of reciprocals to estimate the data error level [96,120,121]. Consistently, the data inversion used an Occam inversion approach, as implemented in the Profile R/R2/R3 software package, accounting for the error level estimated from the data themselves [114]. Results are shown in Figures 9–11.

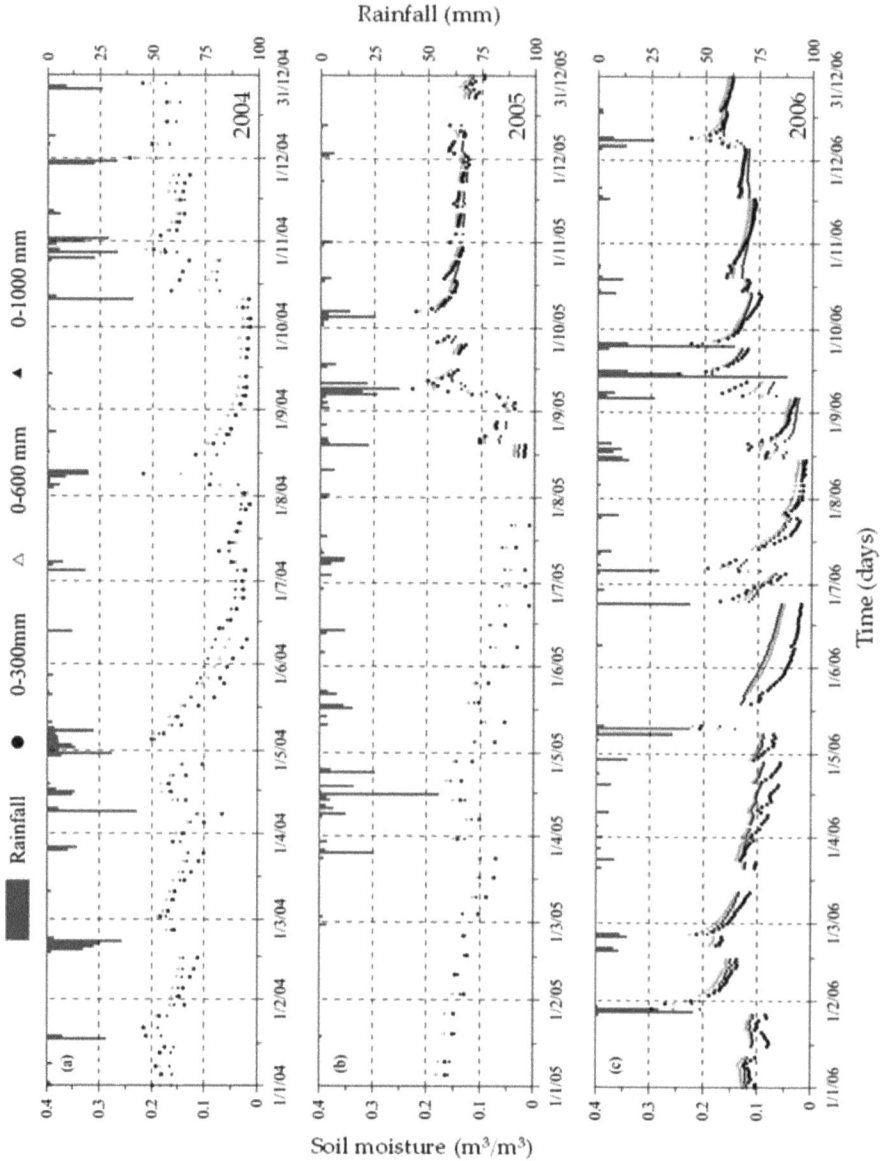

Figure 7. Grugliasco site, 2004–2006. Daily soil water content and rainfall data within the layers 0–0.3, 0–0.6, and 0–1.0 m from 1 January 2004 to 31 December 2006. Since 2005 data have been automatically sampled at hourly step intervals.

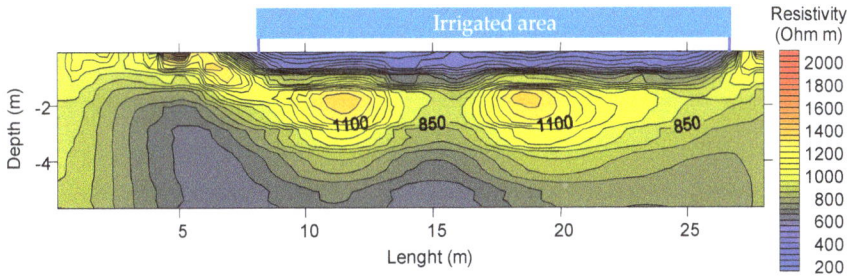

Figure 8. Grugliasco site, 28 August 2009 (hour 17:30). Absolute ERT image at the end of irrigation along the longitudinal ERT line (see Figure 3 for location).

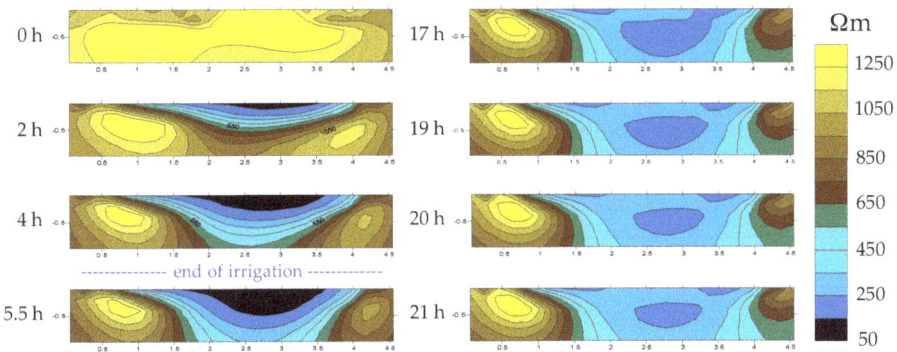

Figure 9. Grugliasco site, August 2009. Absolute ERT images along the transverse line (see Figure 3 for location).

Figure 9 shows the time evolution of absolute resistivity during and after irrigation. The changes in resistivity are particularly clear as the soil profile was initially very dry (at the end of August). The same patterns are apparent, with a more distinctive character, in the relative time-lapse images shown in Figure 10. Here a comparison is also made between a simple approach, where ratios of the absolute images in Figure 9 are taken pixel by pixel, and the more accurate approach based upon ratio inversion, e.g., [96]. In this specific case the two approaches yield practically the same results as, given the geometry of the system and the corresponding resistivity variations over time (the ERT line is transverse to a long stripe where resistivity changes in a fairly homogeneous manner), no 3D effect is present. Note that in general a simple pixel-by-pixel ratio of absolute images does not lead to acceptable results, as the absolute images are 2D inversions in what is actually a 3D space, and ratio or difference inversions shall be used [122].

Finally, Figure 11 shows the estimated soil moisture patterns in the ERT section, derived from converting resistivity into moisture content by means of a laboratory-calibrated Archie's law [86]. The infiltration front estimated by GPR measurements is also depicted (red dotted lines).

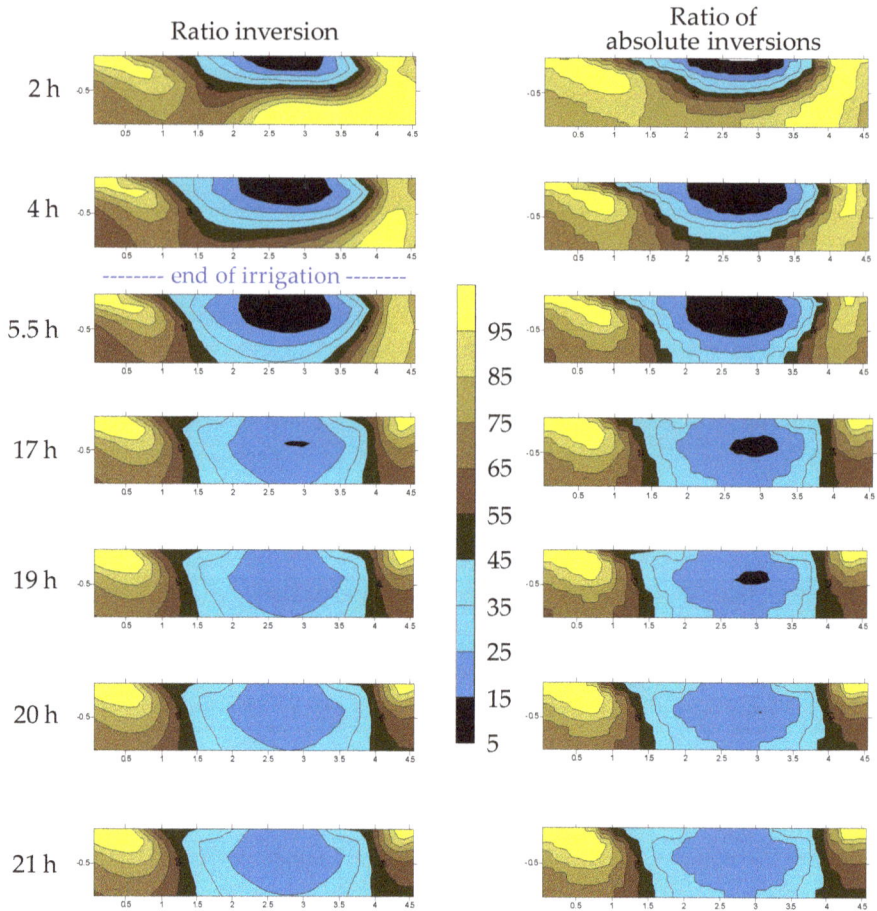

Figure 10. Grugliasco site, August 2009. Relative ERT images along the transverse line (see Figure 3 for location). Both results of ratio inversion and ratios of absolute inversion are shown as percentage of the background (pre-irrigation) values. In this case the two approaches yield very similar results, as the resistivity changes occur homogeneously across the measurement line (no 3D effects).

The performed TDR measurements are depicted in Figure 12 as a function of time and depth. The data are represented as averaged values together with their standard deviations. Initial conditions are dry and homogeneous with soil water contents about 0.08–0.10 m^3/m^3 along all the investigated soil profiles. After the first irrigation step the surface layer immediately reaches values close to saturation (about 0.30 m^3/m^3), but variability is evident between measurements despite the uniform testing conditions. In the following irrigation steps the saturation conditions persist in the surface layer investigated but the variability between measurements progressively decreases. Considering the whole profile, the soil moisture gradually increases along with the irrigation volumes. The following day, starting with the early morning, conditions changed considerably. At the surface, the soil moisture was more uniform and decreased from saturation conditions down to the field capacity (around 0.17 m^3/m^3), while, considering the whole depth investigated, conditions remain the same. All these considerations agree and integrate the information provided by the ERT imaging.

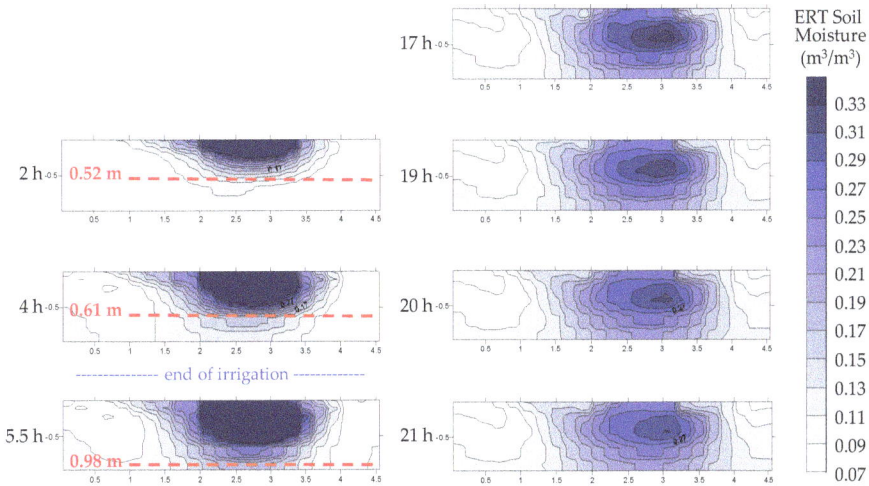

Figure 11. Grugliasco site, August 2009. Moisture content at different times after irrigation, estimated from ERT imaging and laboratory calibrated Archie's law results. A comparison is made against the depth of the infiltration front estimated by GPR measurements (red dotted lines) [32].

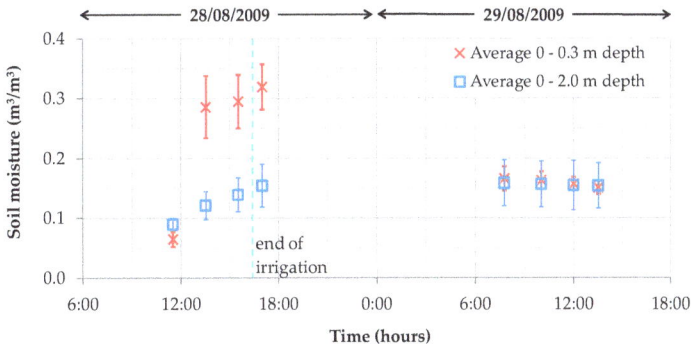

Figure 12. Grugliasco site, August 2009. Moisture contents and their standard deviation at different times after irrigation, computed from TDR measurements.

Usability of TDR for deep soil measurement campaigns. A further interesting application for a low-attenuation porous medium (such as the Grugliasco sandy soil case) is represented by the suitability of TDR technique to investigate very thick layers (down to 4 m). Despite the complications connected with the probes' insertion (where an insertion guide apparatus is essential to prevent geometrical probe deformations), Figure 13 illustrates the waveform detected during a single measurement campaign with soil moisture ranging between 0.11 and 0.14 m^3/m^3 along the profile. Only the surface layer (0–0.5 m depth) shows a higher value, around 0.22 m^3/m^3. Figure 13 highlights that retrieved waveforms are really clear, even for 4 m long probes, and signal reflections are evident and easy toanalyse.

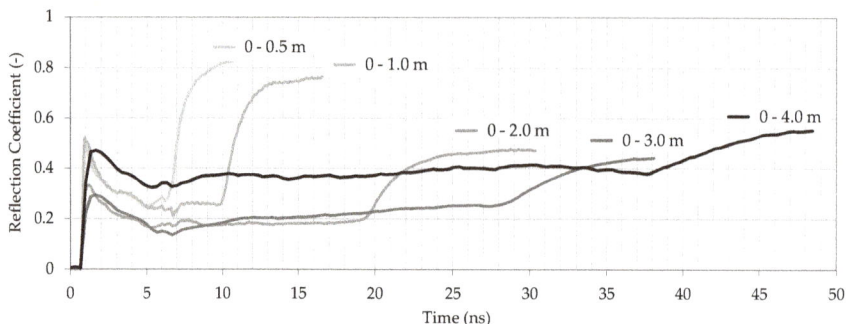

Figure 13. Grugliasco site, 20 May 2010. Waveforms detected with different TDR waveguides lengths vertically inserted into the soil from the surface down to 4 m depth.

4.2. Mountain Permanent Meadow

Long-term soil moisture measurements. In the graph below (Figure 14), the soil volumetric water content—averaged every 30 min—together with rainfall data—collected by the ARPA VdA meteorological station located nearby—are plotted on an hourly base from 1 January to 31 December 2016.

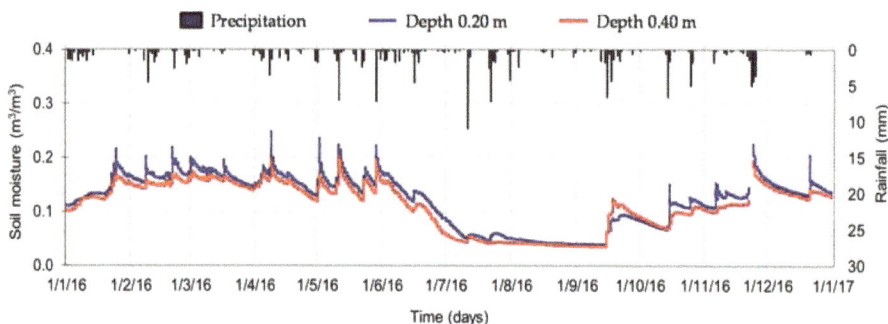

Figure 14. Cogne site, January–December 2016. Precipitation and soil moisture measured at 0.20 and 0.40 m depths.

The volumetric soil moisture collected in the Gimillan site (Figure 4) during the whole year 2016 never exceeded 0.3 m^3/m^3. This threshold can be considered to be in agreement with the literature data for coarse texture soils. Interesting data can be observed at the beginning of the year. In this period the soil is frozen, with a low amount of liquid water content. Then, in the second half of January, a peak occurs without any surveyed rainfall. This behaviour can be ascribed to the soil water (and snow) melting due to the temperature increase, especially considering the south–southeast aspect. Other interesting data can be provided during the summer (in particular from July to the first half of September). In this period, soil moisture is characterized by progressively lower values, with a minimum situated around 0.05 m^3/m^3. In this dry period the correspondence between the occurrence of rainfalls and soil moisture increase is not so evident. This behaviour might be attributed to the effect of the dense vegetation coverage (withered by the water scarcity), driving to interception rather than infiltration into the soil.

With reference to the soil moisture dynamics, fast-acting peaks of soil water increase/decrease can be observed. This attitude can mainly be attributed to the following factors: (1) hydrological properties of the coarse soil texture (e.g., fast infiltration processes; limited water retention/storage);

(2) morphological characteristics of the site, namely the SSE slope aspect, which implies accelerated atmospheric and evaporation dynamics (e.g., solar radiation incidence; mountain/valley breezes). Notwithstanding the abovementioned aspects, it is interesting to highlight that during the surveyed period, the soil moisture values detected at 0.20 m depth were higher than the 0.40 m depth ones, probably due to the dense vegetation and the spatial variability. However, the time dynamic is more enhanced at 20 cm depth.

EMI, ERT and TDR measurement campaigns. EMI surveys at this site are particularly complex given the steep slope and the shrubs (see Figure 4), thus spatial sampling is different at the two dates (as seen in the extent of the mapped areas represented in Figure 15). The most evident feature refers to the strong difference in mean electrical resistivity, with October being much drier than September. Here, the irregular morphology characterized by depressions and bumps, together with the variability of soil and soil-cover vegetation, led to heterogeneous dynamics in time and space, more evident in drier conditions.

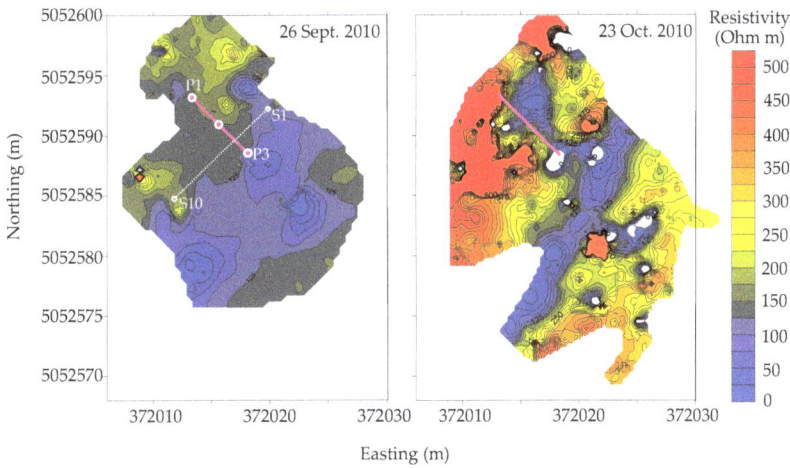

Figure 15. Cogne site, 2010. EMI maps, shown as electrical resistivity values. The strong change between 26 September (more conductive) and 23 ctober (more resistive) is remarkable. The maps represent the mean electrical resistivity from the ground surface to a depth estimated around 1.5 m. The purple line shows the location of the ERT line (results in Figure 16). The white dots represent the two TDR transects: P indicates the progressive 0.75 m depth measurements; S indicates the progressive 0.15 and 0.30 m depth measurements.

It is very informative to compare EMI maps with localized ERT profiles (Figure 16). The latter are taken in reference to the purple line marked in Figure 13. Considering that the values in the EMI maps are obtained in a configuration that gives one value somehow averaged from the ground surface down to 1.5 m depth, i.e., practically involving the entire thickness shown by ERT sections in Figure 16, and given the strong resistivity contrast (both in September and October), this average only has an indicative meaning. Exploring the ERT transects output, note how at both times the shallow soil layer (around 0.5 m thick) is much more resistive than the deeper subsoil. This leads to a very heterogeneous soil profile, so that the EMI maps in Figure 15 only provide a rough mean of the electrical properties averaged over very different values. The ERT images show, consistently with the EMI maps, that in October the site was much drier than in September. More advanced approaches using multi-coil/multi-frequency EMI and depth inversion are, in general, strongly recommended, e.g., [91].

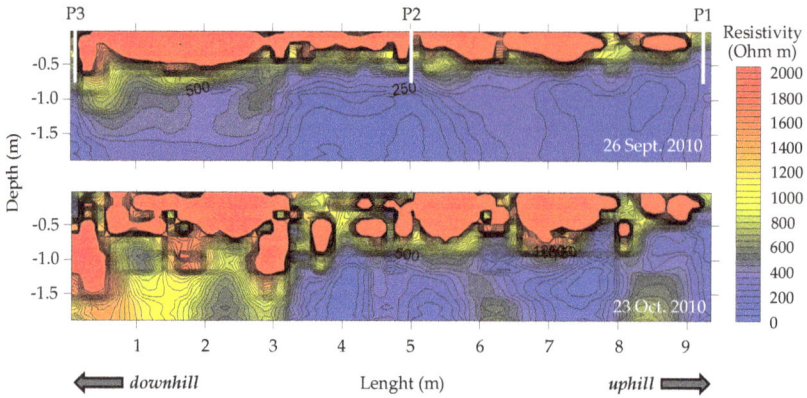

Figure 16. Cogne site, 2010. ERT profiles acquired at the two survey dates along a short profile (see Figure 15 for location). The white lines identified by the P indicate the three 0.75 m depths TDR measurement points. Note that the resistivity scale is much larger in the ERT profiles than in the EMI maps, in order to give a clearer presentation of the variations of resistivity with depth.

Values summarized in Table 2 highlight a general agreement between TDR water content and ERT resistivity behaviour along all the transect ("P" values—see Figure 16 for the exact location). "S" values—see Figure 15 for the transect identification—confirm the water content decrease in the upper layers where the resistivity is higher (as easily seen also in Figure 16) but a strong variability connected with the irregular micro morphology of each measurement point become evident.

Table 2. Cogne site, 2010. TDR soil moisture values measured during the 26 September 2010 EMI and ERT monitoring campaigns. The two TDR transect alignments are depicted in Figure 15.

Probe Code	Transect "P"			Transect "S"											
	P1	P2	P3	S1	S2	S3	S4	S5	S6	S7	S8	S9		S10	
Depth from surface (m)	0.75			0.15	0.15	0.30	0.15	0.15	0.30	0.15	0.15	0.30	0.15	0.15	0.30	0.15	0.15	0.30
Soil moisture (m^3/m^3)	0.41	0.37	0.38	0.33	0.33	0.31	-	0.39	0.22	0.31	0.41	0.40	0.33	0.35	0.34	0.32	0.32	0.27

During the 23 October 2010 EMI and ERT monitoring campaigns, no TDR specific transects were surveyed, however, the CS616 0.10 m depth probe provided the hourly dataset depicted in Figure 15. Despite the data gaps, Figure 17 provides interesting information both on the soil moisture dynamics during the days prior to the measurement, and to the absolute measured soil water content (in agreement with the expected values).

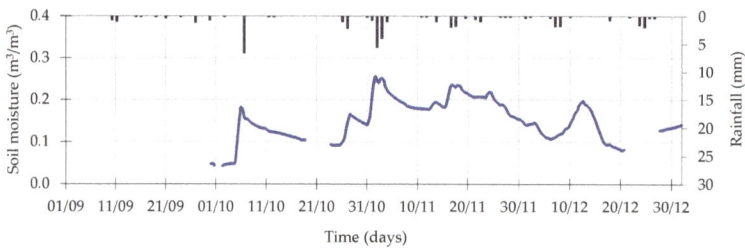

Figure 17. Cogne site, September–December 2010. Precipitation and soil moisture at 0.10 m depth.

4.3. Hilly Vineyard

Long-term soil moisture measurements. A plot of the soil moisture data detected at different depths (namely 0.1, 0.2, 0.3 and 0.4 m below the surface) is depicted in Figure 18, together with the rainfall data.

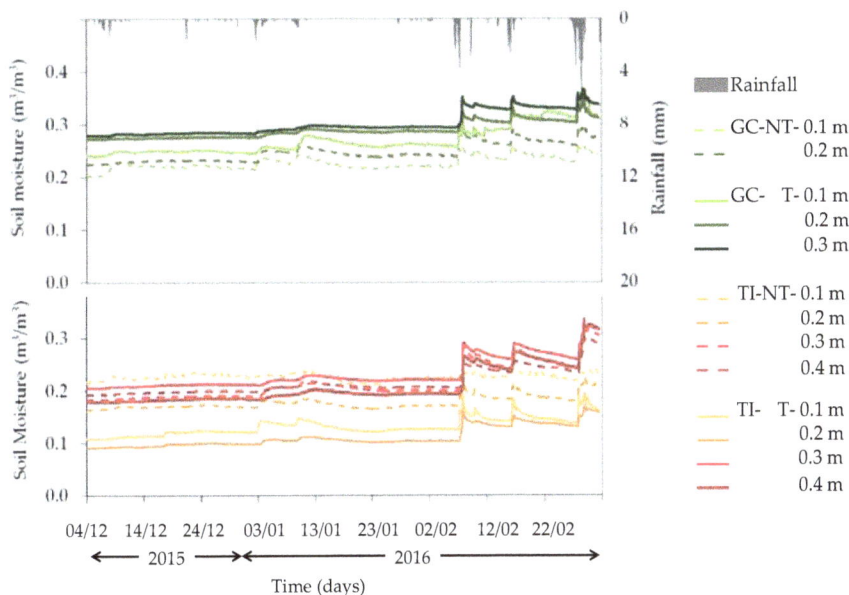

Figure 18. Carpeneto site, winter 2015–2016. Rainfall and soil moisture data collected with an hourly time step (from 0.1 to 0.4 m soil depths). Green lines refer to the Grass Cover condition (GC); red lines refer to the Conventional Tillage condition (TI). Dotted lines refer to the "No Track" condition (NT); solid lines refer to the "Track" condition (T).

The results of the long-term survey (Figure 18) suggest that, under conventional tillage conditions, surface soil moisture in the track position was usually lower than in the central no-track zone. In agreement with Leonard and Andrieux [123] (who found that the major causes in variability of infiltration rates in vineyards are referred to the history of cultivation and the structure of the first soil centimetres), the monitored behaviour can mainly be ascribed to both the agricultural treatment and the soil compaction.

In the grass-covered plot, the soil moisture in the track position was slightly higher than in the central no-track zone along the whole monitoring period. As already observed by Ferrero et al. [124], the tractor traffic in vineyards has a great influence on the spatial variability of soil physical properties, which are strictly related to the topsoil water content. The recurrence of tillage could temporarily decrease this effect, but affects the variability of soil properties over a long period (for more details, see Biddoccu et al. [97]).

EMI and TDR combined campaign. The following Figure 19 show the result obtained the 21 March 2016, from the top of the hill down to the bottom (the last meters of the transects were excluded to avoid measurement distortions due to the presence of a steel grid used to survey the runoff erosion). EMI data were collected in all the rows, starting from the right side (row 18) and moving to the left side (row 0).

In Figure 19, the dots represent the TDR measurement sites along the two transects. The TDR output values are depicted in Figure 20.

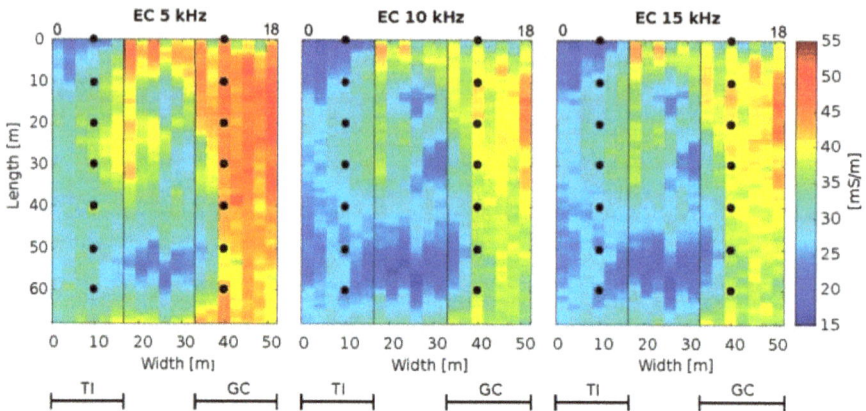

Figure 19. Carpeneto site, 21 March 2016. Electrical conductivity monitored by the EMI equipment and TDR probes locations (black dots) in the testing plot. The three images represent lower depths of investigation, from right to left.

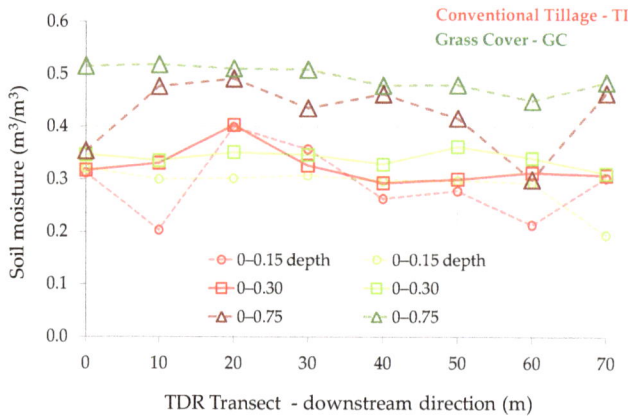

Figure 20. Carpeneto site, 21 March 2016. Soil water content monitored along the two TDR transects (conventional tillage and grass cover) depicted in Figure 19.

The correlation between EC_a and soil moisture is evident (Figures 19 and 20). The results of the EMI measurement campaign show EC_a values along the TI rows lower than those measured along the GC rows. The high value of EC_a along the GC rows can be ascribed to the soil water content, which is higher than in the TI rows, especially in the deeper layers (Figure 20). Moreover, EC_a measured at 5 kHz (highest depth of investigation) is higher than EC_a measured at 10 and 15 kHz, and is in agreement with the soil water content measured by TDR (Figure 20) and also with the long-term measurements (Figure 18). However, the low EC_a values at the bottom of the TI rows (corresponding to the 60 m coordinate) do not completely agree with the TDR water content value: at this point, notwithstanding the lowest TI soil moisture, the EC_a values are not as low as the values detected in similar conditions. Being in the tilled area, this behaviour might be ascribed to the deposition of fine particles eroded from the top toward the bottom.

5. Discussion

Through a comprehensive analysis of the presented case studies, it is possible to evaluate the application pros and cons of the different methods used.

The TDR technique offers wide flexibility in different contexts for both single campaigns and long-term surveys, with excellent temporal and spatial resolution attainable through the automation of measurements. Measurements are rapid, non-destructive and simple to obtain. The higher the number of probes to be acquired simultaneously, the more complex the system will have to be (eventually adopting a multiplexed system). Depending on the monitoring needs, specific experimental layouts can be adopted with vertical and/or horizontal probes of different lengths. In ideal conditions (i.e., low-attenuation porous media with no coarse skeletal material—stones/rocks) TDR probes can be vertically inserted down to unusual depths (e.g., 4 m, as demonstrated in the plain permanent meadow case study) to explore the soil water content/behaviour below the most commonly monitored layers. For these last applications, the adoption of a steel guide during the insertion process is strongly recommended to maintain rods in parallel and prevent geometrical probe deformations. Due to the TDR's inability to detect liquid and ice contents simultaneously (within a bandwidth between a few megahertz and 1.5 GHz), in the case of frozen soils additional measurements with lower frequencies (between 0.1 and 200 kHz [118,119]) would be necessary to avoid underestimations or errors.

Low-frequency sensors directly connected to data loggers combine good accuracy and reliability with a lower cost (and can be used in place of TDR). By their nature they need specific calibration and they suffer from some restrictions related to the constructive features (e.g., probe lengths; more fragility—depending on the types and models), but they are easy to use and efficient in the most common soil moisture investigations.

To determine the spatial distribution of soil moisture, whereas TDR evaluates the specific water content analysing the full signal response (and relating it to the dielectric properties), the ERT system determines the two-dimensional electric resistivity distribution (from surface-based geoelectric measurements and subsequent data inversion). A combination of both methods would lead to better results (e.g., for calibrating the ERT approach). However, determining the soil water content from the resulting electrical resistivity values is more difficult than in the case of the dielectric coefficients using TDR. The inversion procedure can produce mass balance errors, due to a rapid decrease of ERT resolution with depth [40]. This problem may be solved through a coupled hydro-geophysical approach [40,101]. However, if only qualitative assessment of spatial distribution and time variations of soil moisture content are sufficient, an uncoupled approach can be used [32], based on the following steps: (1) inversion of geophysical field data gives the spatial distribution of electrical resistivity; (2) application of a petro-physical relationship to obtain an estimation of moisture content distribution from the electrical resistivity.

It must be noted that electrical resistivity and electrical conductivity, as measured by EMI, are just reciprocal quantities. Thus ERT and EMI measure the same physical parameter, and can be effectively used to complement each other: EMI gives quicker information over large areas, while ERT is more effective at in-depth investigations. In the hilly vineyard test case, for example, EMI measurements allow for identifying differences in EC_a (and thus soil moisture) between the two treatments and in relation to depth, with results that are comparable with the more accurate (and time-consuming) TDR measurements. In the permanent mountain meadow, since EMI measurements are fast (and not invasive), it was possible to map the EC_a spatial variations over the whole field-size plot (also in complex logistical conditions). Some discordances between EMI and TDR measurements could be ascribed to local conditions that can affect TDR local measurements. When possible, a combined EMI and TDR campaign allows us to combine accuracy with wider spatial representativeness of measurements than only TDR campaigns.

Without considering the borehole applications (which have not been analysed in this work), GPR can be used to perform specific campaigns aimed at obtaining soil sub-surface "images" (typically in the range of sub-meters up to tens of meters) referred to the interaction between the transmitted

EM energy and the spatial variation in the complex, frequency-dependent EM properties of the earth materials in the subsurface [125]. For small-scale, site-specific experiments, GPR can be used to monitor the movement of water into the—and through the—subsurface (such as the depth of the infiltration front estimated by GPR measurements in the plain permanent meadow case study). However, the use of GPR systems remains difficult in uneven irregular areas due to the need to drag the instrument to keep it in contact with the soil surface (such as at the permanent mountain meadow testing site). Limitations are also acknowledged in clay-laden soils (such as the hilly vineyard testing site), where the signal penetration into the soil is strongly reduced by the high electrical conductivity.

6. Conclusions

Soil moisture is highly variable at both the spatial and the temporal scale, and soil moisture field measurements are recognized to be fundamental in the integration of in situ, satellite and modelled data. For these reasons, in our work we provide a synoptic view of techniques, supported by case studies, to show that also in very different field conditions (in terms of altitude, land use, and soil type, namely a plain, a mountain meadow and a hilly vineyard) it is possible to estimate time and spatially resolved soil moisture by the same combination of instruments: contact-based methods (i.e., Time Domain Reflectometry—TDR, and two low-frequency probes) for the time resolved, and minimally invasive hydro-geophysical methods (i.e., Electromagnetic Induction—EMI, Ground Penetrating Radar—GPR, and the Electrical Resistivity Tomography—ERT) for the spatially resolved. By doing so, soil moisture dynamics determined by soil's heterogeneity and meteorological events can be efficiently observed and measured. In particular, while plot-scale methods—like EMI, GPR and ERT—provide efficient spatial surveys, local-scale methods—like TDR and other lower-cost soil moisture sensors—provide efficient soil moisture monitoring across time (and space—where properly structured). With less accuracy, and with smaller volume support, the low-frequency sensors can also be used in place of TDR.

An important key future challenge might be represented by the creation of a framework in which the information contained in the local and plot observations is combined with model predictions of soil moisture dynamics, at different spatial scales. Nowadays, this integrated approach would represent the best method to achieve the great potential, especially in hydrological applications, of soil moisture studies.

Acknowledgments: This work has been partially funded by the EU-funded H2020 Project *"ECOPOTENTIAL: Improving Future Ecosystem Benefits through Earth Observations"* (http://www.ecopotential-project.eu) and the Italian National Research Programme 2011–2013 by The Project of Interest *"NextData: a national system for the retrieval, storage, access and diffusion of environmental and climate data from mountain and marine areas"* (http://www.nextdataproject.it/?q=en). The "Agrion—Centro sperimentale per la Vitivinicoltura" (which managed the "Tenuta Cannona" vineyards), is also gratefully acknowledged.

Author Contributions: All co-authors, each within his competence, actively participated in the field campaigns, data processing, and data analysis. Stefano Ferraris is the Grugliasco and Cogne experimental sites scientific coordinator. Eugenio Cavallo coordinated the experiments in Tenuta Cannona. Giulia Raffelli, Stefano Ferraris and Maurizio Previati conceived and designed the paper. The manuscript was written by Maurizio Previati, Giulia Raffelli, Davide Canone, Stefano Ferraris and Giorgio Cassiani. Maurizio Previati, Ivan Bevilacqua, Giulia Raffelli, Giorgio Cassiani and Stefano Ferraris contributed to the figures and the manuscript editing.

Conflicts of Interest: The authors declare no conflict of interest.

References

1. Lin, H. Temporal stability of soil moisture spatial pattern and subsurface preferential flow pathways in the shale hills catchment. *Vadose Zone J.* **2006**, *5*, 317–340. [CrossRef]
2. Tromp-van Meerveld, H.J.; McDonnell, J.J. On the interrelations between topography, soil depth, soil moisture, transpiration rates and species distribution at the hillslope scale. *Adv. Water Resour.* **2006**, *29*, 293–310. [CrossRef]

3. Teuling, A.J.; Hupet, F.; Uijlenhoet, R.; Troch, P.A. Climate variability effects on spatial soil moisture dynamics. *Geophys. Res. Lett.* **2007**, *34*. [CrossRef]

4. Vereecken, H.; Huisman, J.A.; Bogena, H.; Vanderborght, J.; Vrugt, J.A.; Hopmans, J.W. On the value of soil moisture measurements in vadose zone hydrology: A review. *Water Resour. Res.* **2008**, *44*. [CrossRef]

5. Vereecken, H.; Huisman, J.A.; Pachepsky, Y.; Montzka, C.; Van Der Kruk, J.; Bogena, H.; Vanderborght, J. On the spatio-temporal dynamics of soil moisture at the field scale. *J. Hydrol.* **2014**, *516*, 76–96. [CrossRef]

6. Ceaglio, E.; Mitterer, C.; Maggioni, M.; Ferraris, S.; Segor, V.; Freppaz, M. The role of soil volumetric liquid water content during snow gliding processes. *Cold Reg. Sci. Technol.* **2017**, *136*, 17–29. [CrossRef]

7. Bomblies, A. Modeling the role of rainfall patterns in seasonal malaria transmission. *Clim. Chang.* **2012**, *112*, 673–685. [CrossRef]

8. Previati, M.; Bevilacqua, I.; Canone, D.; Ferraris, S.; Haverkamp, R. Evaluation of soil water storage efficiency for rainfall harvesting on hillslope micro-basins built using time domain reflectometry measurements. *Agric. Water Manag.* **2010**, *97*, 449–456. [CrossRef]

9. Canone, D.; Previati, M.; Bevilacqua, I.; Salvai, L.; Ferraris, S. Field measurements based model for surface irrigation efficiency assessment. *Agric. Water Manag.* **2015**, *156*, 30–42. [CrossRef]

10. Canone, D.; Previati, M.; Ferraris, S. Evaluation of Stemflow Effects on the Spatial Distribution of Soil Moisture Using TDR Monitoring and an Infiltration Model. *J. Irrig. Drain. Eng. ASCE* **2017**, *143*, 04016075. [CrossRef]

11. Robinson, D.A.; Campbell, C.S.; Hopmans, J.W.; Hornbuckle, B.K.; Jones, S.B.; Knight, R.; Ogden, F.; Selker, J.; Wendroth, O. Soil moisture measurement for ecological and hydrological watershed-scale observatories: A review. *Vadose Zone J.* **2008**, *7*, 358–389. [CrossRef]

12. Fares, A.; Temimi, M.; Morgan, K.; Kelleners, T.J. In-situ and remote soil moisture sensing technologies for vadose zone hydrology. *Vadose Zone J.* **2013**, *12*. [CrossRef]

13. Brocca, L.; Ciabatta, L.; Massari, C.; Camici, S.; Tarpanelli, A. Soil Moisture for Hydrological Applications: Open Questions and New Opportunities. *Water* **2017**, *9*, 140. [CrossRef]

14. Romano, N. Soil moisture at local scale: Measurements and simulations. *J. Hydrol.* **2014**, *516*, 6–20. [CrossRef]

15. Topp, G.C.; Davis, J.L.; Annan, A.P. Electromagnetic determination of soil water content: Measurements in coaxial transmission lines. *Water Resour. Res.* **1980**, *16*, 574–582. [CrossRef]

16. Topp, G.C. State of the art of measuring soil water content. *Hydrol. Process.* **2003**, *17*, 2993–2996. [CrossRef]

17. Robinson, D.A.; Jones, S.B.; Wraith, J.M.; Or, D.; Friedman, S.P. A review of advances in dielectric and electrical conductivity measurement in soils using time domain reflectometry. *Vadose Zone J.* **2003**, *2*, 444–475. [CrossRef]

18. Robinson, D.A.; Jones, S.B.; Blonquist, J.M.; Friedman, S.P. A physically derived water content/permittivity calibration model for coarse-textured, layered soils. *Soil Sci. Soc. Am. J.* **2005**, *69*, 1372–1378. [CrossRef]

19. Kizito, F.; Campbell, C.S.; Campbell, G.S.; Cobos, D.R.; Teare, B.L.; Carter, B.; Hopmans, J.W. Frequency, electrical conductivity and temperature analysis of a low-cost capacitance soil moisture sensor. *J. Hydrol.* **2008**, *352*, 367–378. [CrossRef]

20. Blonquist, J.M.; Jones, S.B.; Robinson, D.A.; Rasmussen, V.P.; Or, D. Standardizing characterization of electromagnetic water content sensors. *Vadose Zone J.* **2005**, *4*, 1048–1058. [CrossRef]

21. Bogena, H.R.; Huisman, J.A.; Oberdorster, C.; Vereecken, H. Evaluation of a low-cost soil water content sensor for wireless network applications. *J. Hydrol.* **2007**, *344*, 32–42. [CrossRef]

22. Rosenbaum, U.; Huisman, J.A.; Weuthen, A.; Vereecken, H.; Bogena, H.R. Sensor-to-sensor variability of the ECH$_2$O EC-5, TE, and 5TE sensors in dielectric liquids. *Vadose Zone J.* **2010**, *9*, 181–186. [CrossRef]

23. Baker, J.M.; Allmaras, R.R. System for automating and multiplexing soil moisture measurement by time-domain reflectometry. *Soil. Sci. Soc. Am. J.* **1990**, *54*, 1–6. [CrossRef]

24. Heimovaara, T.J.; Bouten, W. A computer-controlled 36-channel time domain reflectometry system for monitoring soil water contents. *Water Resour. Res.* **1990**, *26*, 2311–2316.

25. Dobriyal, P.; Qureshi, A.; Badola, R.; Hussain, S.A. A review of the methods available for estimating soil moisture and its implications for water resource management. *J. Hydrol.* **2012**, *458*, 110–117. [CrossRef]

26. Huisman, J.A.; Hubbard, S.S.; Redman, J.D.; Annan, A.P. Measuring soil water content with ground penetrating radar. *Vadose Zone J.* **2003**, *2*, 476–491. [CrossRef]

27. Serbin, G.; Or, D. Ground-penetrating radar measurement of crop and surface water content dynamics. *Remote Sens. Environ.* **2005**, *96*, 119–134. [CrossRef]

28. Lambot, S.; Weihermüller, L.; Huisman, J.A.; Vereecken, H.; Vanclooster, M.; Slob, E.C. Analysis of air-launched ground-penetrating radar techniques to measure the soil surface water content. *Water Resour. Res.* **2006**, *42*. [CrossRef]

29. Lambot, S.; Slob, E.; Chavarro, D.; Lubczynski, M.; Vereecken, H. Measuring soil surface water content in irrigated areas of southern Tunisia using full-waveform inversion of proximal GPR data. *Near Sur. Geophys.* **2008**, *6*, 403–410. [CrossRef]

30. Busch, S.; Van der Kruk, J.; Bikowski, J.; Vereecken, H. Quantitative conductivity and permittivity estimation using full-waveform inversion of onground GPR data. *Geophysics* **2012**, *77*, H79–H91. [CrossRef]

31. Klotzsche, A.; Van der Kruk, J.; Linde, N.; Doetsch, J.; Vereecken, H. 3-D characterization of high-permeability zones in a gravel aquifer using 2-D crosshole GPR full-waveform inversion and waveguide detection. *Geophys. J. Int.* **2013**, *195*, 932–944. [CrossRef]

32. Rossi, M.; Manoli, G.; Pasetto, D.; Deiana, R.; Ferraris, S.; Strobbia, C.; Cassiani, G. Coupled inverse modeling of a controlled irrigation experiment using multiple hydro-geophysical data. *Adv. Water Resour.* **2015**, *82*, 150–165. [CrossRef]

33. Corwin, D.L.; Lesch, S.M. Apparent soil electrical conductivity measurements in agriculture. *Comput. Electron. Agric.* **2005**, *46*, 11–43. [CrossRef]

34. Robinson, D.A.; Lebron, I.; Kocar, B.; Phan, K.; Sampson, M.; Crook, N.; Fendorf, S. Time-lapse geophysical imaging of soil moisture dynamics in tropical deltaic soils: An aid to interpreting hydrological and geochemical processes. *Water Resour. Res.* **2009**, *45*. [CrossRef]

35. Cassiani, G.; Ursino, N.; Deiana, R.; Vignoli, G.; Boaga, J.; Rossi, M.; Ludwig, R. Non-invasive monitoring of soil static characteristics and dynamic states: A case study highlighting vegetation effects on agricultural land. *Vadose Zone J.* **2012**, *11*. [CrossRef]

36. Brovelli, A.; Cassiani, G. Effective permittivity of porous media: A critical analysis of the complex refractive index model. *Geophys. Prospect.* **2008**, *56*, 715–727. [CrossRef]

37. Brovelli, A.; Cassiani, G. Combined estimation of effective electrical conductivity and permittivity for soil monitoring. *Water Resour. Res.* **2011**, *47*. [CrossRef]

38. Loke, M.H.; Chambers, J.E.; Rucker, D.F.; Kuras, O.; Wilkinson, P.B. Recent developments in the direct-current geoelectrical imaging method. *J. Appl. Geophys.* **2013**, *95*, 135–156. [CrossRef]

39. Vanderborght, J.; Huisman, J.A.; Van der Kruk, J.; Vereecken, H. Geophysical methods for field-scale imaging of root zone properties and processes. In *Soil–Water–Root Processes: Advances in Tomography and Imaging*; Anderson, S.H., Hopmans, J.W., Eds.; SSSA: Madison, WI, USA, 2013; pp. 247–281.

40. Manoli, G.; Rossi, M.; Pasetto, D.; Deiana, R.; Ferraris, S.; Cassiani, G.; Putti, M. An iterative particle filter approach for coupled hydro-geophysical inversion of a controlled infiltration experiment. *J. Comput. Phys.* **2015**, *283*, 37–51. [CrossRef]

41. Hornbuckle, B.K.; England, A.W.; De Roo, R.D.; Fischman, M.A.; Boprie, D.L. Vegetation canopy anisotropy at 1.4 GHz. *IEEE Trans. Geosci. Remote Sens.* **2003**, *41*, 2211–2223. [CrossRef]

42. Saleh, K.; Wigneron, J.P.; Waldteufel, P.; De Rosnay, P.; Schwank, M.; Calvet, J.C.; Kerr, Y.H. Estimates of surface soil moisture under grass covers using L-band radiometry. *Remote Sens. Environ.* **2007**, *109*, 42–53. [CrossRef]

43. Hong, S.; Shin, I. A physically-based inversion algorithm for retrieving soil moisture in passive microwave remote sensing. *J. Hydrol.* **2011**, *405*, 24–30. [CrossRef]

44. Kornelsen, K.C.; Coulibaly, P. Advances in soil moisture retrieval from synthetic aperture radar and hydrological applications. *J. Hydrol.* **2013**, *476*, 460–489. [CrossRef]

45. Reigber, A.; Scheiber, R.; Jager, M.; Prats-Iraola, P.; Hajnsek, I.; Jagdhuber, T.; Horn, R. Very-high-resolution airborne synthetic aperture radar imaging: Signal processing and applications. *Proc. IEEE* **2013**, *101*, 759–783. [CrossRef]

46. Zreda, M.; Desilets, D.; Ferré, T.P.A.; Scott, R.L. Measuring soil moisture content non-invasively at intermediate spatial scale using cosmic-ray neutrons. *Geophys. Res. Lett.* **2008**, *35*. [CrossRef]

47. Zreda, M.; Shuttleworth, W.J.; Zeng, X.; Zweck, C.; Desilets, D.; Franz, T.; Rosolem, R. COSMOS: The COsmic-ray soil moisture observing system. *Hydrol. Earth Syst. Sci.* **2012**, *16*, 4079–4099. [CrossRef]

48. Rivera Villarreyes, C.A.; Baroni, G.; Oswald, S.E. Integral quantification of seasonal soil moisture changes in farmland by cosmic-ray neutrons. *Hydrol. Earth Syst. Sci.* **2011**. [CrossRef]

49. Bogena, H.R.; Huisman, J.A.; Baatz, R.; Hendricks Franssen, H.J.; Vereecken, H. Accuracy of the cosmic-ray soil water content probe in humid forest ecosystems: The worst case scenario. *Water Resour. Res.* **2013**, *49*, 5778–5791. [CrossRef]

50. Zhu, Z.; Tan, L.; Gao, S.; Jiao, Q. Observation on soil moisture of irrigation cropland by cosmic-ray probe. *IEEE Geosci. Remote Sens. Lett.* **2015**, *12*, 472–476. [CrossRef]

51. Shin, Y.; Mohanty, B.P. Development of a deterministic downscaling algorithm for remote sensing soil moisture footprint using soil and vegetation classifications. *Water Resour. Res.* **2013**, *49*, 6208–6228. [CrossRef]

52. Ines, A.V.M.; Mohanty, B.P.; Shin, Y. An unmixing algorithm for remotely sensed soil moisture. *Water Resour. Res.* **2013**, *49*, 408–425. [CrossRef]

53. Peng, J.; Loew, A.; Zhang, S.; Wang, J.; Niesel, J. Spatial downscaling of satellite soil moisture data using vegetation temperature condition index. *IEEE Trans. Geosci. Remote Sens.* **2016**, *54*, 558–566. [CrossRef]

54. Peng, J.; Niesel, J.; Loew, A. Evaluation of soil moisture downscaling using a simple thermal-based proxy: The REMDHUS network (Spain) example. *Hydrol. Earth Syst. Sci.* **2015**, *19*, 4765–4782. [CrossRef]

55. Calvet, J.-C.; Noilhan, J. From near-surface to root-zone soil moisture using year-round data. *J. Hydrometeorol.* **2000**, *1*, 393–411. [CrossRef]

56. Sabater, J.M.; Jarlan, L.; Calvet, J.C.; Bouyssel, F.; De Rosnay, P. From near-surface to root-zone soil moisture using different assimilation techniques. *J. Hydrometeorol.* **2007**, *8*, 194–206. [CrossRef]

57. Albergel, C.; Rudiger, C.; Pellarin, T.; Calvet, J.-C.; Fritz, N.; Froissard, F.; Suquia, D.; Petitpa, A.; Piguet, B.; Martin, E. From near-surface to root-zone soil moisture using an exponential filter: An assessment of the method based on in-situ observations and model simulations. *Hydrol. Earth Syst. Sci.* **2008**, *12*, 1323–1337. [CrossRef]

58. Manfreda, S.; Brocca, L.; Moramarco, T.; Melone, F.; Sheffield, J. A physically based approach for the estimation of root-zone soil moisture from surface measurements. *Hydrol. Earth Syst. Sci.* **2014**, *18*, 1199–1212. [CrossRef]

59. Tebbs, E.; Gerard, F.; Petrie, A.; DeWitte, E. Emerging and Potential Future Applications of satellite-Based Soil Moisture products. In *Satellite Soil Moisture Retrievals: Techniques and Applications*; Petropoulos, G.P., Srivastava, P., Kerr, Y., Eds.; Elsevier: Amsterdam, The Netherlands, 2016; Volume 19, pp. 379–400.

60. Peng, J.; Loew, A.; Merlin, O.; Verhoest, N.E.C. A review of spatial downscaling of satellite remotely sensed soil moisture. *Rev. Geophys.* **2017**, *55*, 341–366. [CrossRef]

61. Mohanty, B.P.; Cosh, M.H.; Lakshmi, V.; Montzka, C. Soil moisture remote sensing: State-of-the-science. *Vadose Zone J.* **2017**, *16*. [CrossRef]

62. Desilets, D.; Zreda, M. Footprint diameter for a cosmic-ray soil moisture probe: Theory and Monte Carlo simulations. *Water Resour. Res.* **2013**, *49*. [CrossRef]

63. Franz, T.E.; Zreda, M.; Ferré, T.P.A.; Rosolem, R.; Zweck, C.; Stillman, S.; Zeng, X.; Shuttleworth, W.J. Measurement depth of the cosmic ray soil moisture probe affected by hydrogen from various sources. *Water Resour. Res.* **2012**, *48*, W08515. [CrossRef]

64. Köhli, M.; Schrön, M.; Zreda, M.; Schmidt, U.; Dietrich, P.; Zacharias, S. Footprint characteristics revised for field-scale soil moisture monitoring with cosmic-ray neutrons. *Water Resour. Res.* **2015**, *51*, 5772–5790. [CrossRef]

65. Hallikainen, M.T.; Ulaby, F.T.; Dobson, M.C.; El-Rayes, M.A.; Wu, L.K. Microwave dielectric behavior of wet soil-part 1: Empirical models and experimental observations. *IEEE Trans. Geosci. Remote Sens.* **1985**, *1*, 25–34. [CrossRef]

66. Evett, S.R. Soil water measurement by time domain reflectometry. In *Encyclopedia of Water Science*; United States Department of Agriculture (USDA): Bushland, TX, USA, 2003; pp. 894–898.

67. Topp, G.C.; Ferré, T.P.A. Methods for measurement of soil water content: Time domain reflectometry. In *Methods of Soil Analysis, Part 4*; Dane, J.H., Topp, G.C., Eds.; SSSA Book Series No. 5; Soil Science Society of America: Madison, WI, USA, 2002; pp. 434–446.

68. Roth, K.; Schulin, R.; Flühler, H.; Attinger, W. Calibration of time domain reflectometry for water content measurement using a composite dielectric approach. *Water Resour. Res.* **1990**, *26*, 2267–2273. [CrossRef]

69. Herkelrath, W.N.; Hamburg, S.P.; Murphy, F. Automatic, real-time monitoring of soil moisture in a remote field area with time domain reflectometry. *Water Resour. Res.* **1991**, *27*, 857–864. [CrossRef]

70. Topp, G.C.; Davis, J.L.; Annan, A.P. Electromagnetic determination of soil water content using TDR: I. Applications to wetting fronts and steep gradients. *Soil Sci. Soc. Am. J.* **1982**, *46*, 672–678. [CrossRef]

71. Topp, G.C.; Davis, J.L.; Annan, A.P. Electromagnetic determination of soil water content using TDR: II. Evaluation of installation and configuration of parallel transmission lines. *Soil Sci. Soc. Am. J.* **1982**, *46*, 678–684. [CrossRef]

72. Zegelin, S.J.; White, I.; Jenkins, D.R. Improved field probes for soil water content and electrical conductivity measurement using time domain reflectometry. *Water Resour. Res.* **1989**, *25*, 2367–2376. [CrossRef]

73. Heimovaara, T.J. Design of triple-wire time domain reflectometry probes in practice and theory. *Soil Sci. Soc. Am. J.* **1993**, *57*, 1410–1417. [CrossRef]

74. Ferré, P.A.; Rudolph, D.L.; Kachanoski, R.G. Spatial averaging of water content by time domain reflectometry: Implications for twin rod probes with and without dielectric coatings. *Water Resour. Res.* **1996**, *32*, 271–279. [CrossRef]

75. Ferré, P.A.; Knight, J.H.; Rudolph, D.L.; Kachanoski, R.G. The sample areas of conventional and alternative time domain reflectometry probes. *Water Resour. Res.* **1998**, *34*, 2971–2979. [CrossRef]

76. Canone, D.; Previati, M.; Ferraris, S.; Haverkamp, R. A new coaxial time domain reflectometry probe for water content measurement in forest floor litter. *Vadose Zone J.* **2009**, *8*, 363–372. [CrossRef]

77. Adelakun, I.A.; Ranjan, R.S. Design of a Multilevel TDR Probe for Measuring Soil Water Content at Different Depths. *Trans. ASABE* **2013**, *56*, 1451–1460. [CrossRef]

78. Nissen, H.H.; Ferré, T.P.A.; Moldrup, P. Sample area of two- and three-rod time domain reflectometry probes. *Water Resour. Res.* **2003**, *39*, 1289. [CrossRef]

79. Vaz, C.M.; Jones, S.; Meding, M.; Tuller, M. Evaluation of standard calibration functions for eight electromagnetic soil moisture sensors. *Vadose Zone J.* **2013**, *12*. [CrossRef]

80. Campbell, C.S.; Campbell, G.S.; Cobos, D.R.; Bissey, L.L. Calibration and Evaluation of an Improved Low-Cost Soil Moisture Sensor. 2009. Available online: http://www.decagon.com (accessed on 12 September 2017).

81. Logsdon, S.D.; Hernandez-Ramirez, G.; Hatfield, J.L.; Sauer, T.J.; Prueger, J.H.; Schilling, K.E. Soil water and shallow groundwater relations in an agricultural hillslope. *Soil Sci. Soc. Am. J.* **2009**, *73*, 1461. [CrossRef]

82. Parsons, L.R.; Bandaranayake, W.M. Performance of a new capacitance soil moisture probe in a sandy soil. *Soil Sci. Soc. Am. J.* **2009**, *73*, 1378–1385. [CrossRef]

83. Evett, S.R.; Schwartz, R.C. Discussion of "Soil Moisture Measurements: Comparison of Instrumentation Performances" by Ventura Francesca, Facini Osvaldo, Piana Stefano, and Rossi Pisa Paola. *J. Irrig. Drain. Res.* **2011**, *137*, 466–468. [CrossRef]

84. Sakaki, T.; Limsuwat, A.; Illangasekare, T.H. A simple method for calibrating dielectric soil moisture sensors: Laboratory validation in sands. *Vadose Zone J.* **2011**, *10*, 526–531. [CrossRef]

85. Friedman, S.P. Soil properties influencing apparent electrical conductivity: A review. *Comput. Electron. Agric.* **2005**, *46*, 45–70. [CrossRef]

86. Archie, G.E. The electrical resistivity log as an aid in determining some reservoir characteristics. *Trans. AIME* **1942**, *146*, 54–62. [CrossRef]

87. Cassiani, G.; Boaga, J.; Rossi, M.; Fadda, G.; Putti, M.; Majone, B.; Bellin, A. Soil-plant interaction monitoring: Small scale example of an apple orchard in Trentino, North-Eastern Italy. *Sci. Total Environ.* **2016**, *543*, 851–861. [CrossRef] [PubMed]

88. Ortuani, B.; Chiaradia, E.A.; Priori, S.; L'Abate, G.; Canone, D.; Comunian, A.; Giudici, M.; Mele, M.; Facchi, A. Mapping soil water capacity through EMI survey to delineate site specific management units within an irrigated field. *Soil Sci.* **2016**, *181*, 252–263. [CrossRef]

89. Deidda, G.P.; Fenu, C.; Rodriguez, G. Regularized solution of a nonlinear problem in electromagnetic sounding. *Inverse Probl.* **2014**, *30*, 125014. [CrossRef]

90. Díaz De Alba, P.; Rodriguez, G. Regularized inversion of multi-frequency EM data in geophysical applications. In *Trends in Differential Equations and Applications*; Ortegón, F., Gallego, M.V., Redondo, N., Rodríguez Galván, J.R., Eds.; SEMA SIMAI Springer Series; Springer: Cham, Switzerland, 2016; Volume 8, pp. 357–369.

91. Shanahan, P.W.; Binley, A.; Whalley, W.R.; Watts, C.W. The Use of Electromagnetic Induction to Monitor Changes in Soil Moisture Profiles beneath Different Wheat Genotypes. *Soil Sci. Soc. Am. J.* **2015**, *79*, 459–466. [CrossRef]

92. Boaga, J. The use of FDEM in hydrogeophysics: A review. *J. Appl. Geophys.* **2017**, *139*, 36–46. [CrossRef]

93. Everett, M.E.; Meju, M.A. Near-surface controlled-source electromagnetic induction: Background and recent advances. In *Hydrogeophysics*; Rubin, Y., Hubbard, S.S., Eds.; Springer: New York, NY, USA, 2005; pp. 157–183.

94. Binley, A.; Kemna, A. DC resistivity and induced polarization methods. In *Hydrogeophysics*; Rubin, Y., Hubbard, S.S., Eds.; Springer: New York, NY, USA, 2005; pp. 129–156.

95. Binley, A.M.; Cassiani, G.; Deiana, R. Hydrogeophysics—Opportunities and Challenges. *Boll. Geofis. Teor. Appl.* **2010**, *51*, 267–284.

96. Cassiani, G.; Bruno, V.; Villa, A.; Fusi, N.; Binley, A.M. A saline trace test monitored via time-lapse surface electrical resistivity tomography. *J. Appl. Geophys.* **2006**, *59*, 244–259. [CrossRef]

97. Constable, S.C.; Parker, R.L.; Constable, C.G. Occam's Inversion: A practical algorithm for generating smooth models from EM sounding data. *Geophysics* **1987**, *52*, 289–300. [CrossRef]

98. La Brecque, D.J.; Heath, G.; Sharpe, R.; Versteeg, R. Autonomous monitoring of fluid movement using 3-D electrical resistivity tomography. *J. Environ. Eng. Geophys.* **2004**, *9*, 167–176. [CrossRef]

99. Cassiani, G.; Godio, A.; Stocco, S.; Villa, A.; Deiana, R.; Frattini, P.; Rossi, M. Monitoring the hydrologic behaviour of a mountain slope via time-lapse electrical resistivity tomography. *Near Sur. Geophys.* **2009**, *7*, 475–486. [CrossRef]

100. Hinnell, A.C.; Ferré, T.P.A.; Vrugt, J.A.; Huisman, J.A.; Moysey, S.; Rings, J.; Kowalsky, M.B. Improved extraction of hydrologic information from geophysical data through coupled hydrogeophysical inversion. *Water Resour. Res.* **2010**, *46*. [CrossRef]

101. Beff, L.; Günther, T.; Vandoorne, B.; Couvreur, V.; Javaux, M. Three-dimensional monitoring of soil water content in a maize field using Electrical Resistivity Tomography. *Hydrol. Earth Syst. Sci.* **2013**, *17*, 595–609. [CrossRef]

102. Garré, S.; Coteur, I.; Wongleecharoen, C.; Kongkaew, T.; Diels, J.; Vanderborght, J. Non-invasive monitoring of soil water dynamics in mixed cropping systems: A case study in Ratchaburi Province, Thailand. *Vadose Zone J.* **2013**, *12*. [CrossRef]

103. Slater, L.; Binley, A.M.; Daily, W.; Johnson, R. Cross-hole electrical imaging of a controlled saline tracer injection. *J. Appl. Geophys.* **2000**, *44*, 85–102. [CrossRef]

104. Cassiani, G.; Binley, A. Modeling unsaturated flow in a layered formation under quasi-steady state conditions using geophysical data constraints. *Adv. Water Resour.* **2005**, *28*, 467–477. [CrossRef]

105. Linde, N.; Binley, A.; Tryggvason, A.; Pedersen, L.B.; Revil, A. Improved hydrogeophysical characterization using joint inversion of cross-hole electrical resistance and ground-penetrating radar traveltime data. *Water Resour. Res.* **2006**, *42*. [CrossRef]

106. Binley, A.; Cassiani, G.; Middleton, R.; Winship, P. Vadose zone flow model parameterisation using cross-borehole radar and resistivity imaging. *J. Hydrol.* **2002**, *267*, 147–159. [CrossRef]

107. Zuecco, G.; Borga, M.; Penna, D.; Canone, D.; Previati, M.; Ferraris, S. Towards improved understanding of land use effect on soil moisture variability: Analysis and modeling at the plot scale. *Procedia Environ. Sci.* **2013**, *19*, 456–464. [CrossRef]

108. Baudena, M.; Bevilacqua, I.; Canone, D.; Ferraris, S.; Previati, M.; Provenzale, A. Soil water dynamics at a midlatitude test site: Field measurements and box modeling approaches. *J. Hydrol.* **2012**, *414*, 329–340. [CrossRef]

109. Jones, S.B.; Wraith, J.M.; Or, D. Time domain reflectometry measurement principles and applications. *Hydrol. Processes* **2002**, *16*, 141–153. [CrossRef]

110. Doolittle, J.A.; Minzenmayer, F.E.; Waltman, S.W.; Benham, E.C.; Tuttle, J.W.; Peaslee, S.D. Ground-penetrating radar soil suitability map of the conterminous United States. *Geoderma* **2007**, *141*, 416–421. [CrossRef]

111. Paniconi, C.; Ferraris, S.; Putti, M.; Pini, G.; Gambolati, G. Three-dimensional numerical codes for simulating groundwater contamination: FLOW3D, flow in saturated and unsaturated porous media. In *Pollution Modeling*; CMP: Boston, MA, USA, 1994; Volume 1, pp. 149–156.

112. Or, D.; Jones, S.B.; VanSchaar, J.R.; Humphries, S.D.; Koberstein, R.L. WinTDR v.6.1: A Windows-based Time Domain Reflectometry Program for Measurement of Soil Water Content and Electrical Conductivity—User Manual. Utah State Univ. Soil Physics Group, Logan. 2004. Available online: https://psc.usu.edu/ou-files/wintdr/Introduction.pdf (accessed on 15 September 2017).

113. Ursino, N.; Cassiani, G.; Deiana, R.; Vignoli, G.; Boaga, J. Measuring and Modelling water related soil—Vegetation feedbacks in a fallow plot. *Hydrol. Earth Syst. Sci.* **2014**, *18*, 1105–1118. [CrossRef]

114. Binley, A. Resistivity Inversion Software. 2013. Available online: http://www.es.lancs.ac.uk/people/amb/ Freeware/freeware.htm (accessed on 12 September 2017).

115. Biddoccu, M.; Ferraris, S.; Opsi, F.; Cavallo, E. Long-term monitoring of soil management effects on runoff and soil erosion in sloping vineyards in Alto Monferrato (North-West Italy). *Soil Tillage Res.* **2016**, *155*, 176–189. [CrossRef]

116. Biddoccu, M.; Ferraris, S.; Pitacco, A.; Cavallo, E. Temporal variability of soil management effects on soil hydrological properties, runoff and erosion at the field scale in a hillslope vineyard, North-West Italy. *Soil Tillage Res.* **2017**, *165*, 46–58. [CrossRef]

117. Starr, J.L.; Paltineanu, I.C. Methods for Measurement of Soil Water Content: Capacitance Devices. In *Methods of Soil Analysis: Part 4 Physical Methods*; Dane, J.H., Topp, G.C., Eds.; Soil Science Society of America, Inc.: Madison, WI, USA, 2002; pp. 463–474.

118. Bittelli, M.; Flury, M.; Roth, K. Use of dielectric spectroscopy to estimate ice content in frozen porous media. *Water Resour. Res.* **2004**, *40*, W04212. [CrossRef]

119. He, H.; Dyck, M. Application of multiphase dielectric mixing models for understanding the effective dielectric permittivity of frozen soils. *Vadose Zone J.* **2013**, *12*, 12. [CrossRef]

120. Binley, A.; Ramirez, A.; Daily, W. Regularised image reconstruction of noisy electrical resistance tomography data. In Proceedings of the 4th Workshop of the European Concerted Action on Process Tomography, Bergen, Norway, 6–8 April 1995; pp. 401–410.

121. Monego, M.; Cassiani, G.; Deiana, R.; Putti, M.; Passadore, G.; Altissimo, L. Tracer test in a shallow heterogeneous aquifer monitored via time-lapse surface ERT. *Geophysics* **2010**, *75*, WA61–WA73. [CrossRef]

122. Daily, W.; Ramirez, A.; LaBrecque, D.; Nitao, J. Electrical resistivity tomography of vadose water movement. *Water Resour. Res.* **1992**, *28*, 1429–1442. [CrossRef]

123. Leonard, J.; Andrieux, P. Infiltration characteristics of soils in Mediterranean vineyards in Southern France. *Catena* **1998**, *32*, 209–223. [CrossRef]

124. Ferrero, A.; Usowicz, B.; Lipiec, J. Effects of tractor traffic on spatial variability of soil strength and water content in grass covered and cultivated sloping vineyard. *Soil Till. Res.* **2005**, *84*, 127–138. [CrossRef]

125. Robinson, D.A.; Binley, A.; Crook, N.; Day-Lewis, F.D.; Ferré, T.P.A.; Grauch, V.J.S.; Knight, R.; Knoll, M.; Lakshmi, V.; Miller, R.; et al. Advancing process-based watershed hydrological research using near-surface geophysics: A vision for, and review of, electrical and magnetic geophysical methods. *Hydrol. Process.* **2008**, *22*, 3604–3635. [CrossRef]

water

MDPI

Article

Spatial Patterns and Influence Factors of Conversion Coefficients between Two Typical Pan Evaporimeters in China

Yanzhong Li [1,2], Changming Liu [1] and Kang Liang [1,*]

[1]   Key Laboratory of Water Cycle and Related Land Surface Processes,
     Institute of Geographic Sciences and Natural Resources Research, Chinese Academy of Sciences,
     Beijing 100101, China; liyz_egi@163.com (Y.L.); liucm@igsnrr.ac.cn (C.L.)
[2]   University of Chinese Academy of Sciences, Beijing 100101, China
*   Correspondence: liangk@igsnrr.ac.cn; Tel.: +86-10-6488-9083

Academic Editors: Tommaso Moramarco and Roberto Ranzi
Received: 18 July 2016; Accepted: 20 September 2016; Published: 27 September 2016

Abstract: Pan measurement is a reliable and efficient method for indicating the evaporative demand of the atmosphere. There are several types of pan evaporimeters worldwide, and the estimation of the conversion coefficients (K_p) between them is necessary in hydrologic research. In China, E601B pans were installed at all meteorological stations beginning in 1998. They replaced the 20 cm pans (ϕ20). To fully use the records from the two pans and obtain long-term pan evaporation, the spatial patterns of K_p between ϕ20 and E601B and the factors that influence K_p are investigated based on records from 573 national meteorological stations from 1998 to 2001. In this study, The results show that higher K_p values are found in southwestern regions and lower values are found in northeastern regions during the warm seasons (from May to September), while K_p values are lower during warm seasons than during cold seasons (from October to April the following year). In addition, net radiation was found to be the dominant climate factor that affects variations in K_p, followed by relative humidity and the vapor pressure deficit. This study can improve the benefit of not only the selection of appropriate evaporimeters by meteorological departments, but also of the study of temporal variability and trends in the evaporative demand.

Keywords: conversion coefficients; pan evaporation; eight climate regions in China; 20 cm diameter pan; E601B (E601); evaporative demand

1. Introduction

Evaporation is a key hydrological process [1]. It is an important nexus between the water cycle and energy budget and can further impact regional and global climate [2–4]. The atmospheric evaporative demand can be evaluated based on potential evapotranspiration [5,6] or evaporimeters [7,8] and is regarded as the upper limit of evaporation [9–11]. The evaporative demand of the atmosphere is controlled by radiative and aerodynamic factors, and it can be calculated using meteorological variables, such as radiation, wind speed, air temperature, and humidity [7,10]. However, these meteorological factors are not always available, especially radiation. Instead, a simple and efficient observational measurement device, the pan evaporimeter, is often adopted to accurately quantify local atmospheric evaporation [8,12–14]. Many types of pan evaporimeters have been installed in different countries to measure the evaporative demand [1,15], including Class A evaporimeters in the US and Australia [16], the GGI-3000 in Russia [17], the MO tank in Britain [18], and the 20 cm pan (ϕ20) and E601B in China [13,19]. These devices are used by meteorological and climate scientists, agricultural scientists, and hydrologists [12,20,21]. Although pan evaporation (E_{pan}) measured from various types of evaporimeters cannot completely represent the actual evaporation [22,23], it can provide in-depth insight into the trends and temporal variations in evaporation in the context of climate change [6,24] and anthropogenic interference [25,26].

Due differences between pan evaporimeters (for example, material type, geometric shape, and installation method), observed evaporation can vary greatly, even for evaporimeters in similar environments [27]. Pan evaporation values measured using different evaporimeters over the same time span, cannot be compared directly. Otherwise, considerable uncertainty would be introduced into the results. A conversion must be performed to make the values comparable. It is necessary to determine the conversion coefficient (K_p) between different measurements before using the values in studies of evaporation trends [28,29] or water resources [8]. A number of previous studies discussed K_p related to pan evaporation at the point scale [15,30], regional scale [31,32], and catchment scale [33]. For example, Hong et al. [30] investigated two types of evaporimeters, one buried in the soil and one exposed to the air, at Nansi Lake Station. The results showed that the buried devices, such as E601 and GGI-3000, had larger K_p values (ratio of the 20 m^2 evaporation tank to these evaporimeters, $K_p > 0.98$) compared to the K_p values ($K_p < 0.88$) of an exposed ones, such as Class A and ϕ20. Fu et al. [15] compared the K_p values of 15 types of evaporimeters to that of a 20 m^2 evaporation tank and found that the K_p values of ϕ20 ($K_p = 0.60$) and E601B ($K_p = 1.07$) were distinctly different at the annual scale. Liang et al. [32] discussed the K_p difference between E_{pan} and reference evaporation (ET_{ref}) in the West Songnen Plain of China and found that it varied significantly in space (0.48–0.68) and time at an annual scale. Xu et al. [33] also studied the K_p difference between E_{pan} and ET_{ref} in the Yangtze River basin. They observed higher K_p values in the central region of the basin, which has a relatively lower vapor pressure deficit, and K_p exhibited monthly variations in the three regions of the catchment. Due to the spatial and temporal variabilities in K_p, considerable uncertainties may be associated with using constant K_p values in large regions when conducting climate change research [28,34]. The spatial patterns of K_p and the dominant factors that control its variation must be determined. Revealing the spatial distribution of K_p and its driving factors is imperative and can improve hydrometeteorological studies.

China is an ideal location to study K_p patterns and the factors that influence K_p. Among the various evaporation pans used in China [15,19,35,36], the two most common are the ϕ20 and E601B pans (Figure 1 and Table 1). In 1998, the E601B pans were first installed at all meteorological stations across China, with the aim of replacing the ϕ20 pan. To maintain comparability between the two evaporimeters, simultaneous observations were collected from 1998 to 2001. By 2002, the ϕ20 pans were successfully replaced by E601B pans at all stations, and E601B became the standard evaporimeter for measuring evaporation in China. To effectively use the records from the two pan evaporation devices in long-term studies of evaporation trends (the ϕ20 pan records date to approximately 1951), the K_p value between them must be determined. In addition, China can be divided into eight climatic regions [13,37] due its large spatial extent. There are significant differences in climate between the regions [38]; for example, the temperature increases from north to south, and precipitation increases from northwest to southeast (Table 1). Ren et al. [35] compared the monthly and annual mean K_p values between the two evaporimeters in different provinces from 1998 to 2001. Additionally, Liu et al. [36] compared the K_p values between E601 (similar to E601B) and ϕ20 in several typical cities from 1986 to 1995. However, few studies have focused on K_p differences in different climatic conditions, which may considerably affect K_p. Mapping the spatial distribution of K_p between the two evaporimeters and determining the driving factors of K_p can allow the pan evaporation records in China to be fully used and provide an understanding of the underlying mechanisms of K_p variability.

Figure 1. Four types of evaporation evaporimeters: Class A from the USA (**a**); GGI-3000 from Russia (**b**); and ϕ20 (**c**) and E601B (**d**) from China.

Table 1. Detailed information regarding the four types of evaporimeters.

Evaporimeter Name	Size	Description
Class-A	Area: 12.56 ft^2 (diameter: 4 ft) Depth: 10 in	Supported by a wood frame and the bottom is 5 cm higher than ground, popular in USA.
GGI-3000	Area: 3000 cm^2 Depth: 60 cm (cylinder) C + 8.7 cm (circular cone)	Buried in the ground with rim about 7.5 cm above the ground level, popular in Russian.
ϕ20	Area: 314 cm^2 (diameter: 20 cm) Depth: 10 cm	Rim is 70 cm from the ground, in every meteorological station in China.
E601B	Same as GGI-3000	Buried like GGI-3000, fiberglass material, surrounded by water and soil circle. Installed in every meteorological station from 1998.

To address these issues, we compare the spatial distribution and temporal variability (monthly for one year) of K_p and investigate the factors that influence K_p between the E601B and ϕ20 pans in China. The main objectives of this paper are as follows: (1) investigate the monthly variation and spatial distribution of K_p for the two pan evaporations during the warm (May–September) and cold seasons (October–April); and (2) determine the contributions of several key climatic factors to the variation in K_p. The results of this study can improve the selection of the appropriate pan in different climate regions and provide information for hydrologic research, especially studies of evaporation trends. This study is structured as follows: in Section 2, the datasets and methods used in our study are described; in Section 3, the spatial distribution of K_p is mapped during the warm and cold seasons. Furthermore, the factors that influence K_p are also investigated in this section. The uncertainties and conclusions are shown in Sections 4 and 5.

This paper reports the initial stage of ongoing research work. The ongoing research and planned studies are as follows: (1) constructing a long-term series of pan evaporation records (from the

1950s–present) and investigating the trends in pan evaporation across the country in the context of climate change [5,13]; (2) performing experiments regarding the pan evaporation of $\phi20$ and E601B and developing some novel approaches or formulations to explain the mechanisms of K_p [20,39]; and (3) investigating and quantifying the effects of various climatic factors on K_p using a modified PenPan model [27] and a partial differential method [6].

2. Data and Methods

2.1. Information and Measurements from Different Evaporimeters

Different types of evaporimeters have been used to measure the evaporative demand of the atmosphere [15]. Two typical pans are recommended by the World Meteorological Organization [40]: Class A from the USA (Figure 1a) and GGI-3000 from Russia (Figure 1b). However, these two evaporimeters are not widely installed in China, but the $\phi20$ and E601B (Figure 1c,d) evaporimeters are commonly used. Additional information regarding the evaporimeters is listed in Table 1.

The $\phi20$ pan with a screen to prevent bird drinking is made of metal and placed at a height of 70 cm (Figure 1c). This evaporimeter is weighed at 20:00 each day using a high-precision weighbridge, and it is then refilled with water to a depth of 20 mm. The daily evaporation rate can be calculated from the following equation:

$$E = \frac{W_1 - W_2}{31.4} + P \tag{1}$$

where E is the pan evaporation rate (mm·day^{-1}); W_1 and W_2 are the pan weights of the previous and current measurements, respectively (g·day^{-1}); P is the total precipitation (mm·day^{-1}, including rain and snow); and 31.4 is the weight of 1 mm of water in the pan (g·mm^{-1}). The E601B is made of fiberglass and has a relatively lower heat transfer to the surrounding area. The evaporation from this evaporimeter should be relatively close to the evaporation recorded from a moderately-sized water body (such as a 20 m^2 evaporation tank) [15,19]. The daily evaporation rate can be calculated using Equation (2):

$$E = P + (H_1 - H_2) \tag{2}$$

where E and P are the same as in Equation (1); and H_1 and H_2 are the water depths of E601B on the previous and current day, respectively. The water depths of H can be read directly from the indicator installed on the stylus holder (Figure 1d).

2.2. Meteorological and Evaporation Data

Daily meteorological data and evaporation data from the evaporimeters at 573 national meteorological stations were obtained from the China Meteorological Administration (CMA, Figure 2) for a four-year period (January 1998–December 2001). The data were quality controlled. Records that were missing for less than three consecutive days were interpolated based on the nearest data. For gaps of more than three days, the missing data were replaced using a simple linear regression based on the nearest stations. Finally, 573 of the 756 meteorological stations with continuous records were selected. Monthly data, which were summed from daily values, were used in the following sections.

To identify the effects of different climatic factors on the K_p values of the two pans, the entire region was divided into eight climatic regions [13,37]: northwest (NW), north center (NC), North China plain (NCP), northeast (NE), east (E), southeast (SE), southwest (SW), and Tibet plateau (TP). According to the aridity index (AI, Table 2), NW, NC, NCP, and NE are humid regions (AI > 1.0), while E, SE, and SW are non-humid regions (AI < 1.0). The climate characteristics of each region are listed in Table 2. Pan evaporation is an integrated process affected by various climate factors, such as the net radiation (R_n), wind speed (U_2), air temperature (T_{mean}), vapor pressure deficit (VPD), relative humidity (RH), and elevation (Elev) [19,21,41]. The U_2 at a height of 2 m was derived from a height of 10 m according to a logarithmic wind speed profile. R_n was calculated from the difference between the incoming net shortwave radiation (R_{ns}) and the outgoing net longwave radiation (R_{nl}) [21].

The coefficients, a_s and b_s, recommended by Allen et al. [21] to calculate the solar radiation (R_s) were not used. The optimized coefficients calibrated using the 116 solar radiation stations were adopted in our paper. These coefficients can significantly reduce the uncertainty associated with radiation [42].

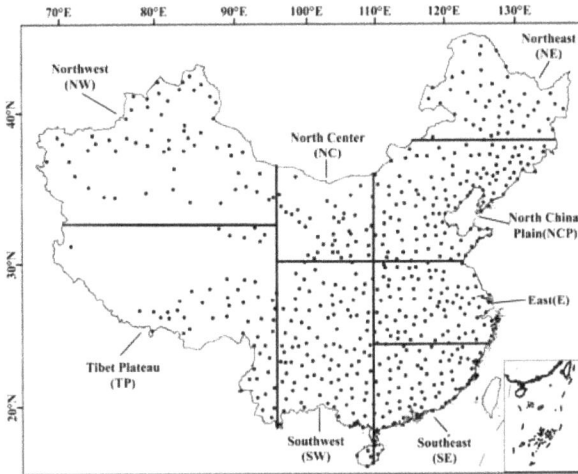

Figure 2. Locations of meteorological stations in China. The eight regional abbreviations are shown in the figure. This map was modified from Liu et al. [13].

Table 2. Climate factor characteristics in the eight climatic regions from 1998 to 2001.

Region	VP (kPa)	T (°C)	U_2 (m/s)	R_s (MJ/m²/Day)	P_{re} (mm/a)	ET_{ref} (mm/a)	AI
NW (n = 57)	0.76	8.45	1.65	15.60	137.03	1078.22	7.87
NC (n = 39)	0.82	9.12	1.72	15.99	265.80	1088.08	4.09
NCP (n = 119)	1.04	9.12	1.91	13.95	494.26	938.91	1.90
NE (n = 48)	0.80	2.52	2.14	13.45	442.05	779.81	1.76
E (n = 83)	1.71	16.43	1.62	11.96	1164.64	902.78	0.78
SE (n = 67)	2.17	20.35	1.30	11.85	1728.42	948.11	0.55
SW (n = 103)	1.70	16.27	1.16	12.02	1179.91	886.25	0.75
TP (n = 57)	1.32	11.78	0.95	15.34	1035.75	941.24	0.91

Notes: VP is the vapor pressure; T is the air mean temperature; and U_2 is the wind speed; All of these parameters are taken at a height of 2 m above the ground. R_s is the solar radiation; P_{re} is the precipitation; ET_{ref} is the reference evapotranspiration calculated using the Penman-Monteith method [21,43]; and AI is the aridity index derived from the ratio of ET_{ref} to P_{re}.

Evaporation data for the E601B pan are not available during the winter in most northern regions in China because the water in the pan freezes. Thus, the year was divided into a warm season (May–September) and a cold season (October–April). This is reasonable because the warm season accounted for most of the annual evaporation (>60%), even in the southern areas; therefore, these months were the most important.

2.3. Calculations and Analysis Method for the K_p

K_p is defined as the ratio of the E601B evaporation to the ϕ20 pan evaporation:

$$K_p = \frac{E_{E601B}}{E_{20}} \tag{3}$$

where E_{E601B} is the monthly evaporation rate of the E601B pan; and E_{20} is the ϕ20 evaporation rate for the same month. Spatial interpolation using the Kriging method was performed in ESRI ArcGIS 10.0

software (Redlands, CA, USA) with the spatial analysis toolbox [6,44] to obtain the spatial distributions of monthly and annual K_p values.

A Pearson's correlation analysis, multiple stepwise regression analysis, and our knowledge of the potential physical driving factors (T_{mean}, RH, VPD, U_2, Elev, and R_n) were used to analyze the potential factors that influence K_p variation. The Pearson's correlation coefficients between K_p and selected climatic factors were calculated using SPSS Statistics 20 (SPSS Inc., Chicago, IL, USA). Stepwise multiple linear regressions were developed between K_p and the potential dominant factors to find the best predictors and the independent explanatory ability of each selected climatic factor based on the spatial variation in K_p.

3. Results and Analysis

3.1. Spatial Patterns of K_p for the Two Pan Evaporimeters

3.1.1. Spatial Distribution of Pan Evaporation during the Warm Season

The spatial distributions of $\phi20$ and E601B evaporation during the warm season from 1998 to 2001 at the national scale are shown in Figure 3a,b, respectively. When the characteristics of the eight climatic regions were combined (Table 2), the differences between the two pan evaporations were spatially related to the variations in meteorological factors. For the warm season from 1998 to 2001, there were statistically significant differences ($p < 0.001$) between the $\phi20$ and E601B pan evaporations, and the annual values were 1014 mm and 597 mm, respectively (Table 3). The spatial distributions of the pan evaporations of $\phi20$ and E601B generally exhibited the same spatial pattern: a clear decreasing trend from northwest to southeast. The highest values were found in the NW region, while the lowest values were found in the TP, NW and NE regions. The $\phi20$ pan evaporation varied from 739 mm to 5573 mm with an average of 1613 mm, and the E601B pan evaporation varied from 470 mm to 2831 mm with an average of 921 mm.

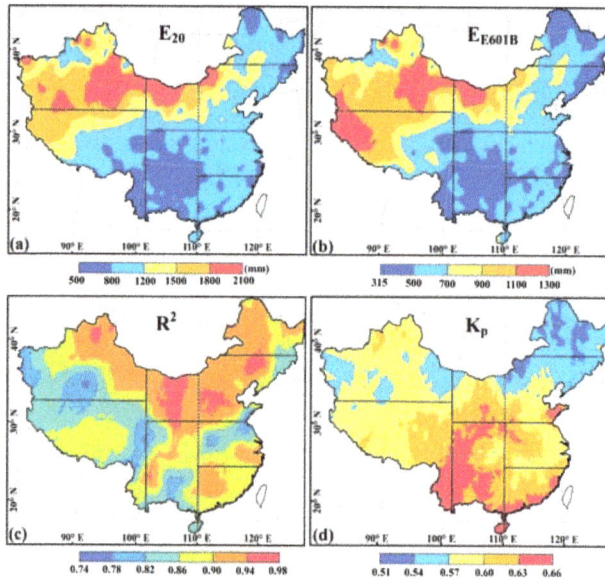

Figure 3. Spatial distributions of $\phi20$ (**a**) and E601B (**b**) evaporation; the determination coefficient R^2 (**c**); and the conversion coefficient K_p (**d**) between the two pan evaporations during the warm season (May–September).

Table 3. Statistical information regarding E601B and ϕ20 pan evaporation (E_{E601B}, E_{20}), the determination coefficient (R^2), and the conversion coefficient (K_p) between the two pans in eight climatic regions during the warm season and three regions in the cold season. The minimum and maximum values are given in parentheses.

Seasons	Regions	E_{20}	E_{E601B}	R^2	K_p
	NW	1613 (739, 5573)	921 (470, 2831)	0.89 (0.67, 0.97)	0.58 (0.50, 0.67)
	NC	1363 (832, 2382)	814 (477, 1436)	0.94 (0.86, 0.99)	0.61 (0.51, 0.70)
	NCP	1123 (648, 2166)	635 (386, 1107)	0.91 (0.52, 0.98)	0.57 (0.47, 0.92)
	NE	1004 (641, 1425)	543 (329, 803)	0.91 (0.68, 0.98)	0.55 (0.46, 0.62)
Warm	E	872 (645, 1236)	524 (401, 792)	0.87 (0.67, 0.97)	0.60 (0.51, 0.75)
	SE	847 (468, 1033)	526 (304, 740)	0.89 (0.74, 0.97)	0.62 (0.53, 0.77)
	SW	752 (498, 1159)	472 (323, 708)	0.87 (0.57, 0.98)	0.63 (0.49, 0.89)
	TP	890 (521, 1766)	620 (328, 1288)	0.84 (0.48, 0.97)	0.62 (0.51, 0.74)
	National	1014 (468, 5573)	597 (304, 2831)	0.89 (0.48, 0.99)	0.60 (0.46, 0.92)
	E	543 (402, 681)	345 (278, 419)	0.93 (0.83, 0.98)	0.66 (0.55, 0.80)
Cold	SE	666 (403, 1224)	444 (250, 819)	0.90 (0.74, 0.97)	0.68 (0.58, 0.80)
	SW	692 (312, 1466)	440 (218, 889)	0.92 (0.76, 0.99)	0.67 (0.49, 0.80)
	Humid	664 (312, 1466)	427 (218, 921)	0.92 (0.74, 0.99)	0.66 (0.49, 0.81)

3.1.2. Spatial Distribution of the Correlation Coefficient during the Warm Season

To quantify the performance of the E601B pan in capturing the spatial distribution and temporal variation of the ϕ20 pan evaporation, the temporal determination coefficients (R^2) of the two pan evaporation series during the warm season were calculated at each station, and the spatial distributions of R^2 were mapped using the Kriging interpolation method [44]. The results showed that the two pan evaporations had high R^2 values in each region (Figure 3c) and varied significantly in one month (Figure 4). The mean value of R^2 in China can reach 0.89, indicating that the E601B values capture the variation in ϕ20 fairly well. This is not surprising because the two pans were influenced by the same meteorological conditions, such as solar radiation, wind speed, temperature, etc. Thus, it may be reasonable to convert ϕ20 evaporation values to E601B evaporation values by multiplying a constant.

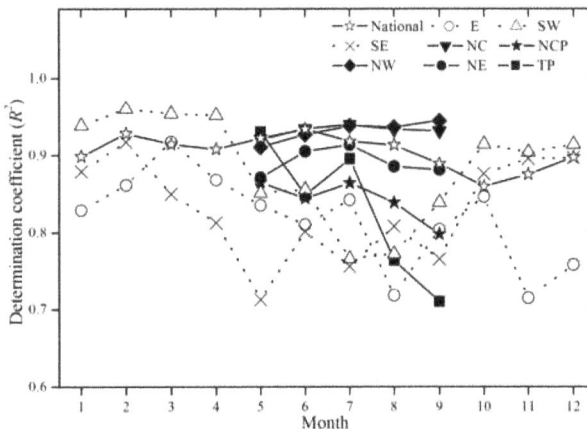

Figure 4. Monthly variation in the coefficient of determination (R^2) values of E601B and ϕ20 pan evaporation in the eight climate regions. The R^2 values in the three humid regions (E, SE, and SW) are illustrated by dotted lines, and the R^2 values in the four non-humid regions and TP region are illustrated by solid lines during the warm season (May–September). The national R^2 is illustrated by the solid black line with open star symbols.

Despite the overall high R^2 between $\phi 20$ and E601B in China, large differences were also found in the eight climatic regions. The R^2 values in the northern regions were generally higher than those in the southern regions, which indicated that the variabilities in the two evaporimeters became more uniform as evaporation increased. The average values of R^2 ranged from 0.89 to 0.94 in the northern regions. R^2 was highest in the NC region, and it varied between 0.86 and 0.99, with an average of approximately 0.94 (Table 3). By contrast, the average values of R^2 in the southern regions (E, SE, and SW) ranged from 0.87 to 0.89 with smaller deviations. The monthly R^2 values in the southern regions were high in the cold season and low in the warm season. The R^2 values in Northern China were high and stable in the warm season. The lowest average R^2 value (0.84) was found in the TP region, and it exhibited a decreasing trend from May to September (Figure 4). This R^2 value in TP may be caused by special climatic conditions. For example, the TP region is the highest plateau in the world with an average elevation of 4000 m, and it is also known as the "Third Pole" of the earth [45].

3.1.3. Spatial Distribution of K_p during the Warm Season

The spatial distribution of K_p during the warm season is shown in Figure 3d. K_p was calculated at each meteorological station independently, and the spatial distribution was obtained using the Kriging interpolation method [44]. The results show that the two pan evaporations exhibited the same spatial distribution (Figure 3a,b), while K_p exhibited significant spatial differences. Overall, K_p varied from 0.46 to 0.92 in China, with an average of approximately 0.60 and a standard deviation of 0.056 (Table 3). K_p was lower in the non-humid northern regions ($K_p < 0.60$) than in the humid southern regions ($K_p > 0.60$), which indicated that the bias in the evaporations of the two pan evaporations was smaller in the humid area during the warm season. Thus, the pan K_p between the two evaporimeters varied substantially, and this variation was larger at low latitudes and smaller at high latitudes. Researchers have documented that the additional heat absorbed by the pan wall has an important effect on K_p [46,47]. Additionally, the $\phi 20$ device can intercept more solar radiation at high latitudes (e.g., NE and NW) than at low latitudes (e.g., SW and SE) due to the solar zenith angle difference. The extra absorbed heat is subsequently transferred into the water through the pan wall, which increases the evaporation rate of the $\phi 20$ pan. The K_p pattern (Figure 3d) generally reflected this process. Additionally, pronounced differences between K_p values in the eight climatic regions were also detected. The smallest K_p values were found in the NE region, and the average K_p ranged from 0.46 to 0.62 with an average of 0.55. The largest values were observed in the SW and TP regions, both with averages of 0.63. The spatial distribution of K_p was similar to that noted by Chen, Gao [19], who compared $\phi 20$ evaporation to reference evapotranspiration in China. Therefore, the regional differences in K_p must be considered to obtain accurate evaporation data series when the evaporation values from the two pans are adopted to determine evaporation trends.

3.1.4. Monthly Variation in K_p during the Warm Season

In addition to the spatial differences, the K_p values of the two evaporimeters also exhibited significant temporal variability. The variation in K_p in eight regions during the warm season is shown in Figure 5. The results are as follows: (1) overall, the monthly average values of K_p in China increased during the warm season from approximately 0.58 in May to 0.63 in September, and the increase was especially rapid from June to August; (2) except in humid regions, K_p varied throughout the warm season with a similar increasing pattern (solid line); (3) in the humid regions, K_p decreased from May to June and then increased until August or September (dotted line), which confirmed the finding of Allen et al. [21] that K_p was high in humid environments; and (4) the lowest values of the entire warm season ($K_p = 0.54$) occurred in the NE region with mean monthly variation from 0.51 to 0.58. The largest values varied from the SW region in May to the TP region in June and July and to SW region again in August and September.

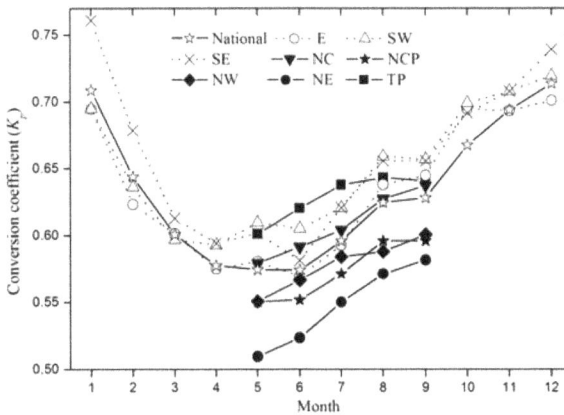

Figure 5. Monthly variation in the conversion coefficient (K_p) of E601B and ϕ20 pan evaporation in the eight climate regions. The K_p values in the three humid regions (E, SE, and SW) are illustrated by dotted lines, and the K_p values in the four non-humid regions and TP region are illustrated by solid lines during the warm season (May–September). The national K_p is illustrated by the solid black line with open star symbols.

3.1.5. Spatial Pattern and Temporal Variability of K_p in Humid Regions during the Cold Season

In humid regions (E, SE, and SW), the annual precipitation is larger than 1000 mm and the aridity index is less than 1.0 (Table 2). Due to the relatively high minimum temperature, there are no frozen periods throughout the whole year. Therefore, the evaporation records of E601B are still available during the cold season (October–April). The spatial pattern of evaporation (E_{20}, E_{E601B}) and the R^2 and K_p between the two pans during the cold seasons are shown in Figure 6. As was the case in the warm season (Figure 3), the evaporation from the two pans exhibited similar spatial patterns during the cold season in humid areas, increasing from the central to marginal regions (Figure 6a,b) and averaging 664 mm and 427 mm annually for the ϕ20 and E601B pans, respectively (Table 3). The R^2 values also increased from the central to marginal regions and were relatively high, with an average of 0.92. This trend indicated that the E601B evaporimeter could capture the variation in ϕ20 better during the cold season than during the warm season.

Although the two pan evaporations had similar spatial patterns, the spatial variations in K_p during the cold season were not constant. The highest values occurred in the northwest and southeast parts of the humid region, indicating larger differences between the two evaporations, while the lowest values occurred in the northeast and southwest areas, indicating smaller differences between the two evaporations. Overall, the K_p values ranged from 0.49 to 0.81 (with a mean of 0.66 ± 0.057) in the humid regions. These values were significantly different from those during the warm season ($p < 0.001$). Of the three regions, the K_p values in the E region were significantly lower than those in the others regions ($p < 0.01$), and there was no significant difference between the values in the SE and SW regions. The highest K_p values appeared in the SW region during the warm season and in the SE region during the cold season.

The K_p between the two pans varied significantly during the cold season (October–April of the following year): (1) the K_p increased from October to the maximum value ($K_p \approx 0.71$) in December or January of the following year, and then decreased until April ($K_p \approx 0.58$); and (2) in the three regions, K_p exhibited the same pattern of variations. The variation of K_p at the annual timescale showed an increase from approximately May to December and then a decrease until April. The monthly variation pattern was most similar to that found by Fu et al. (2009), whose study was conducted at an experimental evaporation station in the E region. The average value of K_p during the cold season (0.66 ± 0.057) was greater than that during the warm season (0.62 ± 0.047), indicating a smaller difference between the two pan evaporations during the cold season. The average value of K_p in all months in the humid regions was 0.64, with a standard deviation of 0.057.

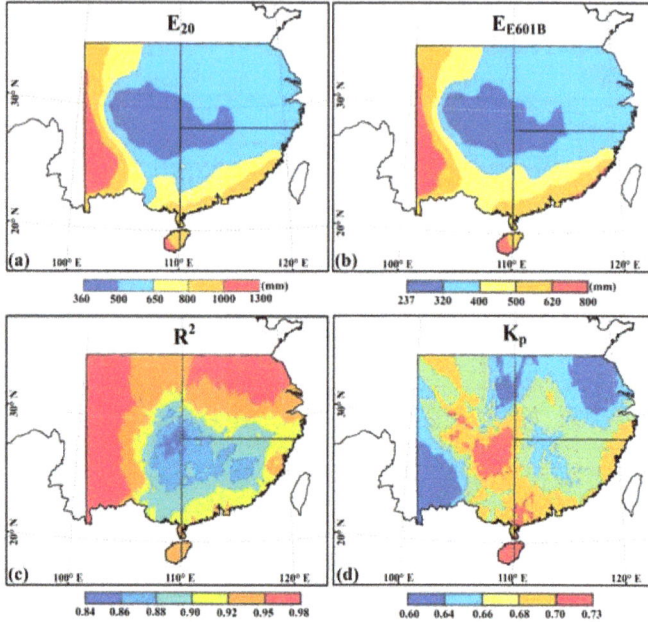

Figure 6. Spatial distributions of $\phi20$ (**a**) and E601B (**b**) evaporation; the correlation coefficient R^2 (**c**); and the conversion coefficient K_p (**d**) between E601B and $\phi20$ evaporation in the cold season (October–April).

As noted above, except for in the TP region, the monthly variability in K_p exhibited a similar pattern in both the warm and cold seasons, which indicated that some specific factors may control the variability in the K_p value. Therefore, meteorological factors were quantitatively explored to identify those that made the greatest contributions to K_p variability.

3.2. Potential Factors That Influence K_p

The Pearson correlation coefficients (r) and significance levels between K_p and several climatic factors are listed in Table 4. K_p was generally negatively correlated with R_n, VPD, U_2, Elev, and T_{mean}. The relationships between K_p and the climatic factors were similar to those noted by Xu et al. [33] in the Yangtze River basin, China. They investigated the K_p of $\phi20$ and the Penman-Monteith reference evapotranspiration; and found that a high conversion coefficient was associated with a relatively high RH and low U_2. At the national scale, K_p was significantly negatively correlated ($p < 0.01$) with R_n ($r = -0.4$) and VPD ($r = -0.33$), and positively correlated with RH ($r = 0.33$) (Table 4). As climate change has occurred in recent decades [48], the pan evaporation paradox [12,13] associated with the variation in solar radiation has gained increasing attention at the global [49,50] and regional scales [51,52]. The change in net radiation (R_n) is expected to strongly affect K_p because of the high correlation between K_p and R_n; therefore, it should be considered in studies of evaporation trends that use different pan measurements. However, at the regional scale, the r values exhibited a significant geographic distribution. In five of the eight climatic regions, NCP, NE, E, SE, and SW, the K_p values had the highest correlation with R_n, followed by RH and VPD. In the other three regions, the K_p values were more associated with RH (especially in the NC and TP regions, with *r* values of 0.46 and 0.42, respectively), and the following factors were found in VPD and T_{mean} (Table 4). Therefore, further analysis must be performed to quantitatively separate the effects of different climate variables on K_p values.

Table 4. Pearson's correlation between the conversion coefficients (K_p) and various climate factors, and the independent explanatory powers of each variable based on K_p variation derived from the stepwise multiple linear regressions (%, in parentheses). T_{mean} is the mean air temperature, RH is the relative humidity, VPD is the vapor pressure deficit, U_2 is the wind speed at a height of 2 m, Elev is the elevation, and R_n is the net radiation. The dominant factors are highlighted in bold.

Variables and Regions	T_{mean}	RH	VPD	U_2	Elev	R_n	Combined
NW (*n* = 1528)	−0.04 (0.3)	**0.34** [a] **(11.2)**	−0.14 [a] (0.5)	−0.04 (0.8)	−0.12 [a] (1.4)	−0.12 [a] (1.2)	- (15.2)
NC (*n* = 984)	−0.21 [a]	**0.46** [a] **(21)**	−0.4 [a] (0.1)	−0.04 (4.5)	0.1 [b]	−0.36 [a] (5.8)	- (32)
NCP (*n* = 3065)	−0.03 (3.3)	0.21 [a] (0.8)	−0.19 [a]	−0.04 [b] (0.4)	−0.08 [a] (0.5)	**−0.24** [a] **(5.5)**	- (10.4)
NE (*n* = 973)	0.04 (8.6)	0.22 [a]	−0.16 [a]	−0.16 [a]	−0.04 (0.6)	**−0.3** [a] **(8.8)**	- (17.7)
E (*n* = 3858)	−0.3 [a] (3)	0.17 [a] (0.9)	−0.33 [a] (2.8)	−0.15 [a]	0.04 [b] (0.4)	**−0.43** [a] **(18.1)**	- (25.2)
SE (*n* = 3207)	−0.33 [a] (0.3)	−0.03 (1.1)	−0.23 [a] (0.3)	0.06 [b] (0.7)	0.01	**−0.35** [a] **(12.2)**	- (14.5)
SW (*n* = 4688)	−0.23 [a] (0.5)	0.25 [a] (3.9)	−0.32 [a] (2.2)	−0.11 [a]	−0.05 [a] (0.3)	**−0.35** [a] **(12.2)**	- (19.1)
TP (*n* = 1408)	0.3 [a]	**0.42** [a] **(17.7)**	−0.21 [a]	−0.06 [b] (1.9)	−0.21 [a] (2.8)	0.06 [b] (0.4)	- (22.6)
National (*n* = 19,848)	−0.17 [a] (0.2)	0.33 [a] (4.4)	−0.33 [a] (1.4)	−0.15 [a] (0.3)	−0.06 [a] (1.3)	**−0.4** [a] **(15.5)**	- (23.1)

Notes: [a] Significant at the 0.01 level; [b] significant at the 0.05 level; - is no value.

The percent contributions of each climate factor to K_p variation were analyzed using multiple stepwise regression with K_p as the dependent variable and the six climatic factors as the independent variables (Table 4). The climate factors together explained approximately one quarter of the K_p variation in China. Of the six climatic factors, R_n had the most explanatory power, with an independent explanation percentage of 15.5%, followed by RH and VPD (4.4% and 1.4%, respectively). The combined explanation percentage was 23.1%, which implied that more complicated mechanisms affect the variations in K_p. Similar to the *r* values, the highest explanatory percentages in five of the eight regions were found associated with R_n (the NCP, NE, E, SE, and SW regions) and ranged from 5.5% (NCP region) to 18.1% (E region), while RH in the other regions ranged from 11.2% (NW region) to 21% (NC region). The greatest combined explanation of the climatic factors was found in the NC region (32%), followed by the E (25.2%) and TP (22.6%) regions, and the weakest explanatory power was found in the NCP region (10.4%).

The two evaporimeters at each station experience the same macroclimate, so the microclimate differences between them can be magnified by the regional difference in climatic factors. For example, the volume and depth of the water in the pans are also important to K_p variability [16]. The difference in the water volume in the two pans caused the water temperature to increase at different rates. Due to the relatively low heat capacity, the water temperature of $\phi 20$ increases faster during the day and also decreases faster at night than that of the E601B pan. This phenomenon may be significant in regions with high temperature differences, such as the NE and NW regions (Table 3). By contrast, the E601B, compared to the $\phi 20$, has no effect because it was buried in the ground and surrounded by water (Figure 1). Similarly, Lim et al. [47], based on an energy balance experiment, found that the K_p between a Class A pan and steady state lake evaporation was mainly dependent on the additional radiation absorbed by the pan wall. The wall of the ϕ20 also had a large area based on the ratio of the wall to the water surface (Figure 1c). Thus, the wall absorbed additional radiation and transferred heat to the water, which increased the evaporation [27]. In addition, on nights without solar radiation, the appreciable storage of heat within the pan wall (Class A or ϕ20) may have caused additional evaporation [27], while E601B had a relatively small evaporation because it was buried in the soil. A negative correlation between K_p and R_n was found in this paper (Table 4), and this finding has also been confirmed by previous experimental data [15,41,53]. In addition, wind speed can decline rapidly with decreasing height due to higher surface roughness and friction near ground [21]. Therefore, a small ϕ20 evaporimeter located at a higher elevation and exposed to the faster wind speed should have more evaporation than the E601B evaporimeter, resulting in a small K_p. This condition can be amplified by regional differences in wind speed. For example, the average wind speed is high in the NE and NCP regions (Table 2); thus, the ϕ20 evaporation could be larger than that of E601, resulting in smaller K_p values than in other regions during the warm season (Table 3).

In conclusion, several factors, including water temperature, vapor pressure, wind speed, turbulence, heat transfer, and heat storage, can affect pan evaporation. The combined influence of various factors produces different spatial patterns and monthly variations in K_p. However, further analysis of the contribution of each factor to K_p requires rigorous experimental investigation of the physics of pan evaporation [47,53] or more reasonable evaporation models [14].

4. Uncertainties

Uncertainties of the spatial distribution and monthly variation of K_p existed in this study, as well as the explanatory powers of the climatic factors. First, although the spatial distribution of K_p is useful for evaporation studies, it should be noted that only larger R^2 values indicate a high confidence of converting the ϕ20 evaporation to E601B evaporation. Therefore, in combination with the spatial distribution of R^2 (Figures 3a and 4), it was reasonable to obtain E601B evaporation by multiplying K_p by the ϕ20 evaporation in the northern part of China, such as in the NC, NCP, and NE regions. However, there are uncertainties in the TP and NW regions due to the relatively sparse distribution of monitoring stations in such large areas (Figure 2); thus, K_p should be used cautiously in these regions. Second, the K_p value was calculated based on the assumption that K_p remains constant over time. This may not be true in some places, and its further testing of the inter-annual variability is required. In addition, the climatic factors that were selected to explain K_p may not be independent. For example, VPD is a function of RH and temperature [21].

Despite these uncertainties, this study presented the spatial distribution of K_p and investigated the dominant climatic factors. Our findings were similar to those of previous studies [19,33,41,53]. Therefore, more attention should be paid to the spatial and temporal variations in K_p before using evaporation data from different evaporimeters. Some researchers have documented that evaporation from Class A (Figure 1a), GGI-3000 (Figure 1b), and ϕ20 (Figure 1c) evaporimeters had relatively small K_p values compared those of 20 m^2 evaporation tanks [8,15,30]. By contrast, E601B evaporation had a high value of K_p and was close to the free water surface evaporation [15]. Researchers have compared various types of evaporimeters at the station scale in the E and NCP regions of China [15,30]. They found that the E601B, compared to other evaporimeters, had a larger conversion coefficient and smaller coefficient of variation with 20 m^2 evaporation tanks, indicating that the evaporation from E601 was much closer to evaporation from the free water surface. The Class A and ϕ20 pans, which are exposed to the air and can absorb additional radiation via the wall, had small conversion coefficients with 20 m^2 evaporation tanks [8,47]. Therefore, it is reasonable to substitute E601B for ϕ20 in China, not only because of its stability in different climate regions but also because of its relatively small difference relative to evaporation from large water bodies.

5. Conclusions and Suggestions

This paper analyzed the spatial distribution, correlation coefficient (R^2), and conversion coefficient (K_p) of evaporation from two typical pans in eight climate regions in China. The main conclusions are as follows:

(1) During the warm season, the spatial evaporation patterns of the two pans were similar and showed increasing trends from the southeastern to northwestern regions of China. The R^2 values were relatively high and ranged from 0.48 to 0.99 with an average of 0.89, which indicates that the E601B pans accurately captured the variation in ϕ20 evaporation. The K_p values showed significant spatial variability across China and varied from 0.46 to 0.92 with a mean of 0.60. The highest and lowest K_p values were found in the southwestern (SW and TP) and NE regions, respectively. Generally, K_p increased from May to September in all of the regions, especially during the summer (June, July and August). In the humid regions (the E, SE, and SW regions), the values of R^2 and K_p were higher during the cold season (means of 0.92 and 0.66, respectively) than during the warm season. The monthly K_p values at the annual scale had a unimodal distribution. They increased from May to December or January the following year, and then decreased until April.

(2) The Pearson correlations and the explanatory powers of the variables were calculated using the multiple stepwise regression method. R_n was the dominant climatic factor for the variation in K_p, exhibiting the best correlation ($r \approx 0.4$, $p < 0.01$) and highest independent explanatory power (approximately 15%) in five of the eight regions, followed by RH and VPD. The combined explanation percentage of all of the variables was 23.1%, and significant differences in explanation percentages were observed in the eight climate regions.

Although the contributions of climate factors to K_p variability were identified by the stepwise multiple linear regression method, the selected factors in this study had relatively weak explanatory powers. Further experiments must be performed to investigate the other potential mechanisms that affect K_p, such as the water vapor pressure at the surface and in the air, pan size, wind speed [53], and the energy balance around evaporimeters [47]. Based on this initial stage of this research, ongoing research and planned experiments, new approaches and evaporation equations [14,41] will be explored using the large and valuable meteorological records available in China. Quantification of the contributions of climatic factors to K_p and determination of the mechanisms of K_p variability can improve the understanding of long-term trends in pan evaporation across China.

Acknowledgments: This research was financially supported by Natural Science Foundation of China (No. 41330529, 41501032, 41571024). We will give many thanks to Wilfried Brutsaert and Guobin Fu for their good suggestions to the manuscript.

Author Contributions: This paper was designed and performed by all the authors. Yanzhong Li wrote the draft of the paper. Changming Liu provided the data and financial support. Kang Liang provided the idea and detailed directed the writing.

Conflicts of Interest: The authors declare no conflict of interest.

References

1. Brutsaert, W. *Evaporation into the Atmosphere: Theory, History and Applications*; Springer: Heidelberg, Germany, 1982; Volume 1.
2. Brutsaert, W.; Parlange, M. Hydrologic cycle explains the evaporation paradox. *Nature* **1998**, *396*, 30. [CrossRef]
3. Hetherington, A.M.; Woodward, F.I. The role of stomata in sensing and driving environmental change. *Nature* **2003**, *424*, 901–908. [CrossRef] [PubMed]
4. Jung, M.; Reichstein, M.; Ciais, P.; Seneviratne, S.I.; Sheffield, J.; Goulden, M.L.; Bonan, G.; Cescatti, A.; Chen, J.; De Jeu, R.; et al. Recent decline in the global land evapotranspiration trend due to limited moisture supply. *Nature* **2010**, *467*, 951–954. [CrossRef] [PubMed]
5. Zhang, Y.; Liu, C.; Tang, Y.; Yang, Y. Trends in pan evaporation and reference and actual evapotranspiration across the Tibetan Plateau. *J. Geophys. Res. Atmos.* **2007**, *112*, 1103–1118. [CrossRef]

6. Li, Y.; Liang, K.; Bai, P.; Feng, A.; Liu, L.; Dong, G. The spatiotemporal variation of reference evapotranspiration and the contribution of its climatic factors in the Loess Plateau, China. *Environ. Earth Sci.* **2016**, *75*, 1–14. [CrossRef]

7. Hobbins, M.; Wood, A.; Streubel, D.; Werner, K. What Drives the Variability of Evaporative Demand across the Conterminous United States? *J. Hydrometeorol.* **2012**, *13*, 1195–1214. [CrossRef]

8. Stanhill, G. Is the Class A evaporation pan still the most practical and accurate meteorological method for determining irrigation water requirements? *Agric. For. Meteorol.* **2002**, *112*, 233–236. [CrossRef]

9. Alkhafaf, S.; Wierenga, P.J.; Williams, B.C. Evaporative Flux from Irrigated Cotton as Related to Leaf Area Index, Soil Water, and Evaporative Demand. *Agron. J.* **1978**, *70*, 912–917. [CrossRef]

10. Azorin-Molina, C.; Vicente-Serrano, S.M.; Sanchez-Lorenzo, A.; McVicar, T.R.; Morán-Tejeda, E.; Revuelto, J.; El Kenawy, A.; Martín-Hernández, N.; Tomas-Burguera, M. Atmospheric evaporative demand observations, estimates and driving factors in Spain (1961–2011). *J. Hydrol.* **2015**, *523*, 262–277. [CrossRef]

11. Donohue, R.J.; McVicar, T.R.; Roderick, M.L. Assessing the ability of potential evaporation formulations to capture the dynamics in evaporative demand within a changing climate. *J. Hydrol.* **2010**, *386*, 186–197. [CrossRef]

12. Roderick, M.L.; Farquhar, G.D. The cause of decreased pan evaporation over the past 50 years. *Science* **2002**, *298*, 1410–1411. [PubMed]

13. Liu, X.; Luo, Y.; Zhang, D.; Zhang, M.; Liu, C. Recent changes in pan-evaporation dynamics in China. *Geophys. Res. Lett.* **2011**, *38*, 142–154. [CrossRef]

14. Singh, V.; Xu, C. Evaluation and generalization of 13 mass-transfer equations for determining free water evaporation. *Hydrol. Process.* **1997**, *11*, 311–323. [CrossRef]

15. Fu, G.; Liu, C.; Chen, S.; Hong, J. Investigating the conversion coefficients for free water surface evaporation of different evaporation pans. *Hydrol. Process.* **2004**, *18*, 2247–2262. [CrossRef]

16. Brouwer, C.; Heibloem, M. *Irrigation Water Management: Irrigation Water Needs. Training Manual*; Food and Agriculture Organization of the United Nations: Rome, Italy, 1986.

17. Golubev, V.S.; Lawrimore, J.H.; Groisman, P.Y.; Speranskaya, N.A.; Zhuravin, S.A.; Menne, M.J.; Peterson, T.C.; Malone, R.W. Evaporation changes over the contiguous United States and the former USSR. *Geophys. Res. Lett.* **2001**, *53*, 323–324. [CrossRef]

18. Symons, G.J. Evaporators and evaporation. *Br. Rainfall* **1867**, *7*, 9–10.

19. Chen, D.; Gao, G.; Xu, C.Y.; Guo, J.; Ren, G. Comparison of the Thornthwaite method and pan data with the standard Penman-Monteith estimates of reference evapotranspiration in China. *Clim. Res.* **2005**, *28*, 123–132. [CrossRef]

20. Brustaert, W. Evaluation of some practical methods of estimating evapotranspiration in arid climates at low latitudes. *Water Resour. Res.* **1965**, *2*, 187–191.

21. Allen, R.G.; Pereira, L.S.; Raes, D.; Smith, M. *Crop Evapotranspiration. Guidelines for Computing Crop Water Requirements*; FAO Irrigation and Drainage Paper 56; Food and Agriculture Organization of the United Nations: Rome, Italy, 1998.

22. Zuo, H.; Chen, B.; Wang, S.; Guo, Y.; Zuo, B.; Wu, L.; Gao, X. Observational study on complementary relationship between pan evaporation and actual evapotranspiration and its variation with pan type. *Agric. For. Meteorol.* **2016**, *222*, 1–9. [CrossRef]

23. Liu, W.; Wang, L.; Zhou, J.; Li, Y.; Sun, F.; Fu, G.; Li, X.; Sang, Y.F. A worldwide evaluation of basin-scale evapotranspiration estimates against the water balance method. *J. Hydrol.* **2016**, *538*, 82–95. [CrossRef]

24. Gao, G.; Xu, C.Y.; Chen, D.; Singh, V.P. Spatial and temporal characteristics of actual evapotranspiration over Haihe River basin in China. *Stoch. Environ. Res. Risk Assess.* **2012**, *26*, 655–669. [CrossRef]

25. Liu, X.; Zhang, D. Trend analysis of reference evapotranspiration in Northwest China: The roles of changing wind speed and surface air temperature. *Hydrol. Process.* **2013**, *27*, 3941–3948. [CrossRef]

26. Mao, D.; Wang, Z.; Li, L.; Song, K.; Jia, M. Quantitative assessment of human-induced impacts on marshes in Northeast China from 2000 to 2011. *Ecol. Eng.* **2014**, *68*, 97–104. [CrossRef]

27. Yang, H.; Yang, D. Climatic factors influencing changing pan evaporation across China from 1961 to 2001. *J. Hydrol.* **2012**, *414–415*, 184–193. [CrossRef]

28. Fu, G.; Charles, S.P.; Yu, J. A critical overview of pan evaporation trends over the last 50 years. *Clim. Chang.* **2009**, *97*, 193–214. [CrossRef]

29. Li, Y.; Liang, K.; Liu, C.; Liu, W.; Bai, P. Evaluation of different evapotranspiration products in the middle Yellow River Basin, China. *Hydrol. Res.* **2016**, *47*. [CrossRef]

30. Hong, J.; Fu, G.; Guo, Z. Experimental research on the water-surface evaporation of Nansi Lake in Shandong Province. *Geogr. Res.* **1996**, *15*, 42–49. (In Chinese)

31. Yang, Y.; Yang, L.; Wang, X.; Liu, J.; Qian, R. Variation Charicteristic and Influence Factors of Pan Evaporation in Xingtai of Hebei Province. *J. Arid Meteorol.* **2013**, *31*, 82–88. (In Chinese)

32. Liang, L.; Li, L.; Liu, Q. Spatio-temporal variations of reference crop evapotranspiration and pan evaporation in the West Songnen Plain of China. *Hydrol. Sci. J.* **2011**, *56*, 1300–1313. [CrossRef]

33. Xu, C.Y.; Gong, L.; Jiang, T.; Chen, D.; Singh, V.P. Analysis of spatial distribution and temporal trend of reference evapotranspiration and pan evaporation in Changjiang (Yangtze River) catchment. *J. Hydrol.* **2006**, *327*, 81–93. [CrossRef]

34. Brutsaert, W. Use of pan evaporation to estimate terrestrial evaporation trends: The case of the Tibetan Plateau. *Water Resour. Res.* **2013**, *49*, 3054–3058. [CrossRef]

35. Ren, Z.; Liu, M.; Zhang, W. Conversion coefficient of small evaporation pan into E-601B pan in China. *J. Appl. Meteorol. Sci.* **2002**, *13*, 508–512. (In Chinese)

36. Liu, X.N.; Wang, S.Q.; Wu, Z.X.; Wang, Y. Comparative analyses on two kinds of observed evaporation data in China. *Q. J. Appl. Meteorol.* **1998**, *9*, 321–328. (In Chinese)

37. Liu, B.; Xu, M.; Henderson, M.; Gong, W. A spatial analysis of pan evaporation trends in China, 1955–2000. *J. Geophys. Res. Atmos.* **2004**, *109*, 1255–1263. [CrossRef]

38. Zhang, Q.; Xu, C.Y.; Chen, Y.D.; Ren, L. Comparison of evapotranspiration variations between the Yellow River and Pearl River basin, China. *Stoch. Environ. Res. Risk Assess.* **2011**, *25*, 139–150. [CrossRef]

39. Brutsaert, W. Equations for vapor flux as a fully turbulent diffusion process under diabatic conditions. *Hydrol. Sci. J.* **1965**, *10*, 11–21. [CrossRef]

40. World Meteorological Organization (WMO). *Guide to Meteorological Instruments and Methods of Observation*, 6th ed.; WMO Rep. 8; WMO: Geneva, Switzerland, 1996.

41. Yu, S.L.; Brutsaert, W. Evaporation from very shallow pans. *J. Appl. Meteorol.* **1967**, *6*, 265–271. [CrossRef]

42. Liu, C.; Zhang, D.; Liu, X.; Zhao, C. Spatial and temporal change in the potential evapotranspiration sensitivity to meteorological factors in China (1960–2007). *J. Geogr. Sci.* **2012**, *22*, 3–14. [CrossRef]

43. Zhang, D.; Liu, X.; Hong, H. Assessing the effect of climate change on reference evapotranspiration in China. *Stoch. Environ. Res. Risk Assess.* **2013**, *27*, 1871–1881. [CrossRef]

44. Stein, M.L. Interpolation of spatial data: Some theory for kriging. *Technometrics* **2015**, *42*, 436–437.

45. Qiu, J. China: The third pole. *Nature* **2008**, *454*, 393–396. [CrossRef] [PubMed]

46. Linacre, E.T. Estimating U.S. Class A Pan Evaporation from Few Climate Data. *Water Int.* **1994**, *19*, 5–14. [CrossRef]

47. Lim, W.H.; Roderick, M.L.; Hobbins, M.T.; Wong, S.C.; Farquhar, G.D. The energy balance of a US Class A evaporation pan. *Agric. For. Meteorol.* **2013**, *182–183*, 314–331. [CrossRef]

48. Intergovernmental Panel on Climate Change (IPCC). *Summary for Policymakes: The Physical Science Basis*; Contribution of Working Group I to the IPCC Fifth Assessment Report Climate Change; IPCC: Geneva, Switzerland, 2013.

49. Wild, M.; Gilgen, H.; Roesch, A.; Ohmura, A.; Long, C.N.; Dutton, E.G.; Forgan, B.; Kallis, A.; Russak, V.; Tsvetkov, A. From dimming to brightening: Decadal changes in surface solar radiation. *Science* **2005**, *308*, 847–850. [CrossRef] [PubMed]

50. Gilgen, H.; Roesch, A.; Wild, M.; Ohmura, A. Decadal changes in shortwave irradiance at the surface in the period from 1960 to 2000 estimated from Global Energy Balance Archive Data. *J. Geophys. Res. Atmos.* **2009**, *114*. [CrossRef]

51. Hayasaka, T.; Kawamoto, K.; Shi, G.; Ohmura, A. Importance of aerosols in satellite-derived estimates of surface shortwave irradiance over China. *Geophys. Res. Lett.* **2006**, *33*, 178–196. [CrossRef]

52. You, Q.; Kang, S.; Flügel, W.A.; Sanchez-Lorenzo, A.; Yan, Y.; Huang, J.; Martin-Vide, J. From brightening to dimming in sunshine duration over the eastern and central Tibetan Plateau (1961–2005). *Theor. Appl. Climatol.* **2010**, *101*, 445–457. [CrossRef]

53. Brutsaert, W.; Yu, S.L. Mass transfer aspects of pan evaporation. *J. Appl. Meteorol.* **1968**, *7*, 563–566. [CrossRef]

water

MDPI

Article

Estimating River Depth from SWOT-Type Observables Obtained by Satellite Altimetry and Imagery

Mohammad J. Tourian *, Omid Elmi, Abolfazl Mohammadnejad and Nico Sneeuw

Institute of Geodesy, University of Stuttgart, 70174 Stuttgart, Germany; elmi@gis.uni-stuttgart.de (O.E.);
abolfazl.mohammadnejad@gmail.com (A.M.); sneeuw@gis.uni-stuttgart.de (N.S.)
* Correspondence: tourian@gis.uni-stuttgart.de; Tel.: +49-711-6858-3474

Received: 14 April 2017; Accepted: 26 September 2017; Published: 30 September 2017

Abstract: The proposed Surface Water and Ocean Topography (SWOT) mission aims to improve spaceborne estimates of river discharge through its measurements of water surface elevation, river width and slope. SWOT, however, will not observe baseflow depth, which limits its value in estimating river discharge especially for those rivers with heterogeneous channel geometry. In this study, we aim to obtain river depths from spaceborne observations together with in situ data of river discharge. We first obtain SWOT-like observables from current satellite techniques. We obtain river water level and slope time series from multi-mission altimetry and effective river width from satellite imagery (MODIS). We then employ a Gauss–Helmert adjustment model to estimate average river depth for 16 defined reaches along the Po River in Italy, for which we use our spaceborne observations in two recognized models for discharge estimation. The average river depth estimates along the Po River are validated against surveyed cross-section information, which shows a generally good agreement in the range of ∼10% relative root mean squared error. Furthermore, we analyzed the sensitivity of error in the estimated river depth to errors of individual parameters. We show that the estimated river depth is less influenced by errors of river width and river discharge, while it is strongly influenced by errors in water level. This result gives a perspective to the SWOT mission to infer river depth by coarse estimates of river width and discharge.

Keywords: river discharge; satellite altimetry; satellite imagery

1. Introduction

River discharge is perhaps the single most important hydrologic quantity representing the amount of available freshwater on landmasses [1]. Despite its importance, the publicly available in situ river discharge database has been declining steadily over the past few years due mainly to economic and political reasons [2–4]. The number of available runoff gauging stations has gone down from about 8000 (pre-1970) to less than 1000 (around the year 2015). Also during this period, the total monitored annual stream flow has dropped accordingly by about 75%. Figure 1 indicates that not only the number of existing stations has decreased, but also in situ measurements over a number of important basins in Africa and South America are no longer available. In fact, most of the active gauges are located over developed countries, and the density of stations is much sparser in the non-industrialized countries [5].

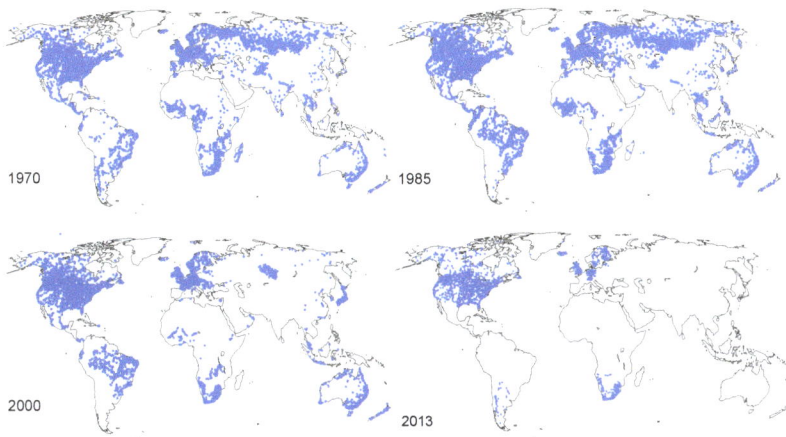

Figure 1. Available stations (updated in March 2017) with discharge data according to the Global Runoff Data Centre (GRDC) database (http://www.bafg.de/GRDC) for different years from 1970–2013.

The potential of spaceborne and airborne techniques for obtaining river discharge estimates has been demonstrated by many studies [1,4,6–11]. The work in [12] found a power law correlation between satellite-derived effective width and discharge using ERS 1 SAR images and simultaneous ground measurements of discharge. Different models are examined in [9] using satellite-derived hydraulic parameters including water surface width, surface velocity, stage and slope. The research in [13] employed TOPEX/Poseidon altimetry data to estimate river discharge based on a relation between water level and discharge. The study in [14] used a data assimilation technique (ensemble Kalman filter) built around a river hydrodynamic model to recover water depth and discharge. The work in [4] introduced a statistical approach to derive discharge from river height through a rating curve based on quantile functions. At-Many-stations Hydraulic Geometry (AMHG) was proposed in [15] where paired coefficients and exponents of At-a-station Hydraulic Geometry (AHG) from many cross-sections of a given river reach are functionally related to one another, following a log-linear relationship. The study in [16] presents the Metropolis–Manning (MetroMan) algorithm to estimate river bathymetry, the roughness coefficient and discharge based on input measurements of river water surface elevation (H) and slope (s) using the Metropolis algorithm in a Bayesian Markov chain Monte Carlo scheme. The research in [17] implemented a form of Manning's equation to retrieve the river low flow bathymetry, roughness and discharge (H_0, K, Q). The Mean-annual Flow and Geomorphology (MFG) algorithm was introduced in [11], which uses the so-called wide-channel approximation leading to a form of Manning's equation that approximates river depth as the difference between water surface elevation and the cross-sectional average river bathymetry. Furthermore, Ref. [11] described the Mean Flow with Constant Roughness (MFCR) approach, which assumes a constant value for the roughness coefficient and uses the Water Balance Model (WBM) mean annual flow estimation. Recently, different methodologies were developed to reconstruct a spatially- and temporally-complete estimate of discharge from a set of observations [18–21]. Moreover, Kalman filters and smoothers have been widely used with river water level data for discharge estimation. The work in [14] assimilated synthetic water elevation observations into a river hydrodynamics model using an ensemble Kalman filter to estimate river discharge. The study in [22] used Kalman filtering to assimilate water level information with simulations from a coupled hydrological and hydrodynamic model to estimate discharge in an ungauged basin scenario. The study in [23] estimated river bathymetry for retrieving river discharge from the Surface Water and Ocean Topography (SWOT) satellite mission using an ensemble Kalman filter-based data assimilation algorithm coupled with a hydrodynamic model. The research in [24] assimilated in situ and radar altimetry data into a large-scale hydrologic-hydrodynamic model for streamflow forecasts in the Amazon.

However, all of these studies are limited to some extent by the characteristics of present-day sensors. The proposed SWOT mission, a joint project between the United States and France, aims to alleviate the limitations and significantly improve spaceborne estimates of river discharge. The SWOT mission will provide measurements of water surface elevation, river width and slope, through which discharge can be estimated for river widths down to 50–100 m [25]. Despite the promising perspective of the mission, one should keep in mind that SWOT will not observe baseflow depth (Figure 2), which limits its value for estimating river discharge especially for those rivers with heterogeneous channel geometry [26]. Therefore, river channel bathymetry is a significant source of uncertainty in estimating discharge from spaceborne measurements [7].

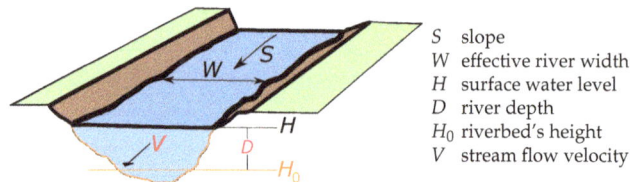

S slope
W effective river width
H surface water level
D river depth
H_0 riverbed's height
V stream flow velocity

Figure 2. A schematic representation of a river with observable parameters (black) and unobservable ones (red, orange) from space. H_0 is the average riverbed's height that does not correspond to the height of the deepest location in the channel section.

Some recent research was devoted to estimate river depth from currently available remote sensing techniques. The work in [27] assessed the feasibility of mapping fluvial systems with passive optical remote sensing and analyzing the physical processes of radiative transfer in shallow stream channels. A spectrally-based approach was suggested in [28] to retrieve depth from passive optical image data. Further, Ref. [29] developed a Forward Image Modeling (FIM) framework for the remote mapping of river morphology via depth retrieval from passive optical images. The research in [30] used a combination of remote sensing imagery and open-channel flow principles to estimate depths for each pixel in an imaged river, which was called the Hydraulically-Assisted Bathymetry (HAB) model. Moreover, Ref. [31] derived mean reach depth from the parameters of the power law establishing the rating curve between water stages from satellite altimetry. A general relation to estimate the bankfull depth was introduced in [32] from observed width and slope, although with significant error in the estimation. The work in [14] used a data assimilation technique (ensemble Kalman filter) built around a river hydrodynamic model to recover water depth and discharge. Similarly, Ref. [7] estimated bathymetric depth and slope using synthetically-generated SWOT measurements and a Data Assimilation (DA) methodology. The study in [33] combined coupled hydrologic/hydrodynamic modeling of an Arctic river with virtual SWOT observations using a local ensemble Kalman smoother to characterize river depth. An algorithm to obtain estimates of river depth and discharge was developed in [34] based on Manning's equation, which reduces the computational expense in contrast to data assimilation techniques. Finally, Ref. [26] proposed "linear" and "slope-break" extrapolation methods that seek to identify optimal locations where there is high correlation between w (cross-sectional flow width) or W (effective width) and H (water surface elevation). As they concluded, the slope-break method can detect fewer optimal locations, even though it shows fewer errors than the linear method. Recently, Ref. [35] estimated discharge and as a by product bathymetry based on variational data assimilation using data of water levels and width.

However, none of the aforementioned studies used real spaceborne observations of both water level and river width at the same time to estimate river depth. In this paper, we aim to obtain the river depth using SWOT-type spaceborne observations: water surface elevation (H), effective river width (W) and slope (S). To this end, we use satellite altimetry data for H and S and satellite imagery for the estimation of W. We use our spaceborne observations in two recognized models for discharge estimation developed by Bjerklie [9] and Dingman and Sharma [36]. Since river depth is an unknown in these models, we employ a Gauss–Helmert adjustment model to estimate average river depth along 16 reaches defined over the Po River in Italy. We validate our results against surveyed cross-section information along the river.

2. Case Study

The Po River, located in the Po valley in Italy, flows 652 km eastward through many important Italian cities, including Turin (Torino), Piacenza and Ferrara (Figure 3). Since the river is subject to heavy flooding, over half its length is controlled with dikes. A major dam is located at the village of Isola Serafini 40 km downstream of Piacenza [37]. For this study, we defined 16 reaches along the river from Bressana ($45.11°$, $9.12°$) up to Berra ($44.98°$, $11.98°$) almost every 20 km (Table 1). In order to obtain two distinct reaches up and downstream of the dam at Isola Serafini, Reach 5 ends at the dam with a length of 11.8 km, and Reach 6 starts from the dam location with a length of 28.2 km. The river reaches are delineated in Figure 3. We selected this river for this study because of the following reasons.

1. The Po River with its narrow width (150–650 m) highlights the limitations of both altimetry and imagery.
2. We have access to a variety of in situ data for the validation purposes.

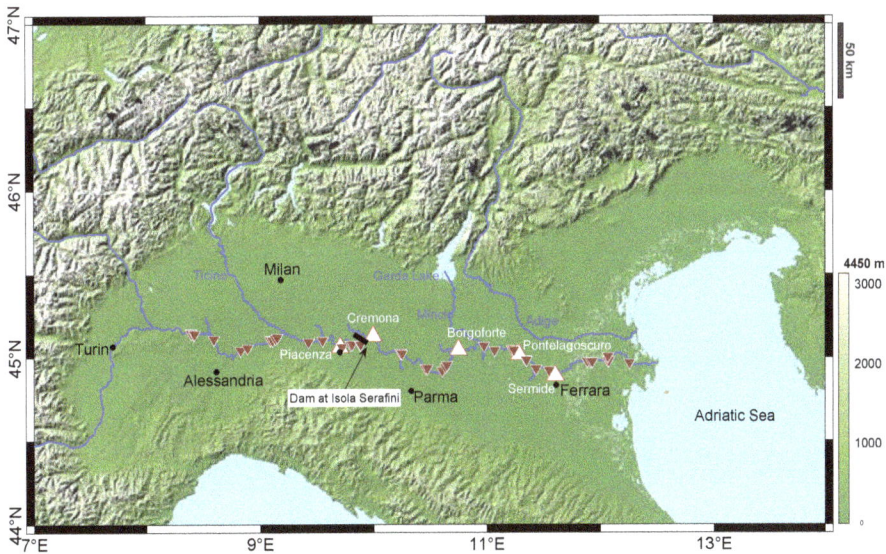

Figure 3. The Po River flowing eastward in Italy. The red triangles indicate the location of virtual stations and the white ones the location of gauging stations from west to east: Piacenza, Cremona, Borgoforte, Sermide and Pontelagoscuro.

Table 1. Selected reaches along the Po River.

Reach Number	Starting Point			Ending Point			Reach Length
	Lat.	Lon.	Chainage	Lat.	Lon.	Chainage	
	(°)	(°)	(km)	(°)	(°)	(km)	(km)
1	45.11	9.12	250.00	45.10	9.34	270.00	20.00
2	45.10	9.34	270.00	45.11	9.55	290.00	20.00
3	45.11	9.55	290.00	45.08	9.66	310.00	20.00
4	45.08	9.66	310.00	45.07	9.82	329.36	19.35
5	45.07	9.82	329.36	45.09	9.90	341.16	11.80
6	45.09	9.90	341.16	45.03	10.07	369.36	28.20
7	45.03	10.07	369.36	45.00	10.28	389.34	19.98
8	45.00	10.28	389.34	44.94	10.46	409.37	20.02
9	44.94	10.46	409.37	44.96	10.66	429.37	20.00
10	44.96	10.66	429.37	45.04	10.79	449.36	20.00
11	45.04	10.79	449.36	45.07	11.00	469.33	19.97
12	45.07	11.00	469.33	45.06	11.21	489.36	20.03
13	45.06	11.21	489.36	44.97	11.39	509.36	20.00
14	44.97	11.39	509.36	44.92	11.57	529.34	19.98
15	44.92	11.57	529.34	44.96	11.75	549.36	20.02
16	44.96	11.75	549.36	44.98	11.98	569.35	19.99

3. Data

In this study, we obtained river water level and width from spaceborne sensors. We use the data from different altimetry missions: TOPEX/Poseidon, ENVISAT, CryoSat-2 and the satellite with ARgos and ALtiKa instruments (SARAL/ALtiKa), to generate water level time series along the river. In order to obtain river width, we use MODIS surface reflectance eight-day composites with 250 m spatial resolution (MOD09Q1). We also use in situ data of river discharge for the estimation of river depth and surveyed cross-section information for the validation. Table 2 lists all the data used in this study.

Table 2. Summary of the datasets used in this study. AIPO, Agenzia Interregionale Fiume Po.

Variable	Dataset	Resolution		Time Period	Source
		Spatial	Temporal		
Effective river width W	MODIS	250 m	8 d	2000–2014	[38]
Variable Surface water level H and slope S	Multi-mission altimetry	–	3 d	2000–2014	
	TOPEX/Poseidon	–	10 d	1992–2002	
	ENVISAT	–	35 d	2002–2010	
	TOPEX/Poseidon XT	–	10 d	2002–2005	[39]
	ENVISAT XT	–	30 d	2010–2012	
	CryoSat-2	–	369 d	2012–2014	
	Jason-2	–	10 d	2008–2016	
River discharge Q	in situ	–	1 d	1995–2013	AIPO
Channel sections	surveyed	every 250 m	–	–	AIPO

3.1. In Situ Data

In situ datasets at five gauging stations along the Po River are available from from 1995–2013. Table 3 lists the location, elevation, average width, average flow velocity and annual discharge at these gauging stations.

Table 3. The location and height of available in situ data along the Po River. The height refers to the mean water level derived from daily time series.

Name	Lat. (°)	Lon. (°)	Elevation (m)	Average Width (km)	Average Flow Velocity (m/s)	Average Discharge (m^3/s)
Piacenza	45.06	9.70	42.37	106	0.61	933
Cremona	45.13	9.99	29.03	157	0.82	1075
Borgoforte	45.05	10.75	14.05	164	0.84	1313
Sermide	45.02	11.29	9.50	329	0.44	1378
Pontelagoscuro	44.89	11.61	3.48	175	3.01	1477

Moreover, we have access to channel section surveys at approximately 250 m intervals between Busca Tornello and the outlet of the river (350 km). These data are provided by Agenzia Interregionale Fiume Po (AIPO).

3.2. Water Level from Satellite Altimetry

With the help of virtual stations and the method defined in [39] (see Table 3 in [39]), individual water level time series along the Po River are densified. That geodetic method allows us to connect hydraulically and statistically all virtual stations of several satellite altimeters and produce water level time series at any location along the river. Here, we generate densified altimetric water level time series at the location of each selected reach. For that, we stacked the selected altimetric measurements by shifting the water level hydrographs of all the virtual stations according to a corresponding time lag. The time lag represents the time that the stream flows from one virtual station to the one downstream. For the time lag estimation, which is necessary for stacking the altimetric measurements, we used average river width $\bar{W}$ obtained from satellite images (explained in Section 3.3) together with the average slope $\bar{S}$ derived from satellite altimetry as inputs for a hydraulic model that estimates average flow velocity from the width and slope of a river [39]:

$$V = 1.48 \cdot \bar{W}^{0.8} \bar{S}^{0.6} , \tag{1}$$

in order to compute the time lag, we then estimate the wave traveling time T_L by [6,40]:

$$T_L = \frac{L}{c} \tag{2}$$

in which L is the distance between virtual stations and c is the celerity. In the case of the kinematic wave equation, the momentum balance can be arranged using the Manning formula, and consequently, the celerity c can be expressed as [41]:

$$c = 5/3V , \tag{3}$$

We derive average flow velocity values at each virtual station using Model (1), the estimated river width using imagery and river slope from the average water level of each virtual station. After estimating the time lag and stacking the measurements, the following algorithm is performed to densify the time series:

1. Normalization of the data between zero and one: The measurements are merged by normalizing the time series according to their statistical characteristics. This step helps to make the range for the distribution of water level variations at different virtual stations consistent. For the merging process, the data of each virtual station is normalized by assigning the third percentile to zero and the 85th percentile to one.

2. Confidence limit definition of 99% of a Student's t-test for a sliding time window: The length of the time window can be experimented with to achieve the best performance. Here, a time

window of one month (15 days before and 15 days after the selected measurement) is chosen as it leads to optimal results in terms of time series behavior and the number of identified outliers.

3. Outlier identification and rejection: The identification and the rejection of the outliers is carried out by an iterative data snooping method and by iteratively updating the confidence limits. The data snooping method searches for the observation (always one observation) with the largest gross error [42].

4. Rescaling of normalized values to their corresponding river water level heights: After removing the outliers, the combined normalized altimetric values are ready to be rescaled back to their true water level values.

5. Constructing the time series: After rescaling, we now have a cloud of measurements with their corresponding uncertainty for the selected location along the river, which is free from outliers. The dense time series of water level can then be obtained by connecting the measurements using a three-point distance weighted moving averaging.

Following the above algorithm, for the densified time series at the all 16 reaches, we obtain an effective temporal resolution of around three days from individual time series with originally a 10- or 35-day sampling interval. Time series in the Figure 4 show the densified water level and effective river width measurements obtained over the reach number 12 (the Borgoforte gauging station).

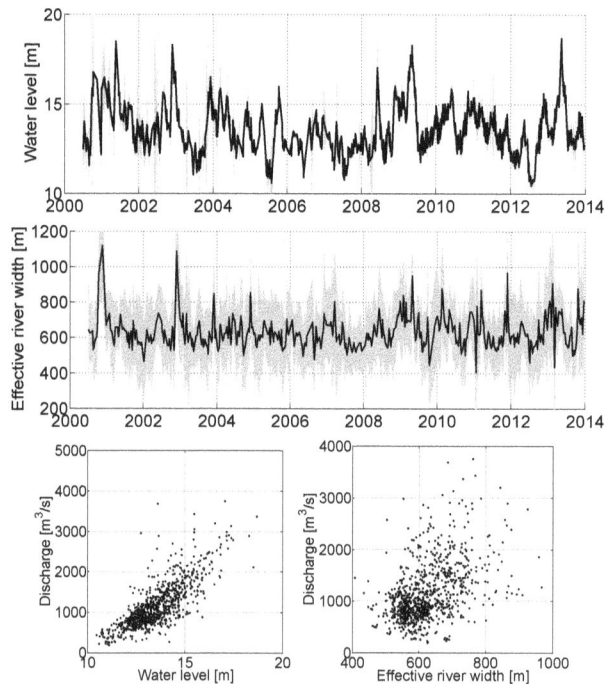

Figure 4. (Top panel) Water level time series from satellite altimetry and effective river width from satellite imagery for Reach 12. (Bottom panel) Scatter plot of river discharge against water level (correlation coefficient = 0.73) and effective river width (correlation coefficient = 0.52).

3.3. River Width from Satellite Imagery

In order to derive the effective width (W) at different river reaches, we apply the algorithm introduced by [38] on MODIS images. The work in [38] developed a Markov Random Fields (MRF) model, which considers spatial correlation between neighboring pixels and the long-term temporal behavior of the river. In this method, to extract the river mask in each image, we define the Maximum A Posteriori (MAP) estimate of the MRF that models the interaction between different constraints and auxiliary sources of information. Based on Bayes' rule, the posterior probability ($\mathrm{P}(d|f)$) of the hypothesis f (labeling structure) given the observations d (pixel values) is proportional to the product of the prior probability ($\mathrm{P}(f)$) and the likelihood ($\mathrm{P}(f|d)$). In this case, the likelihood represents the agreement between the label and value of pixels, and the prior probability describes the priority between different hypotheses. Our aim is to find a realization of the MRF maximizing the posterior probability. In this way, an energy minimization frame is developed regarding the problem and solved by the graph cuts technique, which is a powerful solution for a wide range of image segmentation problems. We followed the approach by [38] and used an undirected graph with two terminals to extract the river mask by finding the max-flow solution on the graph. Figure 5 presents an example of a derived water mask. It shows that the method is able to continuously capture meandering river paths even for a narrow river like the Po.

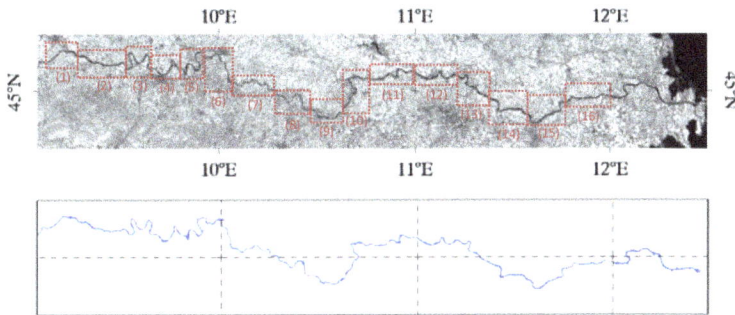

Figure 5. (Top panel) A MODIS image of the Po river with the selected river reaches. (Bottom panel) River mask extracted applying the proposed method by [38]. Date: 28 July 2000.

Although the graph cuts technique is preferred because of its ability to find a globally-optimal solution in polynomial time, it cannot provide any uncertainty measurement associated with the determined river mask. To overcome this deficiency, Ref. [43] introduced a method to measure uncertainty in graph cuts solutions by measuring the max-marginal probability for each pixel. In this way, a probabilistic river mask can also be provided. Following this, a reliable river area time series is generated together with its uncertainty. We retain the assigned labels for pixels with marginal probabilities higher than 10% in both water and land masks. This leaves a third region that contains pixels with marginal probabilities less than 10% in both masks. This is the uncertain region, for which we cannot define a proper label based on the available information. We consider the area of the third region as an error in river area measurement.

Figure 4 shows the time series of effective river width together with the estimated uncertainty for Reach 12. Since the river is very narrow with respect to the MODIS pixel size of 250 m, the magnitude of uncertainty is relatively large. In Figure 4 (bottom panel), we compare the obtained effective river width against in situ discharge for the same reach. An agreement between the behavior of the two time series is expected. The majority of the river section passes through urban areas with river channel contraction at the shoreline leading to the non-natural behavior of the river. Figure 4 (bottom panel) comprises the obtained water level and the effective river width with the in situ discharge data at Borgoforte. The scatter plots clearly demonstrate the challenge of using satellite imagery for discharge estimation, whereas water level from altimetry agrees relatively well.

4. Methodology to Estimate River Depth

In order to estimate the riverbed's height H_0 and, consequently, river depth $H - H_0$, we rely on two models developed for discharge estimation. It should be noted that H_0 is the average riverbed's height that does not correspond to the height of deepest location in the channel section (see Figure 2). Generally, a functional relationship between depth and discharge can be developed from the Manning equation, which is applicable to natural rivers [44]. For a wide rectangular channel with $W > 10(H - H_0)$, discharge can be obtained from the Manning equation as:

$$Q = \frac{1}{n}(H - H_0)^{1.67}W S^{0.5} , \tag{4}$$

in which n is the channel resistance. In practice, the channel resistance cannot be measured directly and often varies considerably with discharge [9]. However, statistical studies by [9,36] have shown that reasonably accurate estimates of discharge Q for within-bank flows can be obtained without resistance as an input variable. The main assumption behind these studies is that the resistance varies with the channel geometry. The works in [9,36] show that discharge can be estimated as:

Bjerklie et al. [9],

$$Q = aW^{1.02}(H - H_0)^{1.74}S^{0.35} \tag{5}$$

Dingman & Sharma [36],

$$Q = aW^{1.17}(H - H_0)^{1.57}S^{0.34} \tag{6}$$

for a wide range of rivers. These models are calibrated using a large database of observed flow measurements from 103 rivers in the United States and New Zealand.

For the estimation of the riverbed's height H_0, we have considered both Models (5) and (6). To this end, we use the estimated river width W from satellite imagery, densified altimetric height H, slope S from satellite altimetry and river discharge Q from in situ data. When all these parameters are available, only the H_0 remains as unknown. However, estimating H_0 in such non-linear models, in which different datasets bring different levels of uncertainty, needs extra care. In fact, considering the stochastics of Figure 4 (bottom left), the estimation of H_0 would mean finding the water level (H) where no water passes through the channel ($Q = 0$) with the added complexity of further stochastic parameters. From the point of view of adjustment, this is an ill-posed extrapolation of data with uncertainties in both axes. Since the data of both axes are provided with uncertainty, standard regression with least squares estimation does not lead to a meaningful H_0. Therefore, we derive the H_0, and consequently river depth, by implementing a Gauss–Helmert Model (GHM). Adjustment with the GHM [45] is a combination of adjustment with condition equations and adjustment with observation equations [46].

In the GHM, we add four terms, e_Q, e_W, e_H, e_S, to Equations (5) and (6), representing the error in discharge, river width, water level and slope, respectively. Hence, Equation (5) would lead to:

$$Q - e_Q - a(W - e_W)^{1.02}(H - e_H - H_0)^{1.74}(S - e_S)^{0.35} = f(a, H_0, e_Q, e_W, e_H, e_S, Q, W, H, S) = 0. \tag{7}$$

We now have an implicit functional relationship $f(.)$ linking unknown parameters $a, H_0, e_Q, e_W, e_H, e_S$ to observations Q, W, H, S. We can linearize the function by splitting up the quantities:

$$H_0 = H_0^0 + \delta H_0, \; a = a^0 + \delta a$$
$$e_W = e_W^0 + \delta e_W, \; e_H = e_H^0 + \delta e_H \tag{8}$$
$$e_S = e_S^0 + \delta e_S, \; e_Q = e_Q^0 + \delta e_Q$$

and choose Taylor points related to both parameters and uncertainties: a^0, H_0^0, e_Q^0, e_H^0 and e_S^0. Linearization around these Taylor points yields:

$$f = f\Big|_0 + \frac{\partial f}{\partial a}\Big|_0 \delta a + \frac{\partial f}{\partial H_0}\Big|_0 \delta H_0 +$$
$$\frac{\partial f}{\partial e_Q}\Big|_0 (e_Q - e_Q^0) + \frac{\partial f}{\partial e_H}\Big|_0 (e_H - e_H^0) + \frac{\partial f}{\partial e_S}\Big|_0 (e_S - e_S^0) + \frac{\partial f}{\partial e_W}\Big|_0 (e_W - e_W^0). \tag{9}$$

After reshaping, one obtains:

$$f = \underbrace{\begin{bmatrix} f^1\Big|_0 - \frac{\partial f^1}{\partial e_Q}\Big|_0 e_Q^0 - \frac{\partial f^1}{\partial e_H}\Big|_0 e_H^0 - \frac{\partial f^1}{\partial e_S}\Big|_0 e_S^0 - \frac{\partial f^1}{\partial e_W}\Big|_0 e_W^0 \\ \cdots \\ f^m\Big|_0 - \frac{\partial f^m}{\partial e_Q}\Big|_0 e_Q^0 - \frac{\partial f^m}{\partial e_H}\Big|_0 e_H^0 - \frac{\partial f^m}{\partial e_S}\Big|_0 e_S^0 - \frac{\partial f^m}{\partial e_W}\Big|_0 e_W^0 \end{bmatrix}}_{\substack{w \\ m \times 1}} + \underbrace{\begin{bmatrix} \frac{\partial f^1}{\partial a}\Big|_0 & \frac{\partial f^1}{\partial H_0}\Big|_0 \\ \cdots & \cdots \\ \frac{\partial f^m}{\partial a}\Big|_0 & \frac{\partial f^m}{\partial H_0}\Big|_0 \end{bmatrix}}_{\substack{A \\ m \times 2}} \underbrace{\begin{bmatrix} \delta a \\ \delta H_0 \end{bmatrix}}_{\substack{\delta x \\ 2 \times 1}} +$$

$$\underbrace{\begin{bmatrix} \frac{\partial f^1}{\partial e_Q}\Big|_0 & \frac{\partial f^1}{\partial e_H}\Big|_0 & \frac{\partial f^1}{\partial e_S}\Big|_0 & \frac{\partial f^1}{\partial e_W}\Big|_0 & 0 & 0 & 0 & 0 & \cdots & 0 & 0 & 0 & 0 \\ 0 & 0 & 0 & 0 & \frac{\partial f^2}{\partial e_Q}\Big|_0 & \frac{\partial f^2}{\partial e_H}\Big|_0 & \frac{\partial f^2}{\partial e_S}\Big|_0 & \frac{\partial f^2}{\partial e_W}\Big|_0 & \cdots & 0 & 0 & 0 & 0 \\ \cdots & \cdots & \cdots & \cdots & \cdots & \cdots & \cdots & \cdots & \cdots & \cdots & \cdots & \cdots & \cdots \\ 0 & 0 & 0 & 0 & 0 & 0 & 0 & 0 & \cdots & \frac{\partial f^m}{\partial e_Q}\Big|_0 & \frac{\partial f^m}{\partial e_H}\Big|_0 & \frac{\partial f^m}{\partial e_S}\Big|_0 & \frac{\partial f^m}{\partial e_W}\Big|_0 \end{bmatrix}}_{\substack{B^\mathsf{T} \\ m \times 4m}} \underbrace{\begin{bmatrix} \delta e_Q^1 \\ \delta e_H^1 \\ \delta e_S^1 \\ \delta e_W^1 \\ \cdots \\ \delta e_Q^m \\ \delta e_H^m \\ \delta e_S^m \\ \delta e_W^m \end{bmatrix}}_{\substack{\delta v \\ 4m \times 1}} = 0 \tag{10}$$

Then, we have the linearized version of (7):

$$w + A\delta x + B^\mathsf{T}\delta v = 0 \tag{11}$$

The upper indices 1, 2, ..., m in matrix B^T and vector δv are an index to the observations of discharge, river width, water level and slope. In this formulation, we have two global parameters a, H_0 and 4-m parameters to be estimated from m epochs of $\{Q^i, H^i, S^i, W^i\}$. In the GHM, the least squares objective function is:

$$\widehat{\delta x} = \operatorname{argmin} \delta v^\mathsf{T} P \delta v, \text{ subject to the Equation (11),} \tag{12}$$

where P is the weight matrix with dimension 4 m × 4 m:

$$P = \operatorname{diag}\left(1/\sigma_{Q_1}^2, 1/\sigma_{H_1}^2, 1/\sigma_{S_1}^2, 1/\sigma_{W_1}^2, ..., 1/\sigma_{Q_m}^2, 1/\sigma_{H_m}^2, 1/\sigma_{S_m}^2, 1/\sigma_{W_m}^2\right). \tag{13}$$

The main benefit of the GHM lies within the weight matrix, which the uncertainties of different measurements carefully take into account. For minimizing Equation (12), the corresponding Lagrangian reads:

$$L(v, \delta x, \lambda) = \frac{1}{2}\delta v^\mathsf{T} P \delta v - \lambda^\mathsf{T}(B^\mathsf{T}\delta v + A\delta x + w), \tag{14}$$

Setting the gradients of the Lagrangian with respect to $\delta v, \delta x, \lambda$ to zero yields the equation system:

$$\begin{bmatrix} P & 0 & B \\ 0 & 0 & A^\mathsf{T} \\ B^\mathsf{T} & A & 0 \end{bmatrix} \begin{bmatrix} \widehat{\delta v} \\ \widehat{\delta x} \\ \widehat{\lambda} \end{bmatrix} = \begin{bmatrix} 0 \\ 0 \\ -w \end{bmatrix} \tag{15}$$

by solving Equation (15), we obtain $\widehat{\delta x}$, $\widehat{\delta v}$ and $\widehat{\lambda}$ as:

$$\widehat{\delta v} = -B(B^\mathsf{T}PB)^{-1}\{w - A[A^\mathsf{T}(B^\mathsf{T}PB)^{-1}A]^{-1}A^\mathsf{T}(B^\mathsf{T}PB)^{-1}w\} \tag{16}$$

$$\widehat{\delta x} = [A^\mathsf{T}(B^\mathsf{T}PB)^{-1}A]^{-1}A^\mathsf{T}(B^\mathsf{T}PB)^{-1}w \tag{17}$$

$$\widehat{\lambda} = (B^\mathsf{T}PB)^{-1}\{w - A[A^\mathsf{T}(B^\mathsf{T}PB)^{-1}A]^{-1}A^\mathsf{T}(B^\mathsf{T}PB)^{-1}w\}, \tag{18}$$

which we can use to update and estimate $\hat{a}$, $\widehat{H_0}$, $\hat{e}_Q$, $\hat{e}_H$, $\hat{e}_S$ and $\hat{e}_W$. We then update unknown parameters and observations and iterate the whole until convergence is achieved for the norm of $\widehat{\delta x}$ and $\widehat{\delta v}$ ($< 10^{-9}$). After the final iteration, we obtain a and H_0 with their uncertainties, for which uncertainties of water level, river width, slope and discharge are taken into consideration.

5. Results, Validation and Discussion

Solving the models by [9,36] with the Gauss–Helmert formalism allows us to estimate two unknown parameters of these models namely a and H_0. For the implementation of the method, we need data of water level (H), slope (S), river width (W) and river discharge (Q). As described in Section 3, we obtain water level time series after applying the densification process [39] over the selected virtual station along the river. Thus, we obtain a water level time series at the middle of each reach, from which we estimate the corresponding time variable slope. The effective river width is obtained from satellite imagery following the method proposed by [38].

In order to fill the weight matrix P, in the Equation (13), we obtain σ_{H_i} from satellite altimetry (error envelope in Figure 4 top); σ_{W_i} are taken from the error estimation of our effective river width estimation (error envelope in Figure 4 middle); and σ_{S_i} are obtained by the error propagation of consecutive height measurements:

$$\sigma_{S_i}^2 = \frac{1}{d^2}\left(\sigma_{H_i^{up}}^2 + \sigma_{H_i^{do}}^2\right), \tag{19}$$

in which d is the distance between the center of two consecutive reaches. For σ_{Q_i}, we consider 10% of Q measurements, i.e., $\sigma_{Q_i} = 0.1 Q_i$.

Unlike H, W and S that are available for each reach, discharge data are not individually available for each reach. We use discharge data at five in situ gauges along the river. However, the question remains which in situ discharge data lead to a better result for the estimation of H_0. The intuitive answer to this question is, of course, the discharge data at the gauge nearest to the selected reach. In order to assess this issue, we first estimated H_0 and a for all the reaches with the help of discharge data at all the gauges. Figure 6 shows the estimated H_0 after the implementation of the GHM for the selected reaches along the Po River. For each reach, we obtain five estimates of $\hat{H}_0$ and $\hat{a}$, for which we use different discharge data. It is interesting to observe that the estimated H_0 for each reach is nearly independent of the choice of discharge data. In principle, we obtain similar values with a discrepancy of ~2% with respect to average riverbed's height with the discharge data of different gauges. Comparing the estimated H_0 with those surveyed along the river (black steps) shows a general good agreement in the range of ~10% relative RMSE.

Such a good agreement is better visible in the scatter plot of estimated H_0 versus in situ H_0 (Figure 7). The scatter is aligned around the diagonal line indicating a good performance of our method for river depth estimation. The results show a marginal difference between the estimated H_0 from the two models, with slightly better performance of the model by [9] (Table 4). As expected, the estimated $\hat{a}$ values, which act like scale factors, are different for both models. The difference is due to the different power for water level, slope and width data. It should be noted that the unit of value a is also different for the two models. In the case of the model by [9], the unit of a is $m^{0.24}\,s^{-1}$, and in the case of the model by [36], it is $m^{0.26}\,s^{-1}$.

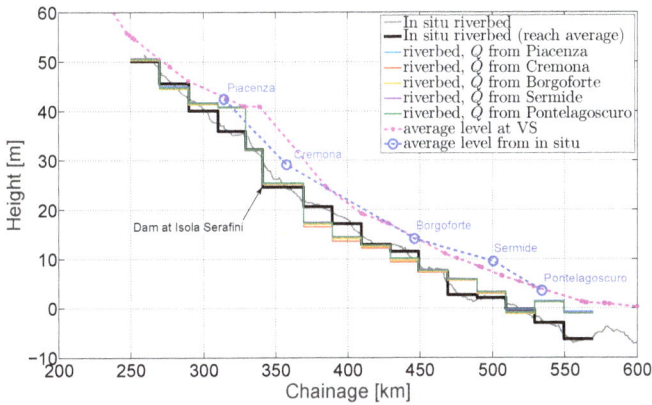

Figure 6. Estimated river bed profile along the Po River for 16 defined reaches. VS refers to virtual station.

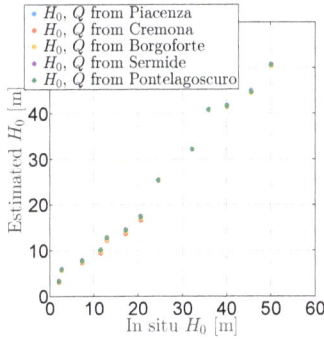

Figure 7. Scatter plot of estimated H_0 using discharge data of different gauging stations along the Po River.

Table 4. Estimated river bed height H_0 using the two selected models, for which discharge data of the nearest gauge are used.

Reach	H_0^{ins} (m)	Model by Bjerklie et al. [9]			Model by Dingman and Sharma [36]		
		$\hat{a}$	$\widehat{H_0}$(m)	$\widehat{H_0} - H_0^{ins}$ (m)	$\hat{a}$	$\widehat{H_0}$ (m)	$\widehat{H_0} - H_0^{ins}$ (m)
1	50.27	0.69	50.37 ± 0.15	0.10	0.39	50.66 ± 0.14	0.39
2	45.62	0.60	44.72 ± 0.15	0.89	0.30	44.96 ± 0.14	0.66
3	40.09	1.36	41.47 ± 0.10	1.38	0.67	41.69 ± 0.09	1.59
4	35.91	2.06	40.73 ± 0.03	4.83	0.88	40.85 ± 0.03	4.94
5	32.20	29.20	32.07 ± 0.01	0.12	11.95	32.13 ± 0.01	0.07
6	24.60	0.80	24.91 ± 0.09	0.31	0.40	25.16 ± 0.08	0.57
7	20.63	0.22	16.06 ± 0.31	4.56	0.13	16.53 ± 0.31	4.10
8	17.18	0.27	13.06 ± 0.26	4.12	0.16	13.58 ± 0.26	3.60
9	12.92	0.74	12.07 ± 0.14	0.85	0.42	12.48 ± 0.13	0.44
10	11.54	0.50	9.37 ± 0.16	2.17	0.27	9.77 ± 0.16	1.77
11	7.38	0.53	7.04 ± 0.16	0.34	0.29	7.48 ± 0.16	0.10
12	2.70	1.40	5.64 ± 0.07	2.94	0.73	5.91 ± 0.07	3.21
13	2.08	1.25	3.00 ± 0.10	0.92	0.66	3.31 ± 0.09	1.23
14	−0.26	0.58	-1.49 ± 0.20	1.23	0.34	-0.87 ± 0.19	0.61
15	−3.00	4.83	1.07 ± 0.04	4.07	2.28	1.23 ± 0.04	4.24
16	−6.29	4.97	-1.32 ± 0.08	4.97	2.64	-1.07 ± 0.07	5.22

In order to investigate the impact of the length of time series on the result, we arbitrarily consider 20%, 40%, 60%, 80% and 100% of data in the time series of water level, width and discharge. Figure 8 shows the estimated H_0 for all reaches, where the results of using different portions of datasets for the estimation of H_0 are very similar to each other and lie within the estimated error. This result shows that the estimation of H_0 is insensitive to the length of the time series.

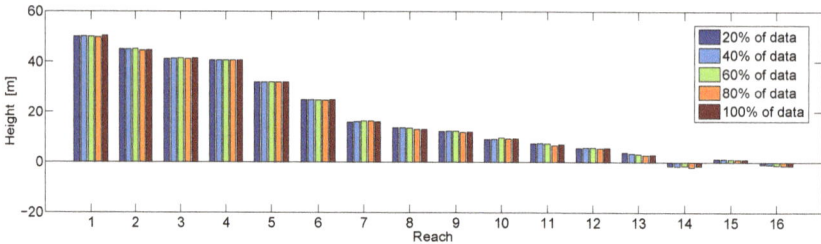

Figure 8. Estimated river bed height by considering 20%, 40%, 60%, 80% and 100% of data in the time series of water level, width and discharge along the Po River for 16 defined reaches.

As a result of GHM, $\widehat{\delta v}$ is also estimated, which comprises the corrections to the original observations. Figure 9 shows estimated residuals for discharge $\widehat{e_Q}$, water level $\widehat{e_H}$ and effective river width $\widehat{e_W}$ data of Reach 1 in the case of using the model by [9]. In fact, the two global unknowns of a and H_0 are estimated by adjusting the observations within the selected model and minimizing the error.

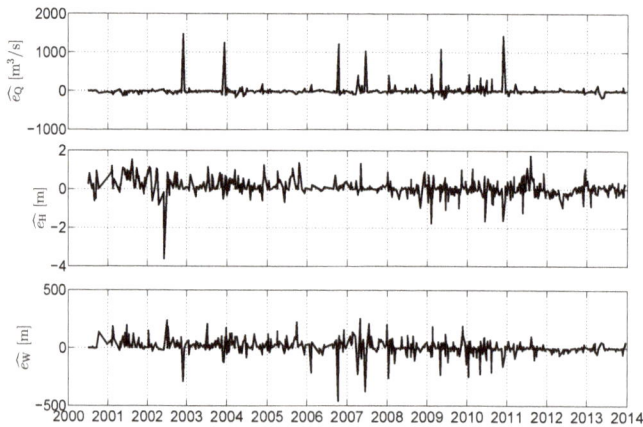

Figure 9. Estimated residual for observations of discharge $\widehat{e_Q}$, water level $\widehat{e_H}$ and effective river width $\widehat{e_W}$ for the first reach along the Po River.

Despite good agreement for most reaches, the estimated H_0 for Reaches 4, 7, 8, 12, 15 and 16 show relatively large errors, which can be better seen in Figure 10.

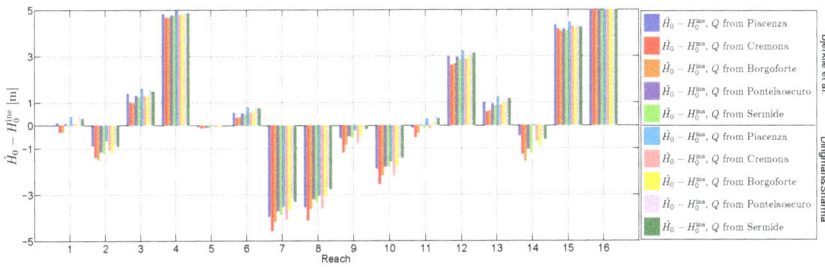

Figure 10. Bar plot of difference between estimated riverbed's height $\widehat{H_0}$ and those from in situ using discharge data of different gauging stations along the Po River.

The errors seem to be independent of the choice of the in situ gauge for discharge data. Table 4 lists the results of H_0 estimation for both of the models [9,36], for which we use the discharge data of the nearest in situ station to the reach. It should be noted that for Reaches 1–5, located upstream of the dam, we use the discharge data at a station also located upstream of the dam. The reason for that was the difference between stream behavior up- and down-stream of the dam, which was discussed in detail by [39]. A possible reason for such large errors is the quality of water level or river width time series at these reaches. The water level time series show generally good correlation (>0.6) with discharge data (Figure A1 in Appendix A), and no specific mismatch is visible at the problematic reaches. On the other hand, the correlation coefficients of effective river width with discharge data for all reaches are generally small especially at Reaches 1, 4 and 6 (Figure A2 in Appendix A). However, again, the reaches with a large error for the estimated H_0 do not show a distinctly low correlation in comparison to other reaches. Based on these results, one can conclude that such large errors in estimated H_0 could be due to the choice of the model. In fact, for heterogeneous reaches with variable channel resistance along the river, a single model does not represent the flow over the entire reach.

The error we obtain for the estimated H_0 originated from different sources. Our error assessment should distinguish between systematic errors, accuracy and the precision of our estimation. The systematic error and accuracy in our case are mainly related to the choice of functional model for depth estimation. This is, of course, testable by using different models and by assessing its effect on the estimation of H_0. In our study, since both models are very similar, we have obtained similar results. In order to assess the sensitivity of estimated H_0 to the precision of W, H and S, we write H_0 as an explicit function of all other parameters (in the case of model by Bjerklie et al.):

$$H_0 = H - \left[\frac{Q}{aW^{1.02}S^{0.35}}\right]^{0.57} , \qquad (20)$$

and obtain its error dH_0 as a function of error in other variables:

$$dH_0 = dH + \frac{\partial H_0}{\partial Q}dQ + \frac{\partial H_0}{\partial W}dW + \frac{\partial H_0}{\partial S}dS , \qquad (21)$$

in which $\partial H_0/\partial H = 1$, which means that any error in H will directly transfer into an equal error in H_0.

The sensitivities $\partial H_0/\partial Q$, $\partial H_0/\partial W$ and $\partial H_0/\partial S$ depend on other parameters that are varying in time. In order to assess the error in H_0 due to the error of water level, width and discharge, here we assume a constant error of 0.7 m for water level time series from altimetry, which would lead to a constant error of 0.05 m/km for the slope of two consecutive reaches with a distance of 20 km (Equation (19)). For the effective river width, we use our time variable uncertainty estimates based on the method by [43]. For discharge, a multiplicative error of 10% is assumed. Figure 11 shows partial errors in H_0 due to the errors of other parameters in Reach 1. The contribution is about 0.2 m of the discharge's precision on the precision of estimated H_0. The average error in H_0 due to the errors of our width estimates is about 0.3 m. Since $\partial H_0/\partial S$ is time variable, a constant error in slope leads to a time variable error contribution with the average value of approximately 0.2 m for reach 1. This result shows that among all the parameters in Models (5) and (6), errors in height will contribute the most in the estimation of H_0 under the assumption that $dH = 0.7$ m.

Figure 12 shows the average of time series for the error contribution of different parameters over all 16 reaches. It is interesting to observe a large sensitivity of H_0 to slope for Reaches 15 and 16, at which we obtain a relatively large error. This means that for these reaches, the error in slope will be amplified with a larger error in the estimation of river depth. Moreover, the effects of errors in discharge and river width for the estimation of river depth are generally smaller than those for water level and slope.

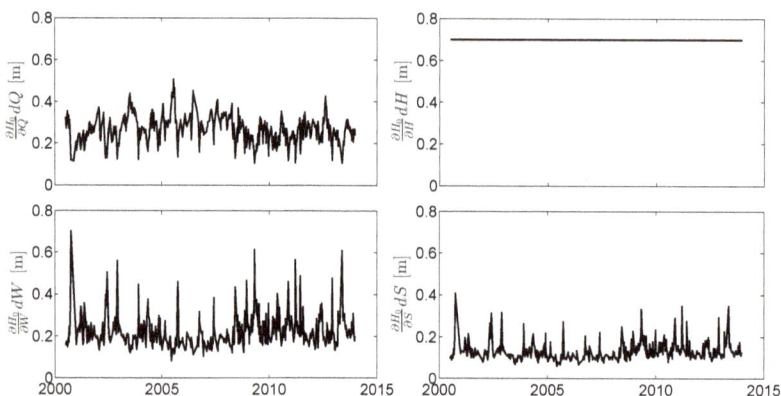

Figure 11. Time series of error in H_0 due to the error in Q, H, W and S for Model (1) over Reach 1.

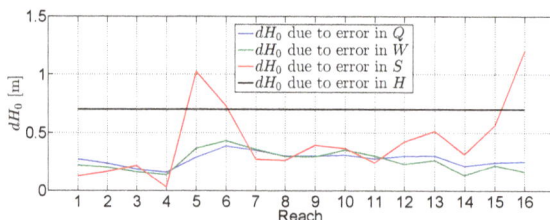

Figure 12. Average error in H_0 due to errors in discharge, water level, effective river width and slope over all 16 reaches.

The direct formulation (21) is also beneficial for investigating why the estimation of H_0 is insensitive to the choice of discharge station. Figure 13 shows that $\partial H_0/\partial Q$ for Reach 1 varies between 3×10^{-4} and 12×10^{-4} with an average value of 6.8×10^{-4}. On the other hand, according to Table 3, average discharge at different stations is 933, 1075, 1313, 1378 and 1477 m^3/s, respectively. This would mean that for the estimation of H_0 of Reach 1, a mischoice of Pontelagoscuro station with an average discharge of 1477 m^3/s instead of Piacenza with an average discharge of 933 m^3/s would lead to only 0.36 m error in H_0 ($6.8 \times 10^{-4} \times (1477 - 933) = 0.36$).

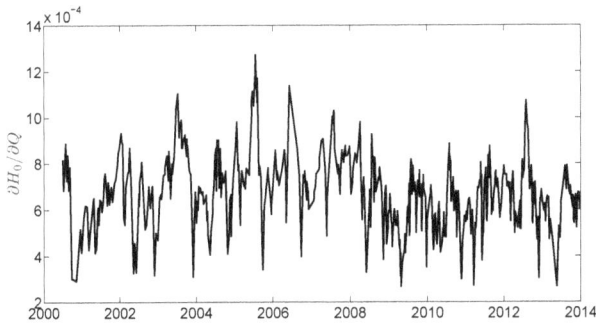

Figure 13. The time variable sensitivity of estimated H_0 to discharge observation $\partial H_0 / \partial Q$.

6. Perspective for the SWOT Mission

In the future, the SWOT mission will provide measurements of water surface elevation, river width and slope, through which discharge can be estimated for river widths down to 50–100 m, which allows us to obtain consistent and coherent information about the spatial distribution of river discharge. However, SWOT does not have the objective of and cannot be a replacement for in situ gauge networks [47]. SWOT will irregularly sample mid-latitude locations approximately three times per its 21-day cycle rather than nearly continuously, as in a gauge estimate. Although irregular sampling is appropriate to monitor the global water cycle, it is inadequate for many local-scale questions on rivers where SWOT may not fully observe temporal dynamics [11,47].

Moreover, SWOT will not observe full river networks. According to the study by [25] given the ability to observe rivers wider than 100 m, SWOT would observe more than 60% of the global sub-basins with an area of 50,000 km^2. In terms of estimation of river depth, recent studies suggest that it may be possible to estimate river depth simultaneously from SWOT observations alone using multiple overpasses over several adjacent river reaches. However, studies are required to improve the understanding of the number of essential observations for a proper estimation of river depth [48].

The results in this study give a perspective of the SWOT mission and its scientific challenges in the following ways:

- Satellite altimetry is put on an operational footing through the Sentinel 3 series of the European Copernicus program. At the same time, research satellites such as CryoSat-2, SARAL/AltiKa or Jason-3 remain in orbit and provide complementary space-time sampling. This constellation addresses many current limitations and opens a significant area of investigation into the operational use of altimetry missions for hydrological purposes. Developing methods to generate dense time series of water level using altimetry data, demonstrated in this study, raise the hope of using operational altimetry together with SWOT data in the future for a better monitoring of temporal dynamics in many rivers around the world.
- Remote sensing techniques have been introduced as viable choices to monitor surface water variations [5]. Optical and SAR satellite imagery missions provide the opportunity to monitor surface water extent repeatedly at appropriate time intervals. Recent missions provide images with better spatial and temporal resolution. Different Landsat missions have been gathering images in various multispectral bands from the Earth since 1980. Recently, Landsat 8 with 30 m spatial resolution has provided monthly images of the Earth's surface. Applications that demand a high temporal resolution preferably make use of the daily snapshots of MODIS imagery with 250-m resolution. Since 2015, Sentinel-2 has provided images with better resolution (10 m, 20 m, 60 m) and also a high revisit time (five days). On the other hand, SAR missions with images from ERS-1, ERS-2 and ENVISAT in C-band are the main sources for spatial water area monitoring in the tropical area, which is cloudy and rainy most of the year [12]. TerraSAR-X in X-band provides

high resolution images with 1-, 2- and-3 meter pixel size every two days. From 2014 onwards, Sentinel-1A has provided continuous imagery (day, night and all weather) in C-band, expanding our understanding of surface water. Therefore, developing reliable algorithms to obtain reliable dynamic river masks, as developed by [38] and implemented in this study, gives the perspective of the SWOT mission that the results from different imagery sensors could be combined with the results from SWOT for a better monitoring of river masks in the future.

- In terms of river depth estimation, our results in Figures 11–13 highlight the weaker dependency of error in river depth estimation to the error of river width and river discharge. This means that a preliminary result of river width and a coarse river discharge would be enough to estimate river depth. This is especially important for discharge algorithms, which work based on Manning's flow resistance equation, and the initial values of variables play an important role in the fast and precise estimation of discharge. Moreover, this result would mean that one could use a rough estimation of river discharge instead of in situ data and still receive acceptable results for river depth.

7. Summary and Conclusions

In this paper, we estimated the average river depth for 16 reaches along the Po River from SWOT-type observables water surface elevation, effective river width and slope using models by [9,36].

In order to obtain dense time series of water level from altimetry, we followed the method suggested by [39]. We connected all the virtual stations of several satellite altimeters along the Po River and produced water level time series at any location along the river. To this end, the selected altimetric measurements are stacked at the center of each reach by shifting the water level hydrographs of all virtual stations according to time lag between different virtual stations. The stacked measurements are then merged by normalizing the time series according to their statistical characteristics. After an outlier identification process, we rescaled the measurements back to their true water level values. For the densified time series at the 16 reaches, we obtained an effective temporal resolution of around three days from individual time series with originally a 10- or 35-day sampling interval. Time series of slope were obtained by differencing the time series of the water level at two consecutive reaches.

In order to derive effective width, we used MODIS images and extracted a dynamic river mask by employing the algorithm developed by [38]. This algorithm uses an MRF model regarding all sources of information available in images: spatial correlation between neighboring pixels and the long-term temporal behavior of the river. We then extracted the river mask in each image by maximizing a posteriori estimation in Markov random fields through minimizing the energy using the graph cuts technique. Although the graph cuts technique is preferred because of its ability to find a globally-optimal solution in polynomial time, it cannot provide any uncertainty measurement associated with the determined river mask. We have used the method by [43] to estimate uncertainty in the graph cuts solutions by measuring the max-marginal probability for each pixel. Following this, we generated a reliable river area time series together with its uncertainty.

With the help of the effective river width from satellite imagery, as well as water level and slope from satellite altimetry, we estimated river depth by implementing a Gauss–Helmert adjustment scheme on two models by [32,36]. Unlike variables observed from the space, discharge data are not available for each reach individually. We use discharge data at five in situ gauges along the river. An interesting result of this work is that the estimated H_0 for each reach only weakly depends on the choice of discharge data, since we obtained similar values for H_0 with a discrepancy of only $\sim$2% with the discharge data of different gauges. This result was explained by the low partials $\partial H_0 / \partial Q$, making the obtained H_0 less sensitive to the choice of discharge gauge.

Water **2017**, *9*, 753

We validated our results against surveyed cross-section information along the Po River. Validation shows general good agreement in the range of ~10% relative RMSE. However, we faced some problematic reaches (4, 7, 8, 12, 15 and 16) for which the estimated H_0 is relatively large. We inconclusively discussed possible sources for such error by analyzing the quality of water level and effective river width time series. We have demonstrated that the water level time series generally correlate well (>0.6) with discharge data, and no specific pattern at the problematic reaches has been found. On the other hand, the correlation coefficients of effective river width with discharge data show generally small values especially at Reaches 1, 4 and 6. However, again, the reaches with a large error for the estimated H_0 do not show a distinctly lower correlation in comparison to other reaches.

The sensitivity analysis in this study especially demonstrates that the the estimated river depth is less influenced by the error of river width and river discharge, while it is strongly influenced by error in water level and slope. These results give a perspective of the SWOT mission, where the estimation of river depth would be possible using coarse estimates of river width and discharge and improved water level and slope time series. Moreover, this result would mean that one could use a rough estimation of river discharge instead of in situ data and still receive acceptable river depth. On the other hand, given the abundance of existing and future altimetry and imagery missions, developing methods to generate dense time series of water level using altimetry data and reliable river width from satellite imagery raises the hope of using operational altimetry and imagery including the Sentinel-series together with SWOT data in the future for a better monitoring of temporal dynamics in many rivers around the world.

Acknowledgments: The authors are grateful to the National Aeronautics and Space Administration (NASA) for providing the MODIS images at https://ladsweb.nascom.nasa.gov and the European Space Agency for providing the ENVISAT data at https://earth.esa.int/web/guest/data-access. Thanks are also due to AVISO for providing SARAL/AltiKa data through ftp://avisoftp.cnes.fr/AVISO/pub/saral/gdr_t/ and the Jason 2 hydrology product at ftp://ftpsedr.cls.fr/pub/oceano/pistach/J2/IGDR/hydro/. The authors would like to thank Angelica Tarpanelli for sharing the data of surveyed cross-sections along the Po River and the Agenzia Interregionale Fiume Po for providing the the in situ data for the Po River, http://www.agenziainterregionalepo.it.

Author Contributions: Mohammad J. Tourian developed the method, conducted the data analysis and wrote the majority of the paper. Omid Elmi generated effective river width time series from satellite imagery and helped with analyzing the results and writing the manuscript. Abolfazl Mohammadnejad helped with analyzing results and writing the manuscript and also helped with the implementation of the Gauss-Helmert adjustment. Nico Sneeuw supervised the research, helped with the discussion of the method and contributed to manuscript writing.

Conflicts of Interest: The authors declare no conflict of interest.

Appendix A. Comparison of Water Level and Effective River Width with Discharge

Figure A1 shows scatter plots of discharge at nearby gauging stations against water level time series for 16 reaches along the Po River. The water level time series show a generally good correlation (>0.6) with discharge data. Figure A2 shows the scatter plots for the effective river width time series for the 16 reaches. The correlation coefficients of effective river width with discharge data for all reaches are generally small especially at Reaches 1, 4 and 6. Since the Po River width is generally narrow for the MODIS pixels with 250 m resolution, such low correlation coefficients for the obtained river width are expected. Moreover, the river passes through many urban regions with managed channels, making the river width and discharge less correlated.

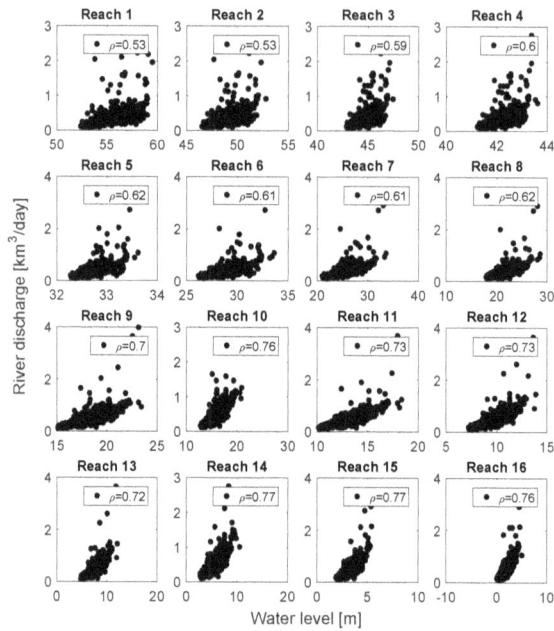

Figure A1. Comparison of water level time series and discharge at nearby gauging stations for 16 reaches along the Po River.

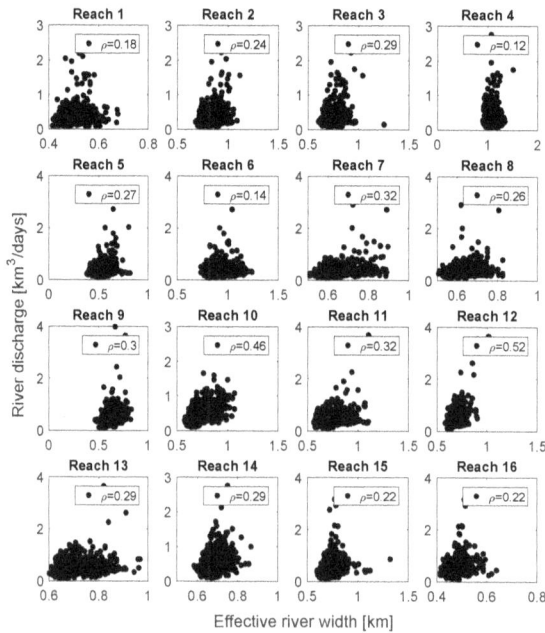

Figure A2. Comparison of effective river width and discharge at close-by gauging stations for 16 reaches along the Po River.

Figure A3 shows the surveyed sections along Reach 4 with the average water level at each section obtained from satellite altimetry. The section at Chainage 312, for instance, would lead to a non-homogeneous dependency between river width and discharge and would consequently lead to an inaccurate estimation of H_0 using the selected models. However, since we consider the averaged parameters of reaches, the effect of an individual problematic section on the final results should be negligible.

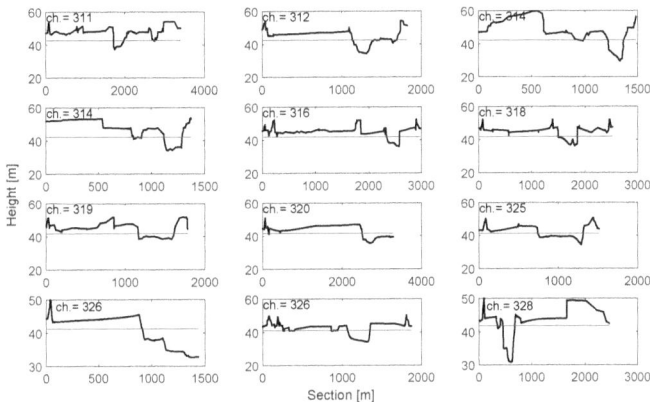

Figure A3. Sections along Reach 4. The red straight line represents the average water level obtained from altimetry.

References

1.   Dingman, S.L.; Bjerklie, D.M. Estimation of River Discharge. In *Encyclopedia of Hydrological Sciences*; John Wiley & Sons, Ltd.: Hoboken, NJ, USA, 2006.

2.   Lorenz, C.; Devaraju, B.; Tourian, M.J.; Riegger, J.; Kunstmann, H.; Sneeuw, N. Large-scale runoff from landmasses: A global assessment of the closure of the hydrological and atmospheric water balances. *J. Hydrometeorol.* **2014**, *15*, 2111–2139.

3.   Fekete, B.M.; Vörösmarty, C.J. The current status of global river discharge monitoring and potential new technologies complementing traditional discharge measurements. In *Predictions in Ungauged Basins: PUB Kick-Off, Proceedings of the PUB Kick-Off Meeting, Brasilia, Brazil, 20–22 November 2002*; IAHS Publication: Wallingford, UK, 2007; Volume 309.

4.   Tourian, M.J.; Sneeuw, N.; Bárdossy, A. A quantile function approach to discharge estimation from satellite altimetry (ENVISAT). *Water Resour. Res.* **2013**, *49*, 1–13.

5.   Alsdorf, D.; Rodriguez, E.; Lettenmaier, D.P. Measuring surface water from space. *Rev. Geophys.* **2007**, *45*, doi:10.1029/2006RG000197.

6.   Tarpanelli, A.; Barbetta, S.; Brocca, L.; Moramarco, T. River Discharge Estimation by Using Altimetry Data and Simplified Flood Routing Modeling. *Remote Sens.* **2013**, *5*, 4145–4162.

7.   Durand, M.; Andreadis, K.M.; Alsdorf, D.E.; Lettenmaier, D.P.; Moller, D.; Wilson, M. Estimation of bathymetric depth and slope from data assimilation of swath altimetry into a hydrodynamic model. *Geophys. Res. Lett.* **2008**, *35*, L20401, doi:10.1029/2008GL034150.

8.   Pavelsky, T.M.; Smith, L.C. RivWidth: A software tool for the calculation of river widths from remotely sensed imagery. *Geosci. Remote Sens. Lett. IEEE* **2008**, *5*, 70–73.

9.   Bjerklie, D.M.; Dingman, S.L.; Vörösmarty, C.J.; Bolster, C.H.; Congalton, R.G. Evaluating the potential for measuring river discharge from space. *J. Hydrol.* **2003**, *278*, 17–38.

10. Elmi, O.; Tourian, M.J.; Sneeuw, N. River discharge estimation using channel width from satellite imagery. In Proceedings of the 2015 IEEE International Geoscience and Remote Sensing Symposium (IGARSS), Milan, Italy, 26–31 July 2015; pp. 727–730.

11. Durand, M.; Gleason, C.J.; Garambois, P.A.; Bjerklie, D.; Smith, L.C.; Roux, H.; Rodriguez, E.; Bates, P.D.; Pavelsky, T.M.; Monnier, J.; et al. An intercomparison of remote sensing river discharge estimation algorithms from measurements of river height, width, and slope. *Water Resour. Res.* **2016**, *52*, 4527–4549.
12. Smith, L.C.; Isacks, B.L.; Bloom, A.L.; Murray, A.B. Estimation of discharge from three braided rivers using synthetic aperture radar satellite imagery: Potential application to ungaged basins. *Water Resour. Res.* **1996**, *32*, 2021–2034.
13. Kouraev, A.V.; Zakharova, E.A.; Samain, O.; Mognard, N.M.; Cazenave, A. Ob' river discharge from TOPEX/Poseidon satellite altimetry (1992–2002). *Remote Sens. Environ.* **2004**, *93*, 238–245.
14. Andreadis, K.M.; Clark, E.A.; Lettenmaier, D.P.; Alsdorf, D.E. Prospects for river discharge and depth estimation through assimilation of swath-altimetry into a raster-based hydrodynamics model. *Geophys. Res. Lett.* **2007**, *34*, 481–496.
15. Gleason, C.J.; Smith, L.C. Toward global mapping of river discharge using satellite images and at-many-stations hydraulic geometry. *Proc. Natl. Acad. Sci. USA* **2014**, *111*, 4788–4791.
16. Durand, M.; Neal, J.; Rodriguez, E.; Andreadis, K.M.; Smith, L.C.; Yoon, Y. Estimating reach-averaged discharge for the River Severn from measurements of river water surface elevation and slope. *J. Hydrol.* **2014**, *511*, 92–104.
17. Garambois, P.A.; Monnier, J. Inference of effective river properties from remotely sensed observations of water surface. *Adv. Water Resour.* **2015**, *79*, 103–120.
18. Paiva, R.C.D.; Durand, M.T.; Hossain, F. Spatiotemporal interpolation of discharge across a river network by using synthetic SWOT satellite data. *Water Resour. Res.* **2015**, *51*, 430–449.
19. Lorenz, C.; Tourian, M.J.; Devaraju, B.; Sneeuw, N.; Kunstmann, H. Basin-scale runoff prediction: An Ensemble Kalman Filter framework based on global hydrometeorological data sets. *Water Resour. Res.* **2015**, *51*, 8450–8475.
20. Pan, M.; Wood, E.F. Inverse streamflow routing. *Hydrol. Earth Syst. Sci.* **2013**, *17*, 4577–4588.
21. Tourian, M.J.; Schwatke, C.; Sneeuw, N. River discharge estimation at daily resolution from satellite altimetry over an entire river basin. *J. Hydrol.* **2017**, *546*, 230–247.
22. Neal, J.; Schumann, G.; Bates, P.; Buytaert, W.; Matgen, P.; Pappenberger, F. A data assimilation approach to discharge estimation from space. *Hydrol. Process.* **2009**, *23*, 3641–3649.
23. Yoon, Y.; Durand, M.; Merry, C.J.; Clark, E.A.; Andreadis, K.M.; Alsdorf, D.E. Estimating river bathymetry from data assimilation of synthetic {SWOT} measurements. *J. Hydrol.* **2012**, *464–465*, 363–375.
24. Paiva, R.C.D.; Collischonn, W.; Bonnet, M.P.; de Gonçalves, L.G.G.; Calmant, S.; Getirana, A.; Santos da Silva, J. Assimilating in situ and radar altimetry data into a large-scale hydrologic-hydrodynamic model for streamflow forecast in the Amazon. *Hydrol. Earth Syst. Sci.* **2013**, *17*, 2929–2946.
25. Pavelsky, T.M.; Durand, M.T.; Andreadis, K.M.; Beighley, R.E.; Paiva, R.C.; Allen, G.H.; Miller, Z.F. Assessing the potential global extent of SWOT river discharge observations. *J. Hydrol.* **2014**, *519 Pt B*, 1516–1525.
26. Mersel, M.K.; Smith, L.C.; Andreadis, K.M.; Durand, M.T. Estimation of river depth from remotely sensed hydraulic relationships. *Water Resour. Res.* **2013**, *49*, 3165–3179.
27. Legleiter, C.J.; Roberts, D.A.; Marcus, W.A.; Fonstad, M.A. Passive optical remote sensing of river channel morphology and in-stream habitat: Physical basis and feasibility. *Remote Sens. Environ.* **2004**, *93*, 493–510.
28. Legleiter, C.J.; Roberts, D.A.; Lawrence, R.L. Spectrally based remote sensing of river bathymetry. *Earth Surf. Proc. Landf.* **2009**, *34*, 1039–1059.
29. Legleiter, C.J.; Roberts, D.A. A forward image model for passive optical remote sensing of river bathymetry. *Remote Sens. Environ.* **2009**, *113*, 1025–1045.
30. Fonstad, M.A.; Marcus, W.A. Remote sensing of stream depths with hydraulically assisted bathymetry (HAB) models. *Geomorphology* **2005**, *72*, 320–339.
31. Leon, J.; Calmant, S.; Seyler, F.; Bonnet, M.P.; Cauhop, M.; Frappart, F.; Filizola, N.; Fraizy, P. Rating curves and estimation of average water depth at the upper Negro River based on satellite altimeter data and modeled discharges. *J. Hydrol.* **2006**, *328*, 481–496.
32. Bjerklie, D.M. Estimating the bankfull velocity and discharge for rivers using remotely-sensed river morphology information. *J. Hydrol.* **2007**, *341*, 144–155.
33. Biancamaria, S.; Durand, M.; Andreadis, K.; Bates, P.; Boone, A.; Mognard, N.; Rodríguez, E.; Alsdorf, D.; Lettenmaier, D.; Clark, E. Assimilation of virtual wide swath altimetry to improve Arctic river modeling. *Remote Sens. Environ.* **2011**, *115*, 373–381.

34. Durand, M.; Rodriguez, E.; Alsdorf, D.E.; Trigg, M. Estimating river depth from remote sensing swath interferometry measurements of river height, slope, and width. *IEEE J. Sel. Top. Appl. Earth Obs. Remote Sens.* **2010**, *3*, 20–31.

35. Gejadze, I.; Malaterre, P.O. Discharge estimation under uncertainty using variational methods with application to the full Saint-Venant hydraulic network model. *Int. J. Numer. Methods Fluids* **2017**, *83*, 405–430.

36. Dingman, S.L.; Sharma, K.P. Statistical development and validation of discharge equations for natural channels. *J. Hydrol.* **1997**, *199*, 13–35.

37. Raggi, M.; Ronchi, D.; Sardonini, L. *Po Basin Case Study Status Report*; Deliverable D32; University of Bologna: Bologna, Italy, 2007.

38. Elmi, O.; Tourian, M.J.; Sneeuw, N. Dynamic River Masks from Multi-Temporal Satellite Imagery: An Automatic Algorithm Using Graph Cuts Optimization. *Remote Sens.* **2016**, *8*, 1005, doi:10.3390/rs8121005.

39. Tourian, M.J.; Tarpanelli, A.; Elmi, O.; Qin, T.; Brocca, L.; Moramarco, T.; Sneeuw, N. Spatiotemporal densification of river water level time series by multimission satellite altimetry. *Water Resour. Res.* **2016**, *52*, 1140–1159.

40. Moramarco, T.; Singh, V.P. Simple Method for Relating Local Stage and Remote Discharge. *J. Hydrol. Eng.* **2001**, *6*, 78–81.

41. Ponce, V.M. *Engineering Hydrology: Principles and Practices*; Prentice Hall: Upper Saddle River, NJ, USA, 1989.

42. Baarda, W. *A Testing Procedure for Use in Geodetic Networks*; Netherlands Geodetic Commission: Delft, The Netherlands, 1968; Volume 2.

43. Kohli, P.; Torr, P. Measuring Uncertainty in Graph Cut Solutions efficiently Computing Min-marginal Energies Using Dynamic Graph Cuts. *Comput. Vis. ECCV* **2006**, *1*, 30–43.

44. Te Chow, V. *Open Channel Hydraulics*; McGraw-Hill Book Company, Inc.: New York, NY, USA, 1959.

45. Helmert, F.R. *Die Ausgleichungsrechnung nach der Methode der Kleinsten Quadrate: Mit Anwendungen auf die Geodäsie und die Theorie der Messinstrumente*; BG Teubner: Leipzig, Germany, 1872; Volume 1.

46. Roese-Koerner, L.R. Convex Optimization for Inequality Constrained Adjustment Problems. Ph.D. Thesis, Universitäts-und Landesbibliothek Bonn, Bonn, Germany, 2015; ISBN 978-3-7696-5171-3.

47. Biancamaria, S.; Lettenmaier, D.; Pavelsky, T. The SWOT Mission and Its Capabilities for Land Hydrology. *Surv. Geophys.* **2016**, *37*, 307–337.

48. Pavelsky, T. Recent Progress in Development of SWOT river Discharge Algorithms. In Proceedings of the 2012 SWOT Discharge Algorithms Workshop, Chapel Hill, NC, USA, 18–20 June 2012; pp. 18–20.

water

MDPI

Article

From Surface Flow Velocity Measurements to Discharge Assessment by the Entropy Theory

Tommaso Moramarco *, Silvia Barbetta and Angelica Tarpanelli

Research Institute for Geo-Hydrological Protection, National Research Council, Via Madonna Alta 126, 06128 Perugia, Italy; s.barbetta@irpi.cnr.it (S.B.); a.tarpanelli@irpi.cnr.it (A.T.)
* Correspondence: t.moramarco@irpi.cnr.it; Tel.: +39-075-50-1404

Academic Editor: Ataur Rahman
Received: 22 December 2016; Accepted: 9 February 2017; Published: 14 February 2017

Abstract: A new methodology for estimating the discharge starting from the monitoring of surface flow velocity, u_{surf}, is proposed. The approach, based on the entropy theory, involves the actual location of maximum flow velocity, u_{max}, which may occur below the water surface (dip phenomena), affecting the shape of velocity profile. The method identifies the two-dimensional velocity distribution in the cross-sectional flow area, just sampling u_{surf} and applying an iterative procedure to estimate both the dip and u_{max}. Five gage sites, for which a large velocity dataset is available, are used as a case study. Results show that the method is accurate in simulating the depth-averaged velocities along the verticals and the mean flow velocity with an error, on average, lower than 12% and 6%, respectively. The comparison with the velocity index method for the estimation of the mean flow velocity using the measured u_{surf}, demonstrates that the method proposed here is more accurate mainly for rivers with a lower aspect ratio where secondary currents are expected. Moreover, the dip assessment is found more representative of the actual location of maximum flow velocity with respect to the one estimated by a different entropy approach. In terms of discharge, the errors do not exceed 3% for high floods, showing the good potentiality of the method to be used for the monitoring of these events.

Keywords: streamflow measurements; surface velocity; maximum velocity; entropy; ADCP; LSPIV; SVR

1. Introduction

The estimate of a reliable discharge is tightly connected to a robust velocity dataset where the mean flow velocity is inferred using velocity measurements carried out for high water levels. This condition is not so usual considering that at hydrometric sites the most common technique used for velocity sampling is based on the current meter from the cableway, and measuring velocity points during high floods in the lower portion of flow area is difficult and dangerous for operators [1]. The stage monitoring is straightforward and relatively inexpensive compared with the cost necessary to carry out flow velocity measurements, which are, however, limited to low flows and constrained by the accessibility of the site. Therefore, the discharge estimation is often obtained by extrapolating the relationship between stage and the discharge obtained for velocity measurements referring to low flows, and, as a consequence, a high uncertainty holds in the assessment [2]. For high flow, the mean flow velocity is hard to estimate, affecting de-facto the reliability of the discharge assessment for extreme events.

To date, ultrasonic devices, such as Acoustic Doppler Current Profilers (ADCP), can be adopted to address the velocity measure for high floods. Indeed, the introduction of ADCP on moving-vessels would allow operators to overcome the aforementioned issues, even if several limitations remain. During high floods, the magnitude of sediment transport may produce a reduction of the signal-noise ratio of acoustic sensors along with the capability of the signal to penetrate through the water [3]. In addition, high flow conditions, mean high velocity, and high turbulence can affect the water depth estimation because of the roll and pitch motions of the vessel [4] as well as vessel tracking [5]. The high velocity and high water levels represent big trouble for the operators considering that branches or trunks of trees can be carried by the flow and could still represent a risk for the instrument as well as the safety of the operators themselves.

On the other hand, the surface flow velocity can be easily monitored by using no-contact technologies as Surface Velocity Radar (SVR) [6,7] and/or Large Scale Particle Image Velocimetry (LSPIV) [8,9]. The radar technology may be affected during low flow when the backscatter of the beam can be blended by high noise, while LSPIV might be not suitable for measurements during scarce seeding, illumination conditions, and/or for any overnight floods [10]. However, the SVR and LSPIV, being no-contact techniques, are suitable for high flows monitoring, without encountering the issues connected to the floating surface material, sediment transport, and dangers for operators.

The information content given by SVR and LSPIV is the spatial distribution of surface flow velocity from which the discharge at a river site might be estimated. In this context, a focal point is how to turn the surface velocity, u_{surf}, measured at a location (vertical) into depth-averaged velocity, u_{vert}, considering the different hydraulic and geometric characteristics of river sites which affect the velocity profile shapes [11]. In the case of maximum flow velocity occurs on the water surface, the two-parameters power law velocity distribution developed by Dingman [12] could be applied to this end.

However, the aspect ratio W/D (with W the channel width and D the flow depth) unfolds a fundamental role on the characteristics of the velocity profile because the presence of the sidewalls may influence the velocity distribution, causing secondary currents that drive the maximum velocity below the water surface, so triggering the well-known dip phenomenon [13].

Overall, the surface velocities are transformed in depth-averaged velocity, assuming a monotonous velocity profile and multiplying the surface velocity, u_{surf}, by a velocity index, $k = u_{vert}/u_{surf}$, equal to 0.85 (e.g., [7] for SVR and [9] for LSPIV). However, the possible occurrence of the dip phenomenon could make the assumption 'weak' for create flows where a monotonous velocity distribution does not take place. Consequently, the velocity index value might be not representative of the two-dimensional spatial velocity distribution anymore, leading to failure of the assessment of discharge.

The entropic velocity distribution developed by [14] can be used to cope with this matter. This approach established a bridge between the probability domain, wherein a probability distribution of the velocity is surmised, and the physical space by deriving the cumulative probability distribution function in terms of curvilinear coordinates in the physical space [15,16]. Therefore, it would be possible to relate the surface velocity measured by SVR and/or LSPIV across the river site, and the maximum velocity occurring below the water surface corresponding to the vertical where the surface velocity is sampled, as shown by [17]. In this context, the identification of the dip, i.e., the location where u_{max} occurs below the water surface, is the main issue. The dip may significantly affect the estimation of the depth-averaged velocity from which, through the velocity area method, the mean flow can be assessed and, hence, the discharge.

This work is aimed to present a new methodology based on the entropy theory of estimating the discharge, starting from the surface flow velocity and taking the dip of the velocity profile into account. Moreover, remarks on the robustness of the velocity index using field data are discussed. The entropy method can be applied regardless of the technique adopted for the surface velocity measurement.

Five gage sites with different geometric and hydraulic characteristics are used as a case study.

The paper is organized as follows. Section 2 provides the theoretical background of the entropy method applied to measured surface flow velocities. Section 3 describes the case study along with the velocity dataset. Section 4 describes the results achieved for the five selected river gage sites and makes a comparison with the velocity index method. Finally, the last section features the conclusions drawn from this investigation.

2. Method

The method is based on the entropy probability density function of velocity, u, such as derived by [18] and simplified by [19] assuming that the entropy velocity profile at the y-axis, where the maximum flow velocity, u_{max}, occurs, holds for all verticals in flow area, yielding:

$$u(x_i, y) = \frac{u_{maxv}(x_i)}{M} \ln\left[1 + \left(e^M - 1\right)\frac{y}{D(x_i) - h(x_i)}\exp\left(1 - \frac{y}{D(x_i) - h(x_i)}\right)\right] \text{ for } i = 1 \dots N_v \quad (1)$$

where $u_{maxv}(x_i)$ is the maximum velocity along the ith vertical, x_i is the position of the ith sampled vertical from the left bank, M is the entropic parameter, $h(x_i)$ is the dip, i.e., the depth of $u_{maxv}(x_i)$ below the water surface, $D(x_i)$ the vertical depth, y is the distance of the velocity point from the bed, and N_v is the number of verticals sampled across the river section. The entropic parameter M, which is a characteristic of the section [20], can be easily estimated through the pairs (u_m, u_{max}) of the available velocity dataset at a gauge site by using the linear entropic relation [14]:

$$u_m = \left(\frac{e^M}{e^M - 1} - \frac{1}{M}\right)u_{max} = \Phi(M)u_{max} \quad (2)$$

where u_m is the mean flow velocity and u_{max} is the maximum flow velocity in flow area.

In addition, Moramarco [21] showed that, at a gauged river site, $\Phi(M)$ is constant for any flow conditions and its estimation can be addressed for ungauged sites as well by leveraging the dependence of $\Phi(M)$ on hydraulic and geometric characteristics.

Even if u_{maxv} occurs below the water surface, it could be estimated for each vertical as a function of surface velocity, u_{surf}, such as proposed by [22]:

$$u_{maxv}(x_i) = \frac{u_{surf}(x_i, D(x_i))}{\frac{1}{M}\ln\left[1 + (e^M - 1)\delta(x_i)e^{1 - \delta(x_i)}\right]} \quad (3)$$

where $\delta(x_i) = \frac{D(x_i)}{D(x_i) - h(x_i)}$. Specifically, if $h(x_i) = 0$, it follows that $\delta(x_i) = 1$ and, hence, that $u_{maxv}(x_i) = u_{surf}(x_i, D(x_i))$. An application of Equation (3) in natural channels was done by [17], who calibrated a hydraulic model by leveraging the surface velocity measurements by SVR at different times.

2.1. Dip Estimate

Considering that the dip, $h(x_i)$, is generally unknown across the flow area, a method is proposed here for its estimation. The approach exploits the experimental outcomes obtained by [13] who investigated the mechanism of the dip phenomenon by leveraging a laboratory channel velocity dataset. Specifically, it was found that for different aspect ratios W/D (with W the channel width and D the flow depth), the location of $u_{maxv}(x_i)$ below the water surface, is mainly linked to the lateral position of the velocity profiles from the sidewalls. Based on this analysis, the dip phenomenon was more significant close to the banks and for large aspect ratio values, in accordance with the following relationship [13]:

$$\delta(x_i) = 1 + 1.3e^{-x_i/D(x_i)} \quad (4)$$

It is worth noting that Equation (4) well represented the measured data obtained in different works dealing with smooth and rough natural channels and we refer to [13] for more details.

Placing Equation (4) into Equation (3) and both into Equation (1), the two-dimensional velocity distribution can be identified by leveraging the surface velocity measured by SVR or LSPIV. In this way, u_m and u_{max} can be estimated using the velocity-area method [23], and, hence, $\Phi(M)$ can be computed by Equation (2) [19]. However, the uncertainty due to the presence of secondary currents lead us to modify Equation (4) in:

$$\delta'_p(x_i) = a_p + \delta(x_i) \tag{5}$$

where a_p is a parameter representing the dip at the y-axis and which is updated through an iterative process, wherein at each iteration, p, a constant value (0.1 m) is added to the initial value $a_1(p=1)$, assumed to be equal to 0.05 m. The iterative parameter refinement process ends by minimizing the objective function when the optimal value of $\Phi_{opt}(M_p)$ is achieved in accordance with:

$$\Phi(M_{obs}) - \Phi(M_p) = \epsilon \tag{6}$$

where ϵ is the threshold error, assumed equal to 0.01.

Steps of the procedure described above are detailed in the schematic diagram as shown in Figure 1. The method can be applied for gage sites in which a robust velocity dataset is available and $\Phi(M_{obs})$ can be easily estimated by the observed pairs (u_m, u_{max}). For ungauged sites, the M parameter can be even estimated by expressing its value in terms of hydraulic and geometric characteristics such as proposed by [21] and/or following [20], who identified a simple way to estimate M from the maximum surface velocity typically located near the middle of the channel [24].

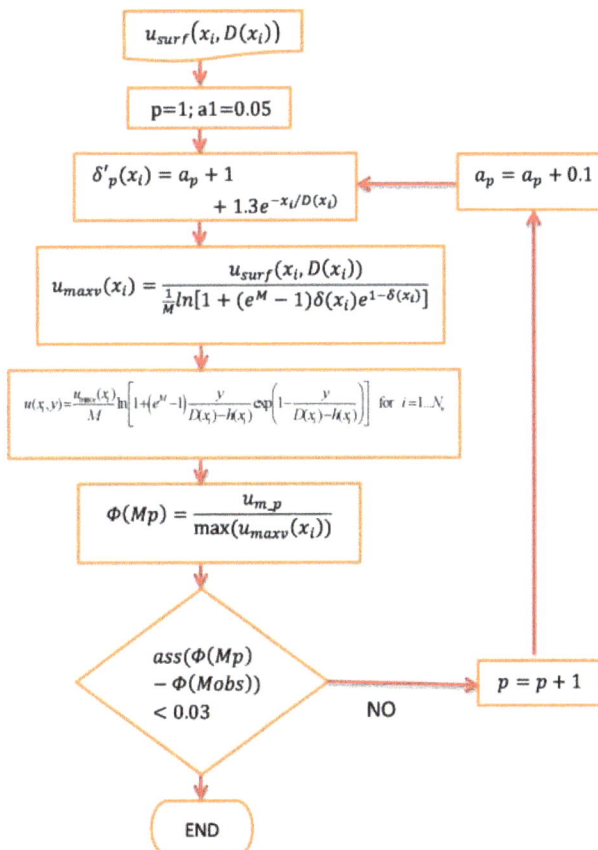

Figure 1. Schematic representation of dip estimation procedure using measured surface velocity, u_{surf} (for symbols see text).

For representing the dip in flow area, the relationship proposed by [25] may be also adopted:

$$\frac{h(x_i)}{D(x_i)} = 0.2ln\frac{58.3M\varnothing(M)}{e^M - 1} \tag{7}$$

Unlike the proposed method, Equation (7) does not need to consider an iterative procedure to estimate the dip along sampled verticals and would be simpler to apply. Specifically, Equation (7) provides $h(x_i)$ that can be used to estimate $\delta(x_i)$. This implies the interest in comparing the two approaches. However, it is worth noting that our target is not to estimate the effectiveness of these relationships in the dip assessment but rather to analyze whether, if given a distribution of dip in flow area, the proposed method is able to provide an accurate estimate of velocity, and considering that Equation (4) was tested with laboratory and field data, it is sufficient for our purpose. However, as stressed above, a comparison with Equation (7) is addressed as well and shown afterwards.

2.2. The Velocity Index

The velocity-area method is applied to estimate the mean flow velocity starting from the depth-averaged velocity, u_{vert}, estimated by velocity points along verticals sampled by using, e.g., current meter across the cross-sectional flow area. In this case, SVR is used for discharge monitoring, the measure of velocity is limited to the surface, and there is the need to turn the surface velocity u_{surf}, into u_{vert}. For that, an approach widely applied is to consider the velocity index, k, defined as:

$$k = u_{vert}/u_{surf} \tag{8}$$

As stressed by [7], k is set to 0.85 for many river gage site configurations (see also [6,9,10]). It has to be pointed out that the choice of k is of paramount importance, considering that Equation (8) refers to a monotonous velocity profile that, however, does not take the dip phenomenon into account. This issue could make the method 'weak' for estimating u_{vert}, in the case of secondary currents taking place.

3. Case Study

Considering that our main purpose is to evaluate the effectiveness of the proposed procedure, the analysis is addressed for the case of gage sites where the velocity measurements are limited only to the surface, e.g., by using SVR. This aspect is of considerable interest for the monitoring of high floods. Therefore, the proposed method is tested using a velocity dataset of five gauged sites. Four are located in the Tiber basin; Santa Lucia (900 km^2), Ponte Felcino (1970 km^2), and Ponte Nuovo (4100 km^2) along the Tiber River and Rosciano (1950 km^2) on the Chiascio River. One section is on the Po River, i.e., Pontelagoscuro (70,000 km^2).

The velocity dataset consists of velocity points data collected during velocity measurements carried out by conventional techniques, i.e., current meter from cableway. The number of velocity points sampled along the verticals is sufficient (more than eight for some verticals) to reconstruct the vertical velocity profile, even in presence of secondary currents. Table 1 shows for all the selected gaged sites the main flow characteristics and, as it can be seen, the measurements cover low and high flow conditions.

Table 1. Flow dataset characteristics: N_m = number of velocity measurements considered, N_v = total number of verticals (from the N_m measurements), $\Phi(M_{obs})$ = observed entropy parameter, Q = measured discharge, D = average flow depth, W = average channel width. The period of sampling is also shown.

River	Site	N_m	N_v	$\Phi(M_{obs})$	Q (m³/s)	D (m)	W (m)	W/D	Period
Tiber	Santa Lucia	16	93	0.66	30–185	2.54	22	8.79	1987–2008
Tiber	Ponte Felcino	8	78	0.60	28–411	3.22	38	11.66	1990–2003
Tiber	Ponte Nuovo	12	77	0.66	10–540	3.98	50	12.30	1986–2005
Chiascio	Rosciano	8	67	0.60	20–378	2.74	35	12.90	1982–2002
Po	Pontelagoscuro	8	89	0.68	380–4000	6.32	270	43.63	1987–1992

4. Results and Discussion

The effectiveness of the proposed procedure in estimating the dip phenomena along with the velocity profile by Equation (1) is discussed here using the velocity dataset of the selected river gage sites. The errors in estimating the depth-averaged velocity along with the discharge are analyzed in-depth. Before that, however, the robustness of the velocity index is investigated using the velocity measurements carried out at gage sites, considering that these measurements contain velocity points well distributed along each vertical sampled in the cross-sectional flow area.

4.1. The Velocity Index Analysis

At a first step, the analysis of the velocity index, k, is addressed here. Unlike [7], who identified the variation of k with relative roughness and flow depth, we investigate how k can vary at each single site and if a dependence on the aspect ratio can be inferred. Considering the velocity dataset in the five gage sites, for each velocity measurement the depth-averaged velocity, u_{vert}, is estimated starting from the measured surface velocity, u_{surf}, i.e., $u_{vert}(x_i, D_i) = k\, u_{surf}(x_i, D_i)$, and compared with the depth-averaged velocity assessed by the sampled velocity points along each vertical.

The velocity at the water surface sampled by the current meter is used to mimic the surface velocity measured by SVR or LSPIV. It is worth noting that the analysis does not aim to show the capabilities of the above two technologies but only to verify the relationship between surface velocity and depth-averaged velocity, which is useful to estimate the mean flow velocity by using the velocity-area method. Figure 2 shows the box-plot of the velocity index values for the five gage sites as a function of the aspect ratio, W/D, in terms of the 5th and 95th percentile along with the median, minimum, and maximum values. As can be seen, the velocity index is quite scattered at each gage site, while the median value is inversely proportional with the aspect ratio, starting from 1.05 at Santa Lucia, the narrowest river section, and dropping to 0.86 at Pontelagoscuro, the widest site. Therefore, the average value of the k index observed for the five river gage sites deviates from the constant value of 0.85 used in the literature (e.g., [6,7]); for the Po River the mean value is close to 0.85. It is worth nothing that $k > 1$ means the dip phenomenon is significant in the velocity profile, causing a consistent curved shape approaching a parabolic one and leading to the depth-averaged velocity being greater than the surface velocity.

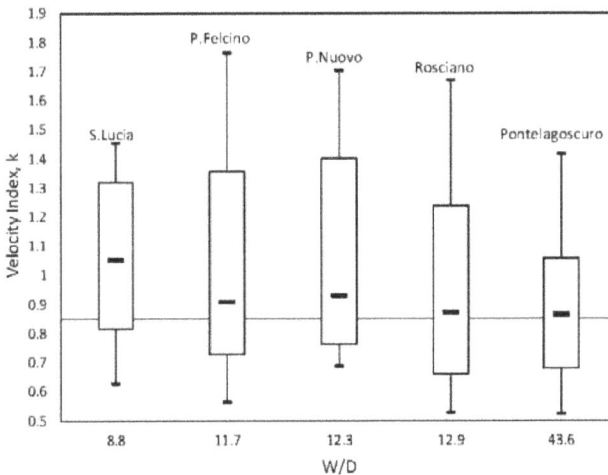

Figure 2. Box-plot of the values of the velocity index, k, as a function of the aspect ratios, W/D, for the five gage sites based on the velocity dataset. The line corresponding to the value of 0.85 is also shown along with the name of the gage sites.

Moreover, inspecting Figure 2, one infers that the velocity index is also influenced by the geometry of the river site, such as expressed by the aspect ratio, and while the median value tends to decrease for large aspect ratios (e.g., Pontelagoscuro site with $W/D = 43.6$), the distribution of the velocity index is always scattered, even in the widest channel. Indeed, as can be inferred from Figure 2, the mean value of k tends to be 0.85 for large values of the aspect ratio, while the variance seems to decrease only for the largest W/D occurring at the Pontelagoscuro site.

It has to be stressed that the limited sample of river gage sites does not allow a general rule for applying the velocity index to any river worldwide to be inferred. However, Figure 3 provides useful insights into understanding why $k = 0.85$ cannot be considered a parameter to be applied to rivers everywhere and for whatever flow condition.

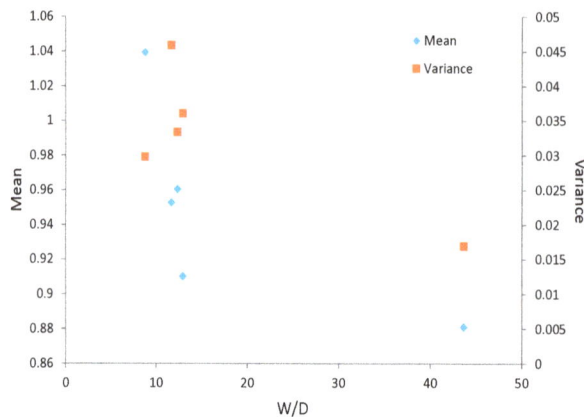

Figure 3. Mean and variance of the velocity index, k, for the selected gage sites as a function of the aspect ratio (W/D).

At a first attempt, Figure 3 could be useful for addressing the uncertainty in estimating the velocity index as a function of W/D. In this case, identifying the relationship expressing k_{mean} and $k_{variance}$ as a function of (W/D) could be of support in understanding the variability of the velocity index and, hence, to evaluate how much the depth-averaged velocity can vary at each vertical sampled in correspondence to the surface velocity measurement. However, more gage sites are necessary to identify a robust relationship that can be leveraged in addressing the uncertainty in the k estimation.

Therefore, the above results indicate that, by using new technologies as SVR and/or LSPIV, the application of a velocity index when dip phenomena occur may affect the accuracy of the mean flow velocity assessment and, hence, of the discharge.

4.2. Comparison of Depth-Averaged Velocities

Based on the previous analysis, it is evident that the assumption of a constant index velocity, i.e., 0.85, might lead to an incorrect estimate of the depth-averaged velocity and, hence, of the discharge. The variability of the index velocity for each vertical sampled in the cross-sectional flow area depends on the presence of dip phenomena due to effect of secondary currents caused by side walls, bridges, and wind effects, as well as the presence of floating material. These effects seem to tend to be smooth for large values of the aspect ratio. Therefore, in the case of streamflow measurements the use of the velocity index method should be applied with care and the reliability verified.

The entropy method proposed here allows the location of the maximum velocity below the water surface (dip) for each vertical sampled in the flow area to be estimated, starting from the measured surface velocity. The procedure illustrated in Figure 1 is applied for all measurements available at the selected gage sites and the results are compared with the ones obtained using the velocity index method with $k = 0.85$.

To this end, the error in magnitude for the depth-averaged velocity is computed for each gage sites and the results are shown in Figure 4 in terms of a box-plot referring to the median, 5th, and 95th percentile. As can be seen and expected, the differences are more noticeable for the narrower gage sites, with a median that does not exceed 6% and 11% for the entropy and the velocity index method, respectively; while for the Pontelagoscuro site, the differences are less evident, even though a slightly larger scatter is always present in the velocity index approach.

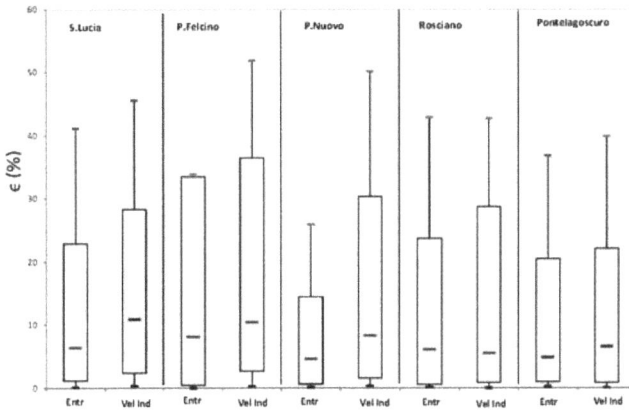

Figure 4. Box-plot of absolute percentage errors (ϵ) in estimating the depth-averaged velocity using the entropy approach (Entr) and the velocity index (Vel Ind).

The entropy approach has been also compared with Chiu and Tung's formulation [24], for which the dip is computed without an iterative procedure. In this case, the dip computed by Equation (7) is replaced in Equation (3), from which $u_{maxv}(x_i)$ is computed and used in Equation (1). By way of example, Figure 5 shows for the Rosciano and Pontelagoscuro gage sites the comparison between the errors in depth-averaged velocity computed using the dip assessment by [25] and the proposed iterative method. As can be seen, the iterative entropy approach performs better than the one based on Chiu's dip, and, as expected, the difference tends to decrease if the aspect ratio increases as for the Pontelagoscuro site. Similar performances obtained at the Rosciano site are for the three sites with lower aspect ratios (W/D). Table 2 shows the mean and standard deviation of the percentage errors in magnitude of estimating the depth-averaged velocity by using Equation (7) and the method proposed here for the dip assessment, showing again that the dip computed by the iterative procedure is more accurate than that based on Equation (7).

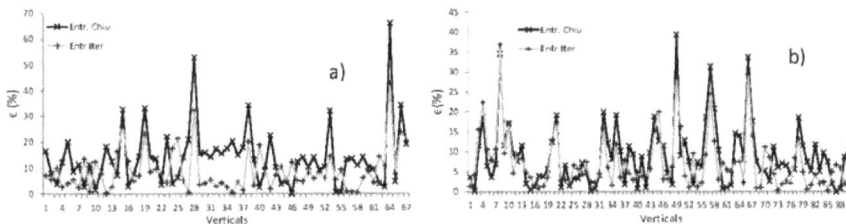

Figure 5. Absolute percentage error in estimating the depth-averaged velocity using the dip assessment by the iterative procedure proposed here (Entr Iter) and Chiu's formula (Entr Chiu) for (**a**) the Rosciano and (**b**) the Pontelagoscuro gage sites.

Table 2. Mean and standard deviation of the absolute percentage error in estimating the depth-averaged velocity using the dip assessment by the iterative procedure proposed here (Entr Iter) and Equation (7) (Entr Chiu) for the investigated gage sites.

Gage Site	Entr Iter		Entr Chiu	
	Mean (%)	Standard Deviation (%)	Mean (%)	Standard Deviation (%)
Santa Lucia	9.8	8.8	12.5	9.1
Ponte Felcino	11	9.6	19	11.1
Ponte Nuovo	8.6	7.9	15.4	10.2
Rosciano	8.9	8.1	15.1	11.4
Pontelagoscuro	7.2	6.9	8.9	7.8

4.3. Comparison of Mean Flow Velocity and Discharge

Once the depth-averaged velocities are computed for each measurement, the cross-sectional mean flow velocity can be estimated using the velocity-area method. Considering that the estimate of dip across the flow area by the proposed approach is more robust than that based on Equation (7), the analysis is here addressed for the proposed procedure and velocity index only.

Figure 6 shows the box-plot of the absolute percentage error in the computation of mean flow velocity for the whole velocity dataset for the five investigated gage sites. As can be seen, the entropy approach performs better than the velocity index one, as is inferable by comparing the median and 95th percentile of errors and the differences in terms of the median and scattering tend to smooth for larger aspect ratios, as for the Rosciano and Pontelagoscuro gage sites. The mean percentage error does not exceed 4.5% and 10% for the entropy and velocity index methods, respectively.

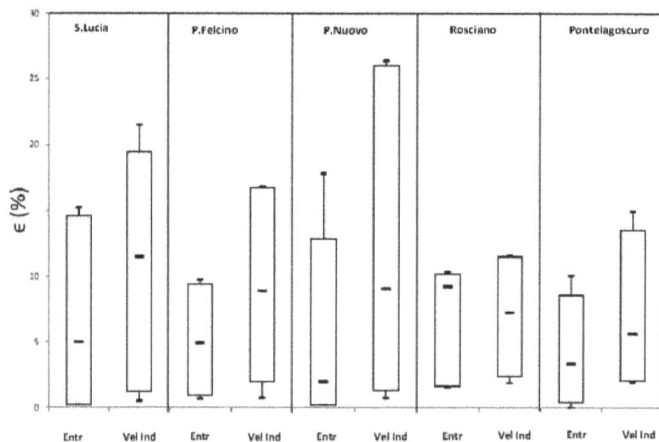

Figure 6. Box-plot of the absolute percentage error (ϵ) in estimating the cross-sectional mean flow velocity using the Entropy approach (Entr) and the velocity index (Vel Ind).

Moreover, considering the performances of the two methods the greatest differences are identified for higher mean flow velocity, as can be seen in Figure 7, where a comparison between the observed discharge and the one computed by the entropy and velocity index approaches is shown for the Santa Lucia, Ponte Nuovo, and Pontelagoscuro gage sites. Similar results are found for Ponte Felcino and Rosciano. Figure 7 demonstrates that, for the velocity index method ($k = 0.85$), the maximum errors shown in the graph in Figure 6 are referred to as the high discharges with an error greater than 26%. This proves that the velocity index method needs to be applied with care if high flood events are of interest. Conversely, for the proposed entropy approach the maximum error in percentage is obtained for the lower flows and does not exceed 14%, while for high floods the errors do not exceed

3%, thus showing the benefit in using the method for the monitoring of high floods starting from the sampling of surface flow velocity.

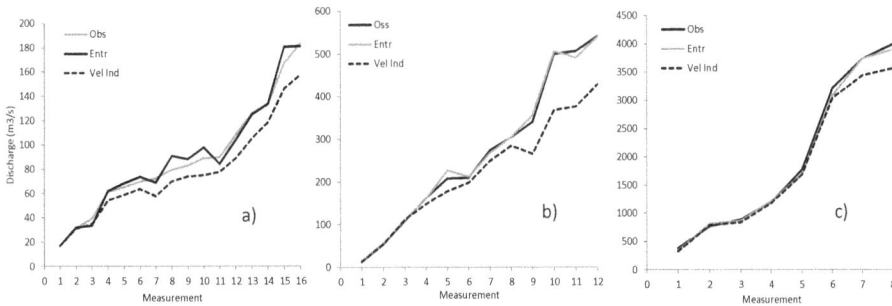

Figure 7. Comparison between observed discharge (Obs) and discharge computed by the Entropy approach (Entr) and the velocity index (Vel Ind) for (**a**) Santa Lucia; (**b**) Ponte Nuovo; and (**c**) Pontelagoscuro.

Finally, the correlation between u_{maxv} values observed and computed by the entropy method for each vertical is found good with the coefficient of determination, R^2, varying from 0.84 for the narrower sites (Santa Lucia and Ponte Felcino) to 0.94 for the wider ones. Moreover, by way of example, Figure 8 shows for the highest flood in the dataset of the Ponte Nuovo gauged site, the comparison between the velocity profile at four different verticals computed by the entropy method, and the velocity points sampled by current meter. It is worth noting that the velocity profiles are obtained just by sampling u_{surf}, and no information is used in terms of dip. It is noticeable how the velocity profile shape computed by Equation (1) is able to embed the observed points and, particularly, u_{maxv}. Similar performances are obtained for the velocity dataset of the other investigated gage sites.

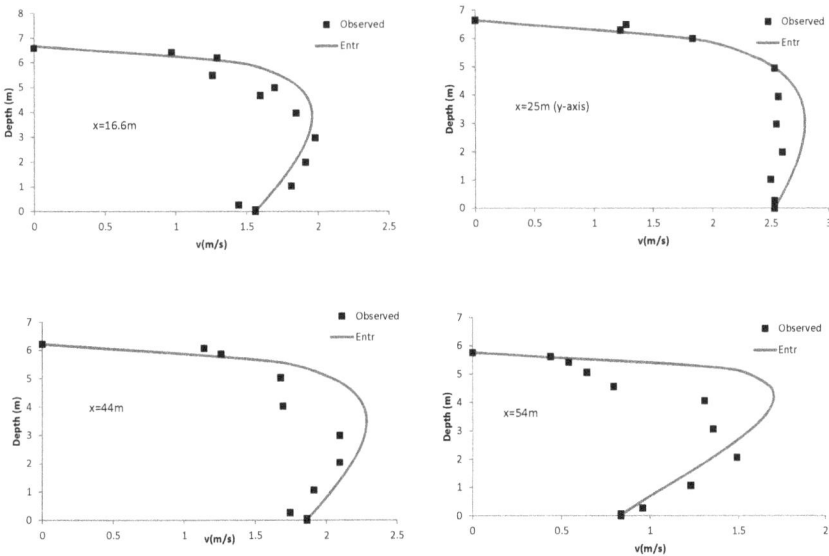

Figure 8. Velocity profiles estimated by entropy method (Equations (1) and (5)) plotted against velocity points sampled by current meter along four verticals during the highest flood of the velocity dataset at the Ponte Nuovo gage site. x is the distance from the left bank, while the y-axis represents the vertical where u_{max} occurs.

95

5. Conclusions

The trialing of emerging no-contact technologies, like SVR and LSPIV, for surface flow velocity measurement in rivers is of considerable interest and represents an effective alternative to conventional monitoring. These new technologies allow the issues related to the limited availability of hydraulic information due to low flow conditions and the difficulties and dangers to operators measuring velocity points in the lower portion of flow area using standard sampling techniques to be overcome.

In this context, this work addresses the crucial matter of turning the surface velocity into depth-averaged velocity, even if secondary current effects occur, for whatever natural channel geometry, providing some useful insights.

Based on the field data used, it has been shown that the proposed entropic algorithm well lends itself to being applied when advanced no-contact technology is used to estimate the discharge by monitoring the surface flow velocity across the river. The dip phenomenon, having a fundamental role in depth-averaged velocity assessment, is accurately identified by the entropy procedure for whatever flow conditions at the investigated gage sites.

The velocity index method, with $k = 0.85$, is of little use for rivers with a lower aspect ratio, with significant errors in estimating high discharges; while for wider rivers, its performance is found quite similar to that of the entropy method, considering that the dip phenomena are quite negligible. The relationship found between the mean and variance of the aspect ration of the k index needs to be further investigated by extending the velocity dataset and the case studies.

Acknowledgments: This work was partly funded by The Project of Interest NextData (MIUR-CNR). The authors wish to thank the Umbria Region Department of Environment, Planning, and Infrastructure, for providing Tiber River basin data. The authors want to thank Cristiano Corradini for the technical support. Finally, the Reviewers are gratefully acknowledged for their useful comments.

Author Contributions: All authors contributed extensively to the work. Tommaso Moramarco developed the methodology, conceived and performed the numerical experiments, and wrote the article. Silvia Barbetta and Angelica Tarpanelli contributed equally to the discussion of methodology, results analysis and the manuscript writing.

Conflicts of Interest: The authors declare no conflict of interest.

References

1. Moramarco, T.; Saltalippi, C.; Singh, V.P. Velocity profiles assessment in natural channel during high floods. *Hydrol. Res.* **2011**, *42*, 162–170. [CrossRef]
2. Di Baldassarre, G.; Montanari, A. Uncertainty in river discharge observations: A quantitative analysis. *Hydrol. Earth Syst. Sci.* **2009**, *13*, 913–921. [CrossRef]
3. Muller, G.; Bruce, T.; Kauppert, K. Particle image velocimetry: A simple technique for complex surface flows. In Proceedings of the International Conference on Fluvial Hydraulics, Balkema, Lisse, The Netherlands, 4–6 September 2002; pp. 1227–1234.
4. Simpson, M.R.; Oltman, R.N. *Discharge Measurement System Using an Acoustic Doppler Current Profiler with Applications to Large Rivers and Estuaries: U.S. Geological Survey Water-Supply Paper 2395*; U.S. Geological Survey: Reston, VA, USA, 1993; 32p.
5. Wagner, C.R.; Mueller, D.S. Comparison of bottom-track to global positioning system referenced discharges measured using an acoustic Doppler current profiler. *J. Hydrol.* **2011**, *401*, 250–258. [CrossRef]
6. Costa, J.E.; Spicer, K.R.; Cheng, R.T.; Haeni, F.P.; Melcher, N.B.; Thurman, E.M. Measuring stream discharge by non-contact methods: A proof-of-concept experiment. *Geophys. Res. Lett.* **2000**, *27*, 553–556. [CrossRef]
7. Welber, M.; Le Coz, J.; Laronne, J.B.; Zolezzi, G.; Zamler, D.; Dramais, G.; Hauet, A.; Salvaro, M. Field assessment of noncontact stream gauging using portable surface velocity radars (SVR). *Water Resour. Res.* **2016**, *52*, 1108–1126. [CrossRef]
8. Fujita, I.; Muste, M.; Kruger, A. Large-scale particle image velocimetry for flow analysis in hydraulic applications. *J. Hydraul. Res.* **1998**, *36*, 397–414. [CrossRef]
9. Tauro, F.; Porfiri, M.; Grimaldi, S. Orienting the camera and firing lasers to enhance large scale particle image velocimetry for streamflow monitoring. *Water Resour. Res.* **2014**, *50*, 7470–7483. [CrossRef]

Water **2017**, *9*, 120

10. Muste, M.; Fujita, I.; Hauet, A. Large-scale particle image velocimetry for measurements in riverine environments. *Water Resour. Res.* **2008**, *44*. [CrossRef]
11. Ferro, V. ADV measurements of velocity distributions in a gravel-bed flume. *Earth Surf. Proccess. Landf.* **2003**, *28*, 707–722. [CrossRef]
12. Dingman, S.L. Probability Distribution of Velocity in Natural Channel Cross Sections. *Water Resour. Res.* **1989**, *25*, 509–518. [CrossRef]
13. Yang, S.Q.; Tan, S.K.; Lim, S.Y. Velocity distribution and dip-phenomenon in smooth uniform open channel flows. *J. Hydraul. Eng.* **2004**, *130*, 1179–1186. [CrossRef]
14. Chiu, C.L. Application of Entropy Concept in Open-Channel Flow Study. *J. Hydraul. Eng.* **1991**, *117*, 615–627. [CrossRef]
15. Singh, V.P.; Fontana, N.; Marini, G. Derivation of 2D power-law velocity distribution using entropy theory. *Entropy* **2013**, *15*, 1221–1231. [CrossRef]
16. Fontana, N.; Marini, G.; De Paola, F. Experimental assessment of a 2-D entropy-based model for velocity distribution in open channel flow. *Entropy* **2013**, *15*, 988–998. [CrossRef]
17. Corato, G.; Moramarco, T.; Tucciarelli, T. Discharge estimation combining flow routing and occasional measurements of velocity. *Hydrol. Earth Syst. Sci.* **2011**, *15*, 2979–2994. [CrossRef]
18. Chiu, C.L. Velocity distribution in open channel flow. *J. Hydraul. Eng.* **1989**, *115*, 576–594. [CrossRef]
19. Moramarco, T.; Saltalippi, C.; Singh, V.P. Estimation of mean velocity in natural channel based on Chiu's velocity distribution equation. *J. Hydrol. Eng.* **2004**, *9*, 42–50. [CrossRef]
20. Farina, G.; Alvisi, S.; Franchini, M.; Moramarco, T. Three methods for estimating the entropy parameter M based on a decreasing number of velocity measurements in a river cross-section. *Entropy* **2014**, *16*, 2512–2529. [CrossRef]
21. Moramarco, T.; Singh, V. Formulation of the entropy parameter based on hydraulic and geometric characteristics of river cross sections. *J. Hydrol. Eng.* **2010**, *15*, 852–858. [CrossRef]
22. Fulton, J.; Ostrowski, J. Measuring real-time streamflow using emerging technologies: Radar, hydroacoustics, and the probability concepts. *J. Hydrol.* **2008**, *357*, 1–10. [CrossRef]
23. Herschy, R.W. *Streamflow Measurement*; Elsevier: London, UK, 1985.
24. Corato, G.; Melone, F.; Moramarco, T.; Singh, V.P. Uncertainty analysis of flow velocity estimation by a simplified entropy model. *Hydrol. Process.* **2014**, *28*, 581–590. [CrossRef]
25. Chiu, C.-L.; Tung, N.-C. Maximum velocity and regularities in open-channel flow. *J. Hydraul. Eng.* **2002**, *128*, 390–398. [CrossRef]

Article

Discharge Measurements of Snowmelt Flood by Space-Time Image Velocimetry during the Night Using Far-Infrared Camera

Ichiro Fujita

Department of Civil Engineering, Graduate School of Engineering, Kobe University, Kobe 657-8501, Japan; ifujita@kobe-u.ac.jp; Tel.: +81-78-803-6439

Academic Editor: Roberto Ranzi
Received: 12 December 2016; Accepted: 7 April 2017; Published: 11 April 2017

Abstract: The space time image velocimetry (STIV) technique is presented and shown to be a useful tool for extracting river flow information non-intrusively simply by taking surface video images. This technique is applied to measure surface velocity distributions on the Uono River on Honshu Island, Japan. At the site, various measurement methods such as a radio-wave velocity meter, an acoustic Doppler current profiler (ADCP) or imaging techniques were implemented. The performance of STIV was examined in various aspects such as a night measurement using a far-infrared-ray (FIR) camera and a comparison to ADCP data for checking measurement accuracy. All the results showed that STIV is capable of providing reliable data for surface velocity and water discharge that agree fairly well with ADCP data. In particular, it was demonstrated that measurements during the night can be conducted without any difficulty using an FIR camera and the STIV technique. In particular, using the FIR camera, the STIV technique can capture water surface features better than conventional cameras even at low resolution. Furthermore, it was demonstrated that measurements during the night can be conducted without any difficulty.

Keywords: river flow measurement; surface flow; image-based technique; Space Time Image Velocimetry (STIV); far-infrared camera; Large Scale Image Velocimetry (LSPIV)

1. Introduction

Intensive and record-breaking rainfall caused severe flood disasters in the world and in Japan in recent years partially due to the global warming [1]. For example, in July 2011, Niigata and Fukushima areas suffered from torrential rain with levee breaches and inundation disasters at many locations due to the maximum flood in recorded history. In September 2011, typhoon No. 12 directly hit Nara and Wakayama Prefectures causing landslide and sediment-related disasters in a wide area. Furthermore, in July 2012, a total of eight first-class river systems in the northern part of Kyushu Island were severely damaged twice by a torrential rain brought by a seasonal rain front. In 2015, the Kinugawa River, a first-class river in Kanto area, suffered from a levee breach, causing a huge inundated area while destroying nearby houses. In the Japanese river management system, 109 rivers are designated as first-class rivers, for which the Japanese government is responsible for the management of major river reach. The other smaller rivers of more than four thousand are classified as second-class rivers, for which the local government takes care of the river reach.

It should also be mentioned that huge flood disasters are taking place every year in the world; e.g., Australia, Brazil, China and Thailand were struck by a record-breaking flooding in the past decades [2]. A specific feature of the recent flood is characterized by the fact that heavy rain with an hourly rainfall intensity of about 100 mm continues for several hours, and this causes a sudden increase of river discharge, which results in an inundation in the worst case. In order to be prepared for such disasters, it is of vital importance to collect and analyze hydrological data such as rainfall intensity, water level and flow discharge. The acquisition of such data is important for the purpose of understanding rainfall run-off processes correctly while tuning various parameters included in a run-off model [3,4]. Among the hydrological data, peak discharge is the most important information because it can determine the flow capacity of a river section in response to a heavy rainfall. However, contrary to the floods in continental large-scale rivers, which last several days or weeks, typical floods in Japan have a duration of several hours; therefore, it becomes difficult to correctly measure the flow at the right timing especially when the peak flow occurs during the night time. In the past, discharge measurements in Japan have been conducted mainly by using a float [5] with its length dependent on the water depth. In Japan, float is made of cardboard pipe in which a certain amount of sand is contained to control its specific gravity close to unity when it floats. According to the Japanese regulation, floats with different lengths are used depending on the water depth; e.g., two-meter float is used when the water depth is greater than 2.6 m and less than 5.2 m.

However, there exist several critical defects in this method because a measurement site of a bridge can be submerged in a huge flood, as actually occurred in the 2011 Niigata Flood. In this case, the measurement itself became impossible and one missed a chance to measure the peak flow. Furthermore, the road to the measurement site was closed due to inundation. In the case of the flood of Kyushu in 2012, the water level rose to the height of the river bank and overflowing flooding actually occurred. Particularly in the dark night time, it is difficult to apply the float method as the measurement might become dangerous in such a condition. In addition, even if measurement is possible, the float may seriously meander under the influence of large-scale river turbulence, which may greatly reduce measurement accuracy.

In the light of such a present status of discharge measurement, the so-called non-intrusive measurement techniques have been paid attention to in recent years, such as a radio-wave velocity meter (RVM) [6,7] or imaging techniques [8–12]. The RVM is usually installed at a bridge to measure streamwise surface velocity continuously at one point by using the effect of the Doppler shift of the surface flow [6,7]. However, the instrument has to be shifted back and forth along the bridge when trying to measure a cross-sectional velocity distribution or a large number of the instruments have to be installed along a bridge with a high cost. An alternative method is the use of surface-flow video images taken from a river bank. The idea behind the use of video images is that surface flow features composed of surface ripples or floating objects are assumed to be advected with the surface velocity as long as the wind effect is negligible. Among the image-based techniques, such as a large-scale image velocimetry (LSPIV) [13–15] or the space-time image velocimetry (STIV) [16–20], STIV seems to be more robust under deteriorated image condition and utilized in the present research. The non-intrusive methods are very useful and effective as they can measure the flow velocity information safely from a remote place. However, the image obtained by a conventional video camera has a shortcoming incapable of recording favorable images during the night. Since peak flows frequently occur in a dark condition in Japan, the existing river monitoring systems' installed conventional cameras are difficult to use for flow measurements. In order to overcome this weak point, we propose a method to use a far-infrared ray (FIR) camera in river flow measurements. It is worth mentioning that traditional invasive instruments such as propeller ammeters, electromagnetic flowmeters or ADCP cannot be applied during high-speed floods. On the other hand, STIV, for example, can measure velocities over 5 m/s [16]. In this paper, we examined the traceability of the surface texture composed of surface ripples without tracers to the surface flow, and discussed the measurement accuracy of river flow rate by using the image obtained by the FIR camera. Furthermore, robustness of the STIV technique to LSPIV in measuring streamwise velocity components when using deteriorated images of low resolution will be shown.

2. Outline of Space-Time Image Velocimetry

2.1. Image Rectification by Camera Calibration and Search Lines

STIV is an image analysis technique to measure streamwise water surface velocity distributions from video images usually taken from a riverbank. This technique assumes that features that appeared on the water surface would follow the surface velocity when the wind effect is negligible. The idea to use surface feature advection is the same as the radio-wave velocity meter or LSPIV. The aspect of STIV different from the other techniques is that STIV pays attention to the time- and space-averaged velocity along a line segment set in the streamwise direction. The line segment is treated as a search line for velocity estimation.

In the first step of STIV, establishing a relation between the screen coordinates (x, y) and the physical coordinates (X, Y, Z), by using the following collinearity equation is required:

$$x = -c \left[\frac{a_{11}(X - X_0) + a_{12}(Y - Y_0) + a_{13}(Z - Z_0)}{a_{31}(X - X_0) + a_{32}(Y - Y_0) + a_{33}(Z - Z_0)} \right] + x_0 \tag{1}$$

$$y = -c \left[\frac{a_{21}(X - X_0) + a_{22}(Y - Y_0) + a_{23}(Z - Z_0)}{a_{31}(X - X_0) + a_{32}(Y - Y_0) + a_{33}(Z - Z_0)} \right] + y_0 \tag{2}$$

where c is the focal distance, a_{ij} with $1 \leq i \leq 3$ and $1 \leq j \leq 3$ are mapping coefficients, (X_0, Y_0, Z_0) is the physical coordinates of the camera, and (x_0, y_0) is the center of the screen image. The coefficients can be determined via a camera calibration procedure by using at least six ground control points. The image rectification can be performed by substituting the water level data H to Z in the above equations. An example of the image rectification for an image obtained by using a normal high definition video camera is shown in Figure 1, together with the search lines set parallel to the streamwise direction. The length of the search line is 17.6 m and the transverse spacing is 5.6 m in this case. The set of search lines covers almost the whole width of the river of 140 m. The direction of search lines is determined by visual inspection after generating a rectified image, i.e., the lines are drawn parallel to the riverbank.

(a)   (b)

Figure 1. Example of an ortho-rectified image (**b**) for an image (**a**) obtained by using a normal high definition video camera, each with search lines.

2.2. Space Time Image

After the set of search lines, space time images (STIs) are generated for each search line by taking brightness distribution in the search line as the horizontal axis and its evolution with time as the vertical axis. Several examples of STI in a transverse direction is shown in Figure 2. The horizontal length is 17.6 m and the downward vertical axis is ten seconds. Since the frame rate of image sampling is 30 frames per second, the vertical length is composed of 300 images. The images are enhanced for a better recognition of surface features by applying the histogram equalization technique. It is clearly seen that an inclined pattern is generated in each STI, indicating that overall surface features are advected along the search line with the flow at almost constant speed, although there are some other noisy patterns created by surface wave motion moving in the direction different from the main flow. The average gradient of the pattern corresponds to the local time- and space-averaged mean velocity along the search line.

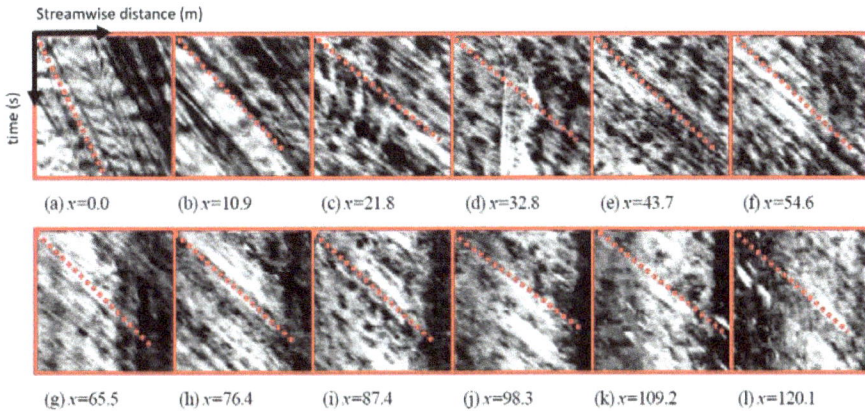

Figure 2. Space time images obtained by normal high definition camera at several transverse location (unit in meter): horizontal scale is 17.6 m and downward vertical scale is ten seconds. Image are enhanced for a better recognition. The dotted line is the main orientation angle of the pattern.

2.3. Measurement of Orientation Angle of the Pattern

In the original STIV, the average orientation angle f of a STI is obtained by dividing the STI into small segments and calculating the local gradient for each segment as indicated in Figure 3. The local orientation angle ϕ can be calculated using the following relations [17]:

$$\tan 2\phi = \frac{2J_{xt}}{J_{tt} - J_{xx}} \tag{3}$$

where

$$J_{xx} = \int_A \frac{\partial g}{\partial x} \frac{\partial g}{\partial x} dxdt \tag{4}$$

$$J_{xt} = \int_A \frac{\partial g}{\partial x} \frac{\partial g}{\partial t} dxdt \tag{5}$$

$$J_{tt} = \int_A \frac{\partial g}{\partial t} \frac{\partial g}{\partial t} dxdt \tag{6}$$

$g(x,t)$ is the grey level intensity in STI and A is the area of the small segment indicated in Figure 3a. Typically, A is a square with a side length of fifteen pixels. Figure 3b provides a distribution of the coherency defined by

$$C_{oh} = \frac{\sqrt{(J_{tt} - J_{xx})^2 + 4J_{xt}^2}}{J_{xx} + J_{tt}} \tag{7}$$

which is a measure of the image pattern coherence and takes a value between zero and one; for an ideal local orientation, its value becomes one, and, for an isotropic gray image, it becomes zero. Therefore, it is possible to calculate the mean orientation angle by preferably obtaining clearer orientation information by taking coherency as a weighting function:

$$\bar{\phi} = \int \phi C_{oh}(\phi) d\phi / \int C_{oh}(\phi) d\phi \tag{8}$$

Figure 3c is a weighted histogram of the orientation angle, showing a distribution with a peak. Since the length and time scales of the STI are given, the mean velocity along the line segment can be calculated by the following relation:

$$U_s = \frac{S_x}{S_t} \tan \overline{\phi}$$ (9)

where S_x is the unit length scale along the search line (m/pixel) and S_t is the unit time scale of the time axis.

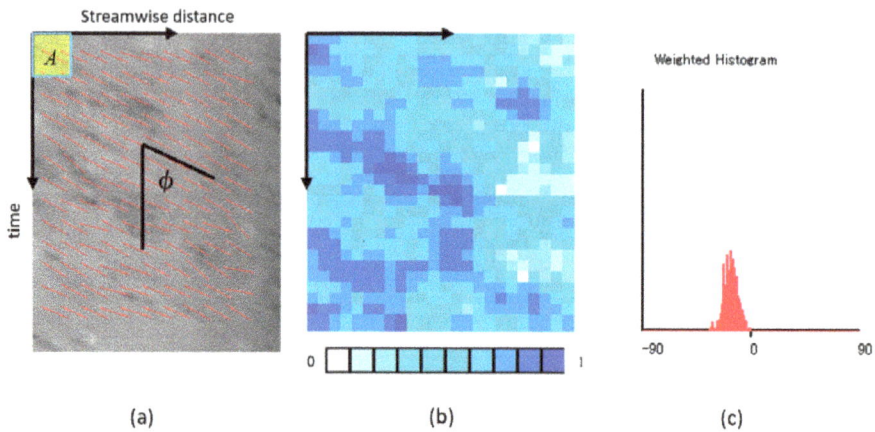

Figure 3. Application of gradient tensor method: (**a**) local angle in segment (ϕ); (**b**) coherency distribution; (**c**) histogram of orientation angles.

3. Outline of the Field Measurement

3.1. Study Area

Field investigations were conducted from 19 April to 22 April in 2012 at the Negoya Bridge of the Uono River, which is a major tributary of the Shinano River, the longest river in Japan. The site is located at Horinouchi in Uonuma City of Niigata Prefecture in the middle of Japan. This location was chosen because daily snowmelt flood occurs frequently in early spring, and we could actually observe the increase of water level of about 40 cm due to the snowmelt during our measurements every day as shown in Figure 4. The water level sensor of a pressure type was temporally installed near the panel R04 (Figure 5). On the measurement day, the water level rose up to the full width of the lower channel with a width of about 150 m, while the bank to bank width is about 220 m. The measurements were mainly conducted at the cross-section downstream of the bridge every hour in order to compare the performance of various measurement methods, i.e., STIV with normal cameras, STIV with a FIR camera, RVM, and an ADCP. Floats were also supplied from the bridge to compare their velocity with the surface velocity measured by the above methods. Figure 5 is the measurement site showing the location of cameras and mark point panels to be used for a camera calibration in STIV measurement. The panels were carefully set so that at least six of them are visible for each camera screen. The measurements were conducted jointly with staffs from International Centre for Water Hazard and Risk Management (ICHARM) under the auspices of UNESCO and several consulting companies.

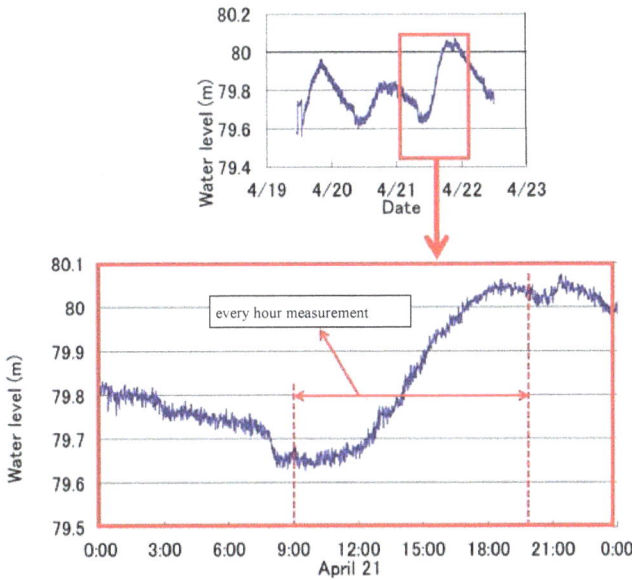

Figure 4. Water level hydrograph at the measurement site.

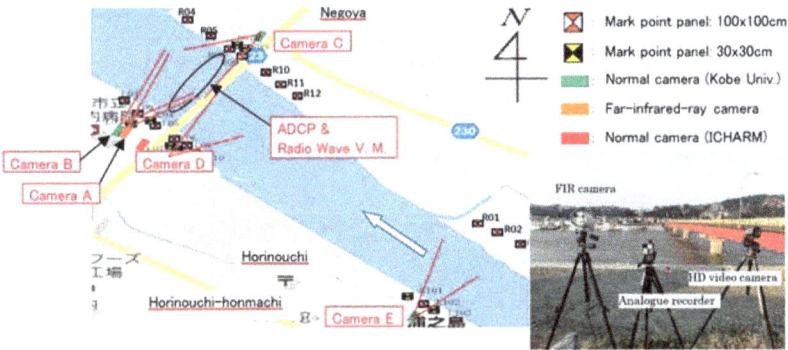

Figure 5. Measurement site description at the Negoya Bridge on the Uono River. L: Panels on the left bank; R: Panels on the right bank; ADCP: acoustic Doppler current profiler; V.M.: velocity meter; FIR: far-infrared-ray; ICHARM: International Centre of Excellence for Water Hazard and Risk Management.

3.2. Image Acquiring Methods

In the present observation, several cameras took images of the water surface from various angles. The cameras are an FIR camera (Camera A in Figure 4) and four normal high-definition video cameras (cameras B to E). Cameras A, B and C viewed the same water surface zone downstream of the bridge where ADCP and radio-wave measurements were executed. Camera D and Camera E took images just upstream and far upstream of the bridge, respectively. Figure 5 shows how the measurement were conducted at the site. It shows that the image-based measurement technique is quite simple, i.e., just take surface images from a riverbank, taking care about the size of the field of view. The important point is to set at least six ground reference points (GRPs) within the field of view for the ortho-rectification of images.

The measurements were repeated every hour for about five minutes from 10:00 a.m. to 8:00 p.m. The FIR camera (FLIR Systems, SR-334, Wilsonville, OR, USA), has a resolution of 320 by 240 pixels and its image sampling frequency is 7.5 fps, while normal high definition (HD) cameras have a resolution of 1920 by 1080 pixels with 30 fps. The video images of the FIR camera were recorded as an NTSC (National Television System Committee) format on an analogue video recorder with a size of 640 by 480 pixels at 30 fps as indicated in Figure 4. Therefore, the recorded images were somewhat blurred with a staggering frame sequence. The relationship between the screen and physical coordinates were determined from the location of mark point panels for each coordinate. Typical panels are made of plywood with a size of 1 m by 1 m. As the resolution of the FIR camera is not high, and, at the same time, it was difficult to detect the panels on the other side of the river about 150 m away from the camera, pocket heating pads were attached on the panel for improving the visibility of panel locations. Since the air temperature is very low (about 5 °C), heated spots were clearly visible in FIR camera screen even from a distant location.

3.3. Comparison of Captured Images

In order to demonstrate the interest of a FIR camera for river flow measurement, images captured in the morning and at night are compared in Figure 6 together with the images captured by a normal camera. It is obvious from the upper images in Figure 6 that the FIR camera can capture the river surface pattern clearly even during the night with low light condition, while a normal camera can recognize only the water surface just beneath street lamps at the bridge as indicated in lower left image in Figure 6. It can be considered that an FIR camera senses only the heat differences of objects, but the intensity of the received infrared ray changes depending on the slope of reflected surface even when the object temperature is the same. It should be noted that the original image size is only 320 by 240 pixels in FIR compared with the normal image size of 1920 by 1080 pixels. Although the surface pattern by FIR camera seems to be rather smoothed out in the night image as seen in Figure 6b, advection of the pattern in the downstream direction was observed in STI, which is a significant advantage over conventional cameras when developing a real-time measurement system using STIV.

(a) 10:00

(b) 20:00

(c) 10:00

(d) 20:00

Figure 6. Comparison of image quality by an FIR camera and a normal high definition video camera: (a) image by FIR camera in the daytime; (b) image by FIR camera at night; (c) image by normal HD camera in the daytime; (d) image by normal HD camera at night.

Figure 7 shows 25 search lines set on oblique and rectified images for the FIR camera A. The search lines were drawn parallel to the river bank at an equal spacing in a similar manner indicated in Figure 1. Its length is 17.8 m and the spacing is 5.7 m, covering the river width of about 140 m. It can be noted in Figure 7b that the image texture within a rectangle region becomes significantly blurred after the ortho-rectification. This is because the pixel image is greatly expanded for the region far from the camera location. As an example, the original spatial resolution along the farthest search line is 33.6 cm/pixel in this case, which is a very large value trying to reproduce the actual water surface roughness pattern. Because of this insufficient information of image texture in that region, it was difficult to apply LSPIV to the whole width since LSPIV uses a pattern matching technique to track the surface texture. As an example, Figure 8 provides the results by LSPIV applied to sequential ortho-rectified images such as shown in Figure 7b. The images ortho-rectified with a pixel size of 0.2 m and a time interval of 0.133 seconds were used for the analysis. The template size for the pattern matching in PIV was set at 33 by 33 pixels, which is 6.6 by 6.6 m in physical scale. In Figure 8, instantaneous vectors are superposed to show variations in data. It is obvious that variation in data increases significantly with the distance from the left bank where the FIR camera is located. Since the actual flow direction is almost unidirectional, the scatter in the data is unreliable. This is because the image is largely stretched in the transverse direction in the process to ortho-rectification. On the other hand, the data nearer to the left bank shows almost uniform variation, indicating that reliable data are obtained only until about half-width, i.e., about 70 m, from the left bank. This feature suggests that the flow rate cannot be analyzed from the LSPIV in the cases where the image resolution is low and the image stretching becomes large in the image transformation, such as in the present situation.

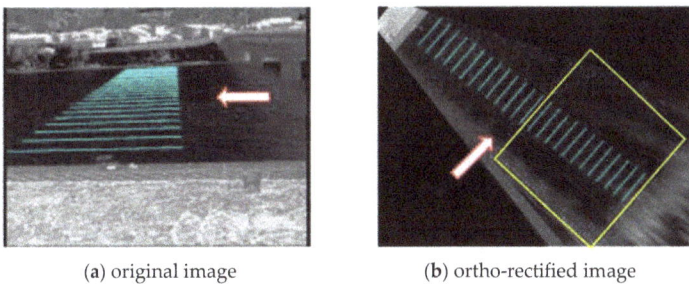

(**a**) original image   (**b**) ortho-rectified image

Figure 7. Images from FIR camera A at 10:00: (**a**) original image taken from the left bank with search lines, (**b**) ortho-rectified image of (**a**).

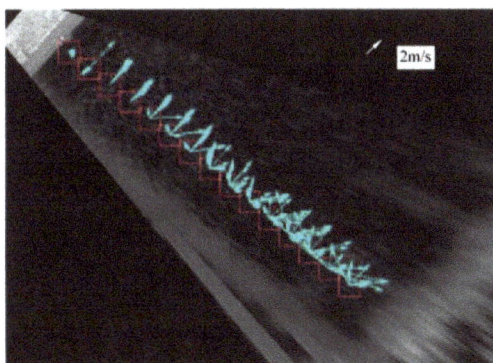

Figure 8. Superposed instantaneous velocity vectors measured by large-scale image velocimetry (LSPIV).

In order to show good sensitivity when using spatiotemporal images, Figure 9 shows the STI obtained by the FIR camera at almost the same position as the STI shown in Figure 2 obtained by a high resolution camera. The images are enhanced by the same technique as for Figure 2. Although the FIR camera used in the measurement have only one sixth of the resolution of the high definition camera, similar patterns appear in the resulting STIs even at the farther locations from the camera location. This feature of STI is the source of robustness in applying STIV to various types of flood flows [16].

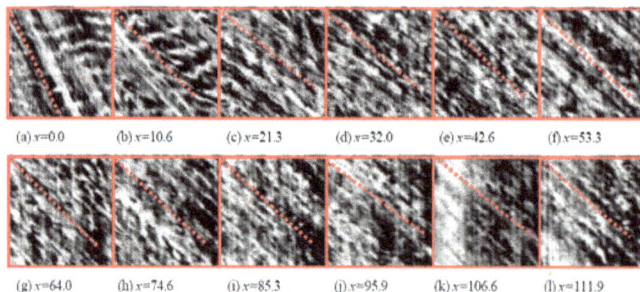

Figure 9. Space time images obtained by an FIR camera at several transverse locations (units in meters): horizontal scale is 17.6 m and downward vertical scale is ten seconds. Images are enhanced for a better recognition. The dotted line is the main orientation angle of the pattern.

3.4. Traceability of Water Surface Features

In using non-contact techniques for surface flow measurement such as RVM, LSPIV or STIV, water surface features generated by turbulence are assumed to be advected with the surface flow velocity as previously mentioned. In order to validate this assumption, an STI for the search line intentionally set along a trajectory of a float captured by camera B is shown in Figure 10 together with an STI by an FIR camera for the same location. The trajectory of the float is visible as a straight line as seen in Figure 10a, indicating that the float moved at a constant speed along the search line, from which its speed can be easily obtained from the slope of the trajectory. In Figure 10a, there appears a texture almost parallel to the float trajectory. An STI at the same location obtained by an FIR camera is shown in Figure 10b. The important point is that the gradient of the oblique patterns obtained by the two cameras is almost the same as that of the float trajectory. Referring to the four straight lines of the same gradient superposed in the figures, a good agreement of the speeds can be visually recognized, indicating the traceability of surface features to the surface flow.

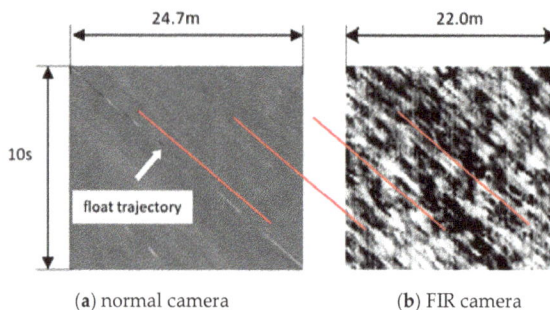

(**a**) normal camera   (**b**) FIR camera

Figure 10. Comparison of space time image (STI) captured downstream of the bridge in the daytime: (**a**) STI captured by normal HD camera for a search line along a float trajectory; (**b**) STI captured by FIR camera for the same search line as (**a**).

Another example for examining traceability of surface features is shown in Figure 11, which compares a trajectory of a piece of driftwood and surface features. Since these STIs are obtained from the camera D located upstream of the bridge, surface features displays straight uniform patterns. This is because the water surface was not disturbed by the pier of the bridge as in the downstream region. The trajectory of driftwood coincides with the surface pattern gradient so that it seems to be embedded in the pattern. This suggests that measurements by STIV should be conducted on the upstream side of the bridge, without disturbance by the bridge pier

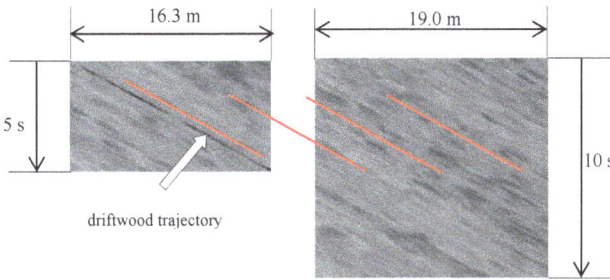

Figure 11. Comparison of STIs with and without a driftwood obtained at the upstream of the bridge from camera D shown in Figure 5.

Finally, Figure 12 compares the trajectory of a luminous float on the STI captured by a normal camera and STI by an FIR camera obtained at night. The luminous float is used conventionally in a dark condition in Japan. It has a fluorescent part at the top of the normal float so that an observer can easily detect the light from the river bank. Again, due to the low spatial resolution of the FIR camera used in the present study, oblique patterns displayed in Figure 12b do not have sharp outlines, but it is possible to determine the slope of the general pattern even in such a deteriorated image condition. Referring to the parallel lines drawn in the figure, the gradient of the light trajectory agrees quite well with the pattern slope by the FIR camera. Therefore, the conventional measurement during the night by using a luminous float can be replaced by STIV using an FIR camera. In this respect, STIV is much more robust than LSPIV in which a reliable matching of image pattern has to be established in a sub-pixel level. Alternatively, STI containing a float trajectory, such as indicated in Figure 12a, can be used for STIV measurement when an FIR camera is not available. The gradient of the pattern can be obtained either manually or automatically with the aid of an STIV algorithm [17].

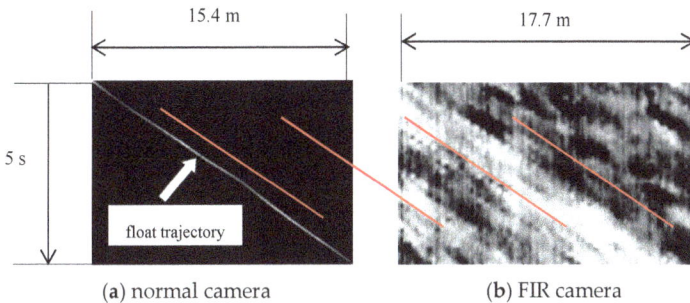

(a) normal camera   (b) FIR camera

Figure 12. Comparison of STI obtained at night: (**a**) STI captured by normal HD camera for a search line along a luminous float trajectory; (**b**) STI captured by FIR camera for the same search line as (**a**).

4. Results and Discussion

4.1. Comparison with Other Measurement Methods

As mentioned previously, the measurements were conducted every hour from 10:00 a.m. to 8:00 p.m. During the measurements, the water level rose about 50 cm due to the melting of snow from the mountainous region. The water level repeated up and down changes due to the snowmelt in the daytime as has been indicated in Figure 5. Figure 13 shows surface velocity distributions downstream of the Negoya Bridge at 1:00 p.m. and 8:00 p.m. measured by various methods, i.e., STIV by the far infrared camera on the left bank, STIVs by normal camcorders on the left and the right banks, a radio-wave velocity meter, and an ADCP. The cross-sectional bottom shape measured by ADCP was also indicated in the figure.

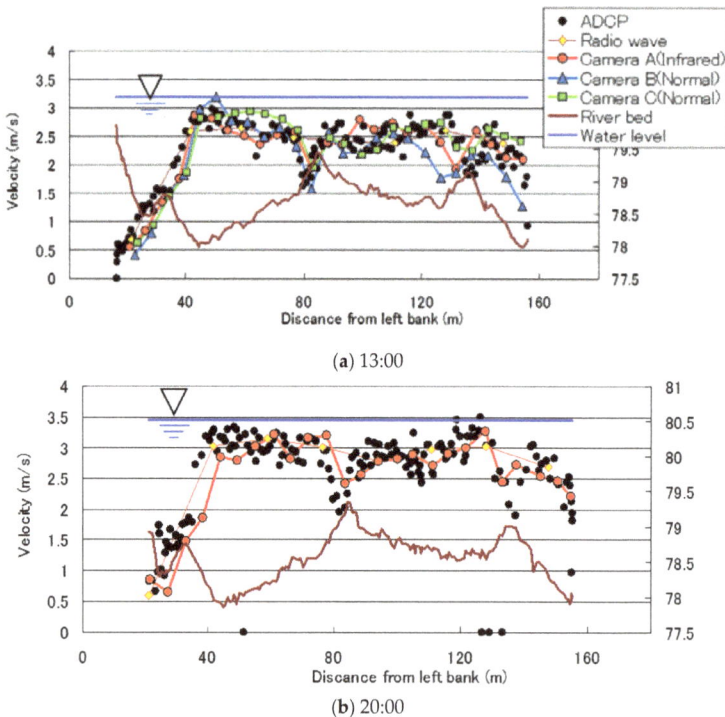

(a) 13:00

(b) 20:00

Figure 13. Comparison of surface velocity distributions by various techniques.

As for the ADCP data, the streamwise velocity component closest to the water surface, about 30 cm below it, were compared as representing the surface velocity. Since the ADCP measurement is conducted by tagging the boat-mounted sensor manually in a transverse direction, the data indicated are almost instantaneous values and have a great scatter especially in the region downstream of the bridge piers, where large-scale unsteady wake flow is generated. On the other hand, measurements by a radio-wave velocity meter were conducted by shifting its position step by step manually along the bridge. It measures a locally averaged velocity within a circular spot with a diameter of a few meters on the water surface by utilizing the Doppler shift effect. The obtained data falls into a scattered range of ADCP data.

Regarding the measurements by STIV, plots by the normal video cameras and the FIR camera show a large scatter within the wake flow region with the maximum relative error more than 0.5 m/s as shown in Figure 13a. Compared to the ADCP data, the FIR camera produced a favorable velocity distribution with a relative error of about 10%, which also agreed well with the RVM data. Since normal cameras are impossible to use at night, STIV by the FIR camera are compared with ADCP and RVM as shown in Figure 13b. Keeping in mind the scatter in ADCP data, data by the FIR camera agrees quite with the other data with a relative error of about 10%.

Comparing the STIV data obtained by normal video cameras and the FIR camera, normal cameras record the light reflection from the water surface roughness while the FIR camera mainly detects the temperature difference as well as the radiation intensity that varies with the roughness shape. Since the spatial resolution along a search line decreases with the distance from the camera, there should be some limitation in terms of the width of the river that STIV can be applicable. According to the author's experience, the present case with a width of about 150 m might be the maximum distance at which STIV can yield reliable data.

Regarding the transverse bed profile at the section downstream of the bridge shown in Figure 13, the river bed is deeper closer to the left bank and a ridge-like deposition occurs downstream of a bridge pier. Another lower deposition can be found closer to the right bank just after the other bridge pier. Generally speaking, water surface velocity distributions measured by the above methods agree well with each other in the entire cross-section with its distribution influenced by the bottom variation, i.e., faster in a deeper zone and slower in a shallower part. The maximum velocity occurred near the left bank increased from about 2.8 m/s at 1:00 p.m. to about 3.3 m/s at 8:00 p.m. due to the increase of discharge.

In order to examine the effect of the bridge piers that disturb the surface features, time variation of velocity distributions measured by STIV with Camera D just upstream of the Negoya Bridge is shown in Figure 14. The location of the Camera D is indicated in Figure 4. As can be seen from Figure 10 obtained at the upstream section, the oblique patterns that appeared on the STI show continuous straight features with little disturbance, indicating that surface features are advected steadily at a constant speed without the influence of hydraulic structures. As a result, the transverse velocity distributions are much smoother than that shown in Figure 13. This feature is important for improving the measurement accuracy by STIV. Unfortunately, we could not calculate the flow rate because the riverbed shape of this section was unknown, but once the topographic data is available, more reliable discharge data could be obtained.

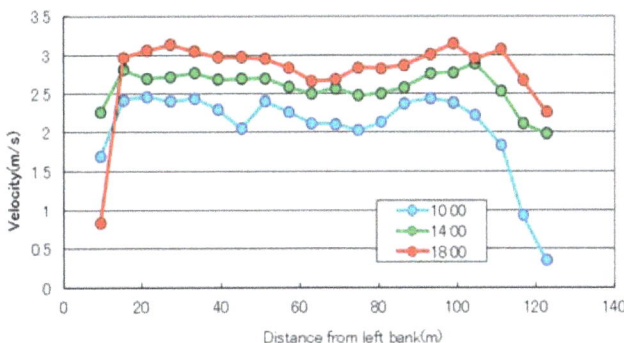

Figure 14. Surface velocity distribution measured by space time image velocimetry (STIV) at the upstream section of the bridge.

4.2. Discharge Measurement Accuracy

Figure 15 compares a discharge directly measured by ADCP and estimated discharges from surface velocity distributions by STIV and a radio-wave velocity meter. In integrating the velocity distribution measured by ADCP, the surface velocity U_s was assumed to be the same as the data 30 cm below the water surface. In using STIV, discharge was obtained by dividing the river width by the interval of the search line Δy_i, multiplying the flow velocity at the search line by a surface velocity ratio α, the local flow depth h_i, and finally summing the piecewise discharges. Therefore, the discharge can be calculated by the following formula:

$$Q_{ADCP/RVM} = \sum_{i=1}^{n} \alpha U_{si} h_i \Delta y_i. \tag{10}$$

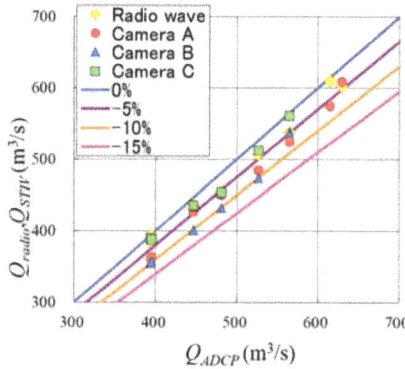

Figure 15. Comparison of measured discharges.

Similar flow rate calculation is also performed by RVM, but the number of measurement cross-sections by STIV was much larger than RVM; i.e., 25 sections in STIV and eight sections in RVM. In STIV, the number of sections can be increased easily, but it becomes physically difficult to conduct a simultaneous measurement in using RVM. In the present discharge calculation shown in Figure 14, the conventional value of 0.85 was used for α, and, as a result, underestimated values were obtained when compared to ADCP data. Therefore, using a value of 0.9 for α would increase the degree of coincidence at the section downstream of the bridge. As shown in Figure 14, even with a value of 0.85 for α, the relative error is less than ten percent, which is acceptable in the practical usage of STIV.

Figure 16 compares the discharge hydrographs obtained by the above methods. The discharge increased about 200 m^3/s due to the snowmelt. The discharge began to increase just after noon, reached its peak at 7:00 p.m. and decreased again after that, which indicates a typical feature of snowmelt flood in Japan.

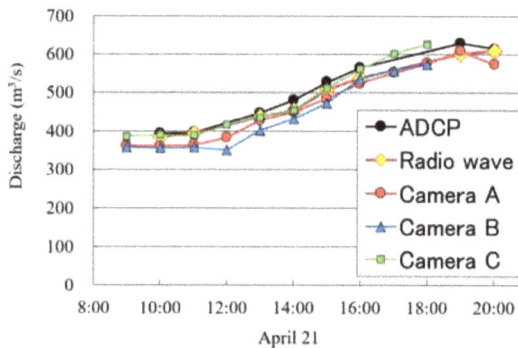

Figure 16. Discharge hydrographs obtained by various techniques.

5. Conclusions

Velocity fields of the snowmelt flood that occurred in the Uono River were measured by various cutting-edge methods such as imaging techniques, ADCP and radio-wave velocity meter. Among them, imaging techniques used to be considered difficult to apply during the night. However, this shortcoming was overcome by introducing a far-infrared ray camera in the present study for the first time for flood flow measurement with surface velocities more than 3 m/s. Puelo et al. [21] used a thermal infrared camera to measure small stream flows, but the velocity level was low with the maximum velocity less than 0.9 m/s, and it was not a flood measurement. Regarding the traceability of surface features, the assumption that surface features are advected with the surface velocity was verified by directly comparing it with the velocities of floating objects such as driftwood or floats in a space-time image. It was also made clear that the measurement accuracy of discharge by STIV was comparable to the direct measurement by ADCP with a relative error less than ten percent. In the actual situation of discharge measurements, measurement sections should have to be located upstream of a bridge because surface features at the downstream section can be destroyed by wake flows generated at the piers. Regarding the use of LSPIV, another widely-used image-based technique, it yielded erroneous values where the stretching of the image in the transverse direction is remarkable. On the other hand, STIV yielded stable results even where the image resolution and quality is low farther from the camera location. In that sense, STIV is a promising technique for measuring river discharges safely from a remote place even for an overtopping flood flow. However, the conventional surveillance camera such as closed-circuit television (CCTV) is difficult to use at night with little surrounding light. Therefore, an FIR camera is recommended for establishing a real-time measurement system that works all the time. Further investigation is required to examine the effects of strong wind or heavy rain on the accuracy of STIV measurements.

Acknowledgments: The study was supported by a research grant from the Foundation of River and Basin Integrated Communications (FRICS) Foundation in collaboration with ICHARM. The data other than imaging data were provided through ICHARM. I am grateful to their support of the present research.

Conflicts of Interest: The author declares no conflict of interest.

Abbreviations

The following abbreviations are used in this manuscript:

ADCP	Acoustic Doppler Current Profiler
RVM	Radio-wave Velocity Meter
FIR	Far-Infrared Ray
ICHARM	International Centre for Water Hazard and Risk Management
LSPIV	Large Scale Particle Image Velocimetry
STIV	Space Time Image Velocimetry
STI	Space time image
CCTV	Closed-circuit television

References

1. Schiermeier, Q. Increased flood risk linked to global warming: Likelihood of extreme rainfall may have been doubled by rising greenhouse-gas levels. *Nature* **2011**, *470*, 316. [CrossRef] [PubMed]
2. Okazumi, T.; Nakasu, T. Lessons learned from two unprecedented disasters in 2011—Great East Japan Earthquake and Tsunami in Japan and Chao Phraya River flood in Thailand. *Int. J. Disaster Risk Reduct.* **2015**, *13*, 200–206. [CrossRef]
3. Xu, G.-Y. Climate change and hydrologic models: A review of existing gaps and recent research developments. *Water Resour. Manag.* **1999**, *13*, 369–382. [CrossRef]
4. Moradkhani, H.; Sorooshian, S. General review of rainfall-runoff modeling: Model calibration, data assimilation, and uncertainty analysis. *Hydrol. Model. Water Cycle* **2009**, *63*, 1–24.

5. Turnipseed, D.P.; Sauer, V.B. *Discharge Measurements at Gaging Stations. U.S. Geological Survey Techniques and Methods Book 3*; USGS: Reston, VA, USA, 2010; Chapter 8, p. 87.

6. Fukami, K.; Yamaguchi, T.; Imamura, H.; Tashiro, Y. Current status of river discharge observation using non-contact current meter for operational use in Japan. *World Environ. Water Resour. Congr.* **2008**, *2008*, 1–10.

7. Welber, M.; Le Coz, J.; Laronne, J.B.; Zolezzi, G.; Zamler, D.; Dramais, G.; Hauet, A.; Salvaro, M. Field assessment of non-contact stream gauging using portable surface velocity radars (SVR). *Water Resour. Res.* **2016**, *52*, 1108–1126. [CrossRef]

8. Fujita, I.; Komura, S. Application of video image analysis for measurements of river surface flows. *Proc. Hydraul. Eng. JSCE* **1994**, *38*, 733–738. (In Japanese). [CrossRef]

9. Fujita, I.; Muste, M.; Kruger, A. Large-scale particle image velocimetry for flow analysis in hydraulic engineering applications. *J. Hydraul. Res.* **1998**, *36*, 397–414. [CrossRef]

10. Muste, M.; Fujita, I.; Hauet, A. Large-scale particle image velocimetry for measurements in riverine environments. *Water Resour. Res.* **2008**, *44*, W00D19. [CrossRef]

11. Muste, M.; Hauet, A.; Fujita, I.; Legout, C.; Ho, H.-C. Capabilities of Large-scale Particle Image Velocimetry to characterize shallow free-surface flows. *Adv. Water Res.* **2014**, *70*, 160–171. [CrossRef]

12. Hauet, A.; Creutin, J.-D.; Belleudy, P. Sensitivity study of large-scale particle image velocimetry measurement of river discharge using numerical simulations. *J. Hydrol.* **2008**, *349*, 178–190. [CrossRef]

13. Jodeau, M.; Hauet, A.; Paquier, A.; Le Coz, J.; Dramais, G. Application and evaluation of LS-PIV technique for the monitoring of river surface velocities in high flow conditions. *Flow Meas. Instrum.* **2008**, *19*, 117–127. [CrossRef]

14. Dramais, G.; Le Coz, J.; Camenen, B.; Hauet, A. Advantages of a mobile LSPIV method for measuring flood discharges and improving stage–discharge curves. *J. Hydro-Environ. Res.* **2011**, *5*, 301–312. [CrossRef]

15. Le Coz, J.; Hauet, A.; Pierrefeu, G.; Dramais, G.; Camenen, B. Performance of image-based velocimetry (LSPIV) applied to flash-flood discharge measurements in Mediterranean rivers. *J. Hydrol.* **2010**, *394*, 42–52. [CrossRef]

16. Fujita, I.; Kumano, G.; Asami, K. Evaluation of 2D river flow simulation with the aid of image-based field velocity measurement techniques. In *River Flow 2014*; Schleiss, A.J., Cesare, G.D., Franca, M.J., Pfister, M., Eds.; Taylor & Francis Group: London, UK, 2014; pp. 1969–1977.

17. Fujita, I.; Watanabe, H.; Tsubaki, R. Development of a non-intrusive and efficient flow monitoring technique: The space time image velocimetry (STIV). *Int. J. River Basin Manag.* **2007**, *5*, 105–114. [CrossRef]

18. Tsubaki, R.; Fujita, I.; Tsutsumi, S. Measurement of the flood discharge of a small-sized river using an existing digital video recording system. *J. Hydro-Environ. Res.* **2011**, *5*, 313–321. [CrossRef]

19. Fujita, I.; Hara, H. Development of space time image velocimetry introduced fast Fourier transform for improving robustness in river surface flow measurement. *J. Hydrosci. Hydraul. Eng.* **2011**, *29*, 123–135.

20. Fujita, I.; Kosaka, Y.; Yorozuya, A.; Motonaga, Y. Surface flow measurement of snow melt flood using a far infrared camera. *J. Jpn. Soc. Civ. Eng. Ser. B1 (Hydraul. Eng.)* **2013**, *69*, I_703–I_708. (In Japanese) [CrossRef]

21. Puleo, J.A.; McKenna, T.E.; Holland, K.T.; Calanton, J. Quantifying riverine surface currents from time sequences of thermal infrared imagery. *Water Resour. Res.* **2012**, *48*. [CrossRef]

water

MDPI

Article

Daily Precipitation Changes over Large River Basins in China, 1960–2013

Bo Qu [1,2], Aifeng Lv [1,*], Shaofeng Jia [1] and Wenbin Zhu [1]

[1] Key Laboratory of Water Cycle and Related Land Surface Processes, Institute of Geographic Sciences and Natural Resources Research, CAS, Beijing 100101, China; geo_qb@163.com (B.Q.); jiasf@igsnrr.ac.cn (S.J.); bfdh198612@163.com (W.Z.)

[2] University of Chinese Academy of Sciences, Beijing 100049, China

* Correspondence: lvaf@163.com; Tel.: +86-10-6485-6539; Fax: +86-10-6485-6539

Academic Editors: Tommaso Moramarco and Roberto Ranzi
Received: 23 February 2016; Accepted: 26 April 2016; Published: 2 May 2016

Abstract: Based on a high-quality dataset of 713 daily precipitation series, changes in daily precipitation events during 1960–2013 were observed in China's ten largest river basins. Specifically, the amount of precipitation in four categories defined by fixed thresholds and their proportion on total precipitation were analyzed on annual and seasonal time scales. Results showed annual precipitation increased by 1.10 mm/10yr in China, but with obvious spatial differences. Regionally, annual precipitation increased significantly in northwestern rivers, upstream areas of the Yangtze River, the Yellow River, southwestern rivers (due to increase in light and moderate precipitation); and in southeastern rivers, downstream areas of the Yangtze River, and the Pearl River (due to increase in heavy and extreme precipitation). Annual precipitation decreased significantly in the mid-Yangtze River and upstream Pearl River (due to decrease in light, moderate, and heavy precipitation). Seasonally, precipitation decreased only in autumn; this was attributable to a decrease in light and moderate precipitation. Results show that the distribution of precipitation intensity over China has shifted to intense categories since the 1960s, there has been an increase in moderate precipitation in Northwestern and Northern China, and an increase in extreme precipitation in Southeastern China. This shift was detected in all seasons, especially in summer. Precipitation extremes were investigated in the categories of extreme precipitation and results show that the risk of flood has been exacerbated over the past half-century in the Huaihe River, the mid- and lower Yangtze River, the Pearl River, and southeastern rivers.

Keywords: daily precipitation; precipitation distribution; precipitation extremes; river basin

1. Introduction

As global climate experiences significant change with increasing concentrations of greenhouse gasses [1], assessing changes in climatic variables has become a key issue for researchers around the world [2]. Although it is widely recognized that global temperature is increasing significantly, there are large uncertainties concerning precipitation change because of variabilities in time and space [3]. Revealing changes in precipitation receives great interest because of its importance to agriculture, water supply, and natural hazards, such as droughts and floods.

Precipitation can be observed and analyzed in terms of long-term mean state and the intensity and frequency of precipitation events. Global land precipitation has increased by about 2% since 1900 [4], but this increase has been neither spatially nor temporally uniform [3].

A study of the US also showed a precipitation increase during the 20th century; the increase was most pronounced during the warm season [5]. To detect changes in the frequency and intensity of precipitation, daily precipitation is usually categorized using two methods: percentile thresholds and fixed thresholds. They are both widely used in many studies [1,5–7]. Brunetti *et al.* observed changes in daily precipitation within percentile categories across Italy and found that the significant decrease in frequency of precipitation was due to a decrease in low-intensity precipitation events [6]. Using fixed thresholds, Karl *et al.* observed that the proportion of total precipitation contributed by extreme precipitation had increased significantly within the US [8]. In the hydrological cycle, the frequency and intensity of precipitation directly affect rainfall-runoff processes and further determine regional water resources. For example, in forest stands, the proportion of interception loss could be up to 100% in light precipitation, while the loss could be 25% in heavy precipitation [9]. Percentile categories are generally defined by specific quantiles of the precipitation distribution at a single station, such as the top 5% for extreme precipitation [6] and bottom 50% for light precipitation [7]. Li *et al.* showed extreme events by the top 5% of precipitation distribution was about 10~20mm/d in Western China [10]. Additionally, the mean intensity of precipitation in the same percentile at different stations, therefore, could be different. Although percentile categories benefit the detection of precipitation changes across regions and seasons [1,6], fixed categories could be more useful in hydrological models, as rainfall-runoff processes generally derive from the amount of precipitation events across a river basin.

In recent years, changes in precipitation characteristics have also been widely discussed in China. Liu *et al.* found that the frequency of precipitation decreased by a significant 10% over the same period for most of China, except the northwest [1]. Precipitation extremes have also been studied extensively in China. A more recent study by Wu *et al.* found that heavy precipitation events had higher rates of change than did mean precipitation [7]. The river basin is an important unit in hydrological processes. Water resource management is principally practiced at the basin scale [11]. It is, therefore, valuable to identify precipitation changes at that scale. Precipitation extremes are one focus in studies over basins in China. Chen *et al.* studied precipitation extremes in large basins across China [12] and, similarly, studies were also performed in the Yangtze River [13], Pearl River [14,15], Lancang River [16], and Yellow River [17]. Zhang and Cong analyzed the mean intensity and frequency of precipitation events in large river basins across China from 1956 to 2005. Their results indicated that precipitation intensity had significantly increased, while precipitation frequency had decreased in all basins except for rivers in the northwest [11]. Changes in the distribution of precipitation intensity, however, are rarely explored in river basins across China. In hydrology, if the distribution of precipitation intensity shifts, even when there is no change in total precipitation, runoff quantities would also significantly change due to changes in rainfall-runoff processes. A relative study is, therefore, urgently needed in China.

Climatic variability research, however, is not an easy task. One of the greatest problems in examining climate change is a lack of high-quality long-term data [6] because of non-climatic noise in climate signals [18]. Although researchers have understood that developing reliable datasets is essential in climate-change studies, checking prospective datasets and adjusting for biases has received little attention around the world and in China.

This study focused on changes in daily precipitation of different intensities over large river basins in China and attempted to reveal changes in the distribution of precipitation intensity in these basins. The research first considered issues of quality control, homogeneity, and completeness and a high-quality dataset of daily precipitation was produced. Second, changes in daily precipitation were analyzed on annual and seasonal time scales during the past half-century. Specifically, fixed thresholds were used to categorize precipitation events. The amount and proportion of precipitation by category and annual or seasonal precipitation were calculated for every station each year, and then the trends were analyzed across China and its ten largest river basins. This paper is organized as follows. Section 2 describes the data, data-processing methods, and analysis methods used in this study. Results and analysis are presented in Section 3, and the discussion and conclusion are given in Section 4.

2. Data and Methodology

2.1. Data and Pre-Processing

The observed daily precipitation dataset obtained from the Chinese National Meteorological Center was used in our study. It contained 819 stations for the period from 1 January 1951 to 31 December 2013. This dataset is well distributed across China, including the Tibetan plateau [1], and is widely used in all kinds of studies [19–24]. In the 1950s, however, there were only about 160 to 400 stations with available data [25]. We, therefore, restricted our study to the period 1960–2013.

2.1.1. Quality Control

Quality control is an essential part of climatic variability research [26]. The dataset used in this study was compiled with quality control procedures including the extreme value test, consistency check, and spatial outliers test [11]. A detailed description of these procedures can be found in the work of Feng *et al.* and Qian and Lin. The study by Feng *et al.* applied quality control tests to daily meteorological data for 1951–2000 from the same source we used, and only 0.02% stations showed low quality [22]. We, therefore, considered the precipitation data to be of good quality.

2.1.2. Data Homogeneity

The homogeneity of meteorological data has been highlighted in many studies; climatic variability research is not possible without clear knowledge about it [18,27–33]. A number of non-climatic factors, including station relocation and changes in instruments, observing regulations, and algorithms for the calculation of means, can introduce bias into climatological time series [6]. These biases have disturbed the record of actual climatic variations over time [30].

A variety of direct and indirect methods has been developed to identify and adjust for inhomogeneities [18]. A comprehensive review of such methods was given by Peterson *et al.* [30]. In China, only 60 stations have metadata available [22], which the direct correction techniques for inhomogeneities require. Relatively indirect methods were, therefore, chosen for this study. Most of the indirect methods are based on the comparison of a candidate series with a reference series formed from nearby or correlative stations; these include Potter's ratio test [34], the standard normal homogeneity test (SNHT) [32], Bayesian procedures [20], and multiple linear regression [31].

No complete protocol, however, is available for processing daily precipitation datasets [27] and most methods were developed for monthly or annual data [30]. For daily precipitation, precipitation frequency is an important factor that affects the homogeneity of a time series [6,29]. Brunetti *et al.* attempted to check the homogeneity of daily precipitation records. They tested two kinds of heterogeneities in daily precipitation records: (1) the precipitation amount and (2) the number of rainy days [6]. This study followed their methods in that both precipitation amount and rainy days per month were used to evaluate the homogeneity in dataset. One of the most-used methods to identify inhomogeneities, the standard normal homogeneity test (SNHT) developed by Alexandersson was mostly used [27,35]. In China, Feng *et al.* also applied the SNHT method to identify inhomogeneity in daily meteorological data. Therefore, SNHT was used in this study.

After identifying inhomogeneities of our dataset, 11 stations showed homogeneity problems in precipitation amount and more stations, 57 stations, displayed homogeneity problems in rainy days. Among them, four stations were overlapped. At the end, a set of 755 stations constitute the newly-homogenized dataset in this study.

2.1.3. Missing Values

Missing values could also introduce bias in a climatological time series, so we attempted to fill in any missing values before proceeding. At first, we filtered the dataset to exclude stations with more than five missing years (nearly 10% of study period) between 1960 and 2013. Specifically, following Lucie *et al.*, the monthly and, consequently, annual, values were considered to be missing when data for more than three consecutive days or more than five random days within a month were missing [36]. In the end, 713 of 755 stations with more than 48 years of complete data remained in our study (Figure 1).

Each candidate station subjected to the SNHT was compared to a reference series created to identify its inhomogeneities. The reference series is a weighted average of a number of the nearest available stations [30]. The SNHT reference series was used to fill in the missing values from the 713 stations.

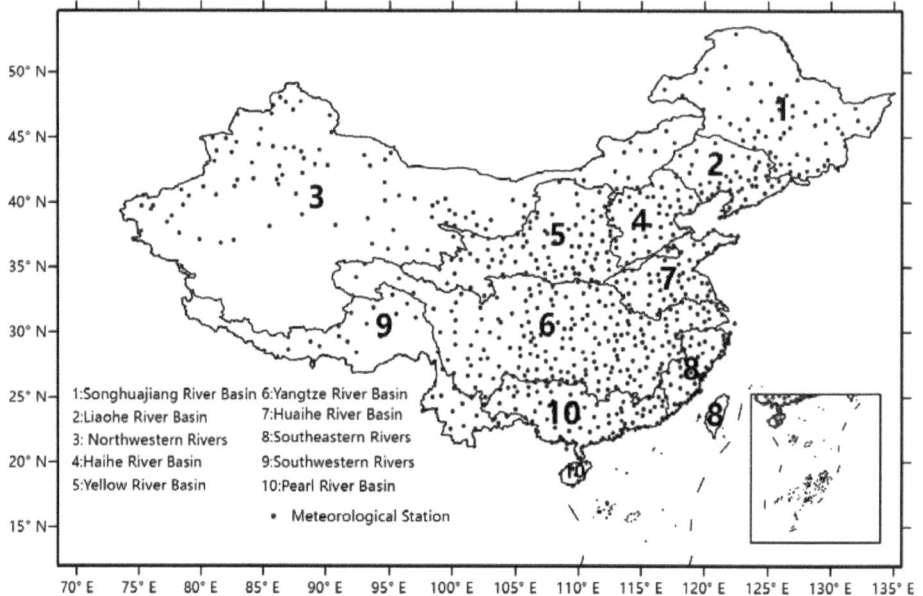

Figure 1. Large river basins and the locations of the 713 meteorological stations evaluated in China.

2.2. Methodology

The China Meteorological Administration (CMA) categorizes daily precipitation (P) into five intensity groups: $P < 10$ mm/d, $10 \leqslant P < 25$ mm/d, $25 \leqslant P < 50$ mm/d, $50 \leqslant P < 100$ mm/d, and $P \geqslant 100$ mm/d. Since the $P \geqslant 100$ mm/d group occurs infrequently in Northern China [13], following Gong *et al.*, we combined the last two groups and defined four categories: light precipitation (<10 mm/d), moderate precipitation ($10–25$ mm/d), heavy precipitation ($25–50$ mm/d), and extreme precipitation ($\geqslant 50$ mm/d). The amount and proportion of daily precipitation falling into these four categories were calculated for each station. We then analyzed the trends over China and in its ten large river basins (Figure 1).

2.2.1. Trends Detection

In this study, the trends were analyzed using the Mann-Kendall test, which is well suited to non-normally distributed data [37,38] and has been widely used in many previous studies [6,7]. To remove lag-1 autocorrelation in a time series, which could overestimate significance of the Mann-Kendall test, the trend-free pre-whitening (TFPW) method developed by Yue *et al.* was adopted before applying the Mann-Kendall test [38].

The Mann-Kendall test represents a ranked-based approach. The test statistic S is given as follows:

$$S = \sum_{i=1}^{n-1} \sum_{j=i+1}^{n} sgn\left(x_j - x_i\right) \tag{1}$$

where:

$$\text{sgn}(x) = \begin{cases} 1, & \textit{for } x > 0 \\ 0, & \textit{for } x = 0 \\ -1, & \textit{for } x < 0 \end{cases} \tag{2}$$

x_i and x_j are the sequential data values, and n is the length of the dataset.

A positive value of S indicates an upward trend and a negative value of S indicates a downward trend. The test is conducted using the normal distribution, with the expectation (E) and variance (V) as follows:

$$E(S) = 0 \tag{3}$$

$$V(S) = \frac{1}{18}\left(n(n-1)(2n+5) - \sum_{k=1}^{h} t_k(t_k-1)(2t_k+5)\right) \tag{4}$$

where t_k is the number of data points in the kth tied group and h is the number of tied groups in the dataset.

The Mann-Kendall Z is then calculated by:

$$Z = \begin{cases} \frac{S-1}{\sqrt{V(s)}} & \textit{for } S > 0 \\ 0 & \textit{for } S = 0 \\ \frac{S+1}{\sqrt{V(s)}} & \textit{for } S < 0 \end{cases} \tag{5}$$

The Z value can be related to a p-value of a specific trend. In a two-sided test for trend, the null hypothesis of no trend H_0 is accepted if $-Z_{1-\alpha/2} \leqslant Z \leqslant Z_{1-\alpha/2}$, where α is the significance level [39]. The null hypothesis of no trend is rejected if $|Z| > 1.65$, $|Z| > 1.96$ and $|Z| > 2.57$ at the 10%, 5%, and 1% significance levels, respectively.

The magnitudes of tendency were estimated using Theil-Sen approach (TSA), which is also a non-parametric statistical technique [39,40]. In this study, the TSA slope in precipitation was multiplied by 10 to express trends per decade. The TSA slope β is given by:

$$\beta = \text{median}\left[\frac{x_j - x_i}{j - i}\right] \textit{ for all } i < j \tag{6}$$

2.2.2. Spatial Interpolation

Meteorological stations can provide reliable and accurate precipitation data, but their spatial distribution is generally uneven in China [7]. In order to explore the temporal variation of ten large river basins, Kriging interpolation was chosen to generate grid data due to its good performance with geographic data [41]. The spherical semi-variogram model was used as the weighting function in interpolation [11]. Trends across the whole country and each basin were then derived from regional averaged data from 1960 to 2013.

3. Results and Analysis

3.1. Annual Character of Precipitation Tendencies

The Mann-Kendall trends for annual precipitation during 1960–2013 are shown in Figure 2. Stations with a significance level of 0.05 are labeled by pluses (positive) or triangles (negative). The averaged trends over China and its ten large river basins are showed in Table 1.

Over the past half-century, annual precipitation increased in China at a magnitude of 1.10 mm/10yr, but this trend was not statistically significant (Table 1). Regionally, significantly increasing trends occurred in the eastern coastal region, northwestern China, and the Tibetan Plateau, while significantly decreasing trends occurred in southwestern and northern China (Figure 2). In the ten large river basins, significant uptrend was observed in southwestern rivers. Large downtrends were observed in the Huaihe, Liaohe, and Haihe Rivers, while large uptrends were observed in southeastern rivers and northwestern rivers, although these findings were not statistically significant (Table 1).

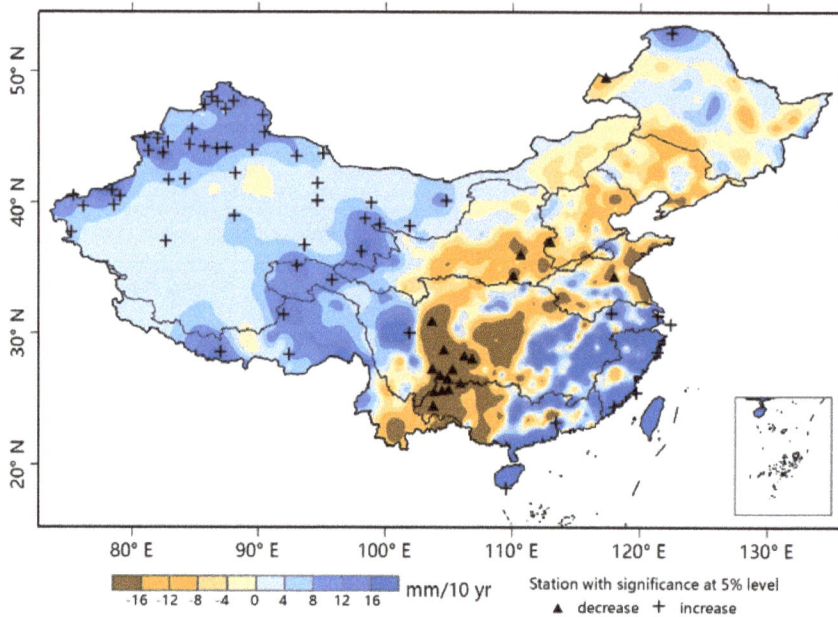

Figure 2. Trends in annual precipitation for 1960–2013. A plus/triangle shows a positive/negative trend for each station with statistical significance at the 5% level.

Table 1. Trends in amounts and proportion of precipitation in each category in the large river basins in China during 1960–2013.

Basin	Amount (mm/10yr)					Proportion (%/10yr)			
	Light Rain	Moderate Rain	Heavy Rain	Extreme Rain	Total	Light Rain	Moderate Rain	Heavy Rain	Extreme Rain
Songhuajiang River	−0.01	−1.65	−0.57	0.20	−2.89	0.25	−0.10	−0.06	0.00
Liaohe River	−1.68	−1.48	−1.80	−2.48	−8.60	0.13	0.12	0.05	−0.29
Northwestern Rivers	1.44[a]	0.99[a]	0.41	0.32	3.69[a]	−0.68[a]	0.37[b]	0.13	0.02
Haihe River	−1.65	0.86	−0.45	−4.39[b]	−5.59	−0.05	0.53[b]	0.08	−0.56[b]
Yellow River	−1.73	−1.44	0.61	−0.31	−2.69	−0.45	0.12	0.33[b]	−0.05
Yangtze River	−2.28[b]	−3.30[b]	0.61	3.88[b]	−1.13	−0.22[b]	−0.28[a]	0.08	0.45[a]
Huaihe River	−3.08[a]	−1.24	−1.80	−1.10	−7.20	−0.31[a]	−0.01	0.04	0.47
Southeastern Rivers	−3.94	−3.71	9.35[b]	15.63[a]	22.21	−0.48[a]	−0.52	0.34[a]	0.71[a]
Southwestern Rivers	1.44	1.04	2.14	1.84[b]	5.29	−0.43[a]	0.19	0.12	0.13[a]
Pearl River	−5.44[a]	−4.48	0.39	5.00	−5.71	−0.23[a]	−0.24[a]	0.04	0.34[b]
All China	−0.41	−0.65	0.47	1.76[b]	1.10	−0.39[a]	0.06	0.15[a]	0.15[a]

[a] Significance at 95% level; [b] Significance at 90% level; Unit "mm/10yr" and "%/10yr" means mm and % per decade.

Light precipitation showed a slight decrease (−1.20 mm/10yr) during 1960–2013 (Table 1), but a contrasting pattern of trends emerged across China (Figure 3a). Northwestern rivers, the headwaters of the Yangtze River and Yellow River, and southwestern rivers showed a significant increase, whereas other regions showed a significant decrease (Figure 3a). In the ten river basins, northwestern rivers and southwestern rivers showed an increase in light precipitation over the past half-century, and the trend in northwestern rivers was statistically significant (Table 1). Light precipitation in the Yangtze, Huaihe, and Pearl Rivers showed significant decreases. Data shown in Table 1 further suggest decreases in light precipitation largely contributes to the decrease in annual precipitation in these regions.

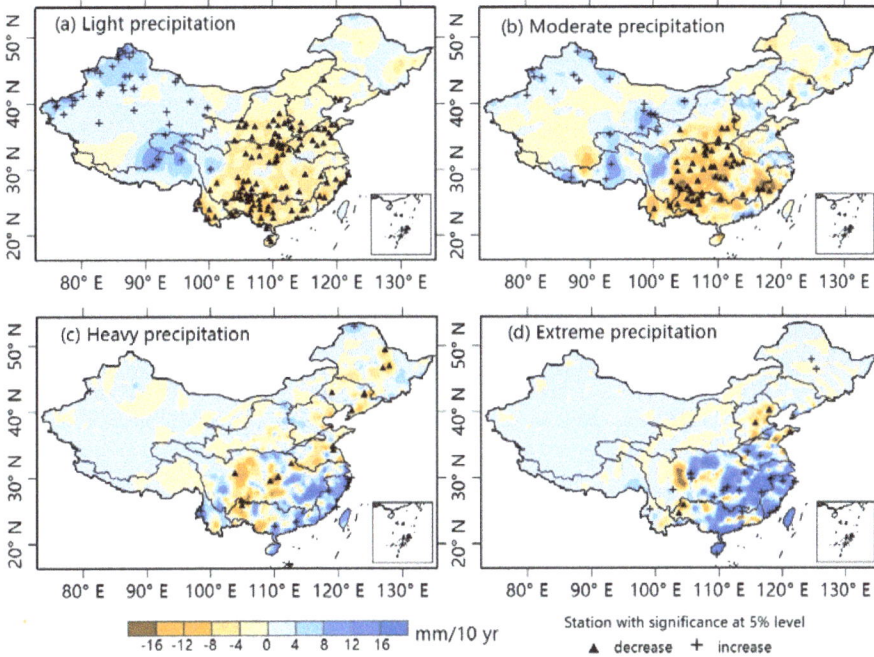

Figure 3. Trends in precipitation of each category for 1960–2013. (**a**) Light precipitation; (**b**) moderate precipitation; (**c**) heavy precipitation; and (**d**) extreme precipitation.

A decreasing trend for moderate precipitation was also detected across China (−0.65 mm/10yr, Table 1). Significant decreases occurred from Central to Southern China, including the mid-Yellow River, mid- and downstream Yangtze River, Pearl River, and southeastern rivers (Figure 3b). In the Yangtze River, moderate precipitation showed a significant decrease at a 0.1 significance level (−3.30 mm/10yr), which had a strong influence on the downtrend of annual precipitation (Table 1). Southeastern rivers and Pearl River also displayed large decreases in moderate precipitation (−3.71 mm/10yr and −4.48 mm/10yr, respectively). Significant increases mainly occurred in Northern and Western China (Figure 3b). Northwestern rivers were dominated by a significant increasing trend (0.99 mm/10yr). In the Haihe River, only moderate precipitation showed a positive trend (0.86 mm/10yr), which would partly alleviate the drier conditions over the past half-century.

Heavy precipitation was mainly detected in monsoon regions of China, and stations with significant increases were mostly located in the lower reaches of the Yangtze and Pearl Rivers and the southeastern rivers (Figure 3c). Table 1 shows that heavy precipitation significantly increased in northwestern rivers (0.82 mm/10yr) and southeastern rivers (9.35 mm/10yr), and fluctuated largely in Liaohe River, Huaihe River, and southwestern rivers, although these trends were not statistically significant. Extreme precipitation was mainly detected in the eastern and southeastern coastal regions, and stations with significant increases stretched from southeastern coastal areas to the middle reaches of the Yangtze and Huaihe Rivers (Figure 3d). In the whole country, on average, extreme precipitation showed a significant increase from 1960 to 2013 (1.76 mm/10yr). Regionally, in northwestern rivers, Yangtze River, southeastern rivers, and southwestern rivers, extreme precipitation significantly increased, while in Haihe River a significant decrease was detected (Table 1). The southeastern rivers displayed a large increase in total precipitation (22.21 mm/10yr), which was primarily the result of an increase in extreme precipitation (15.63mm/10yr).

Karl *et al.* noted that even when there is no change in the total precipitation, the distribution of precipitation events might shift significantly [8]. In this study, the proportion of precipitation falling in each category was sensitive to changes in the distribution of precipitation. Results showed the proportion of light precipitation decreased significantly in nearly whole China over the past half-century (Figure 4a), while moderate precipitation in Northwestern China (Figure 4b) and extreme precipitation in Southeastern China (Figure 4d) significantly increased. On basin scale, a decrease in proportion of light precipitation was detected in most basins except Songhuajiang River and Liaohe River, even though in the northwestern rivers and southwestern rivers, where light precipitation increased from 1960 to 2013 (Table 1). In the Yangtze River and Pearl River, the proportion of moderate precipitation also decreased significantly, mainly resulting from larger percentage of extreme precipitation. The proportion of heavy precipitation increased in nearly all basins except the Songhuajiang River, and the trend was significant in the northwestern rivers, Yellow River, and the southeastern rivers (Table 1). The proportion of extreme precipitation increased in basins locating in Southern China, but decreased significantly in Haihe River, where the distribution of precipitation shifted toward the moderate category (Table 1). Similarly, in Yellow River, the proportion of extreme precipitation also showed decreasing trend, although without significance, and the distribution of precipitation shifted toward the heavy category (Table 1). There was no marked change in the distribution of precipitation in the Songhuajiang River and Liaohe River.

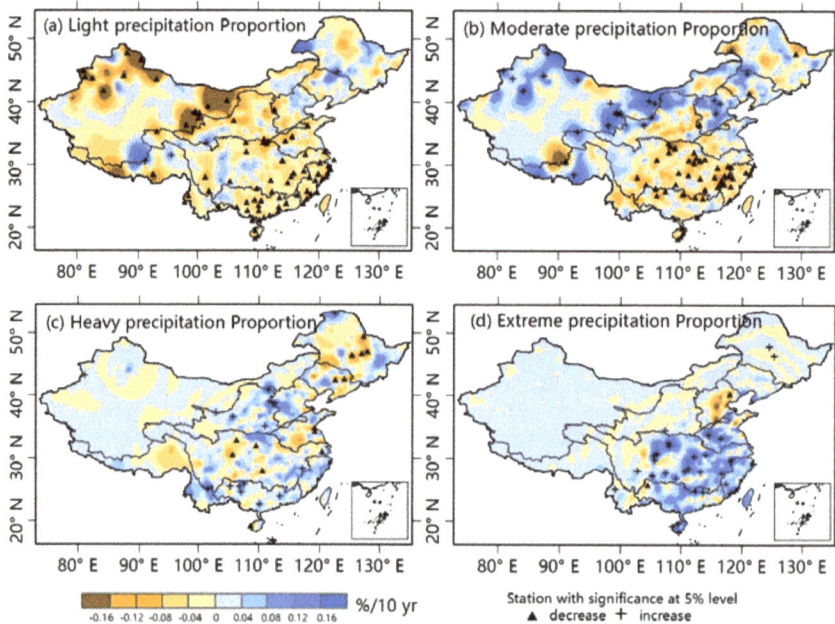

Figure 4. Trends in the proportion of each precipitation category relative to annual precipitation for 1960–2013. (**a**) Light precipitation; (**b**) moderate precipitation; (**c**) heavy precipitation; and (**d**) extreme precipitation.

3.2. Seasonal Characteristics of Precipitation Tendencies

Changes in seasonal precipitation over China during 1960–2013 are shown in Figure 5 and the averaged trends in each season over China and its ten large river basins are given in Table 2. In spring (Figure 5a), precipitation increased in Northern and Western China, especially over the southwestern rivers and the headwaters of the Yellow River and Yangtze River. Significant decreases were observed over the middle and lower reaches of the Yangtze River and the middle reaches of the Yellow River (Figure 5a).

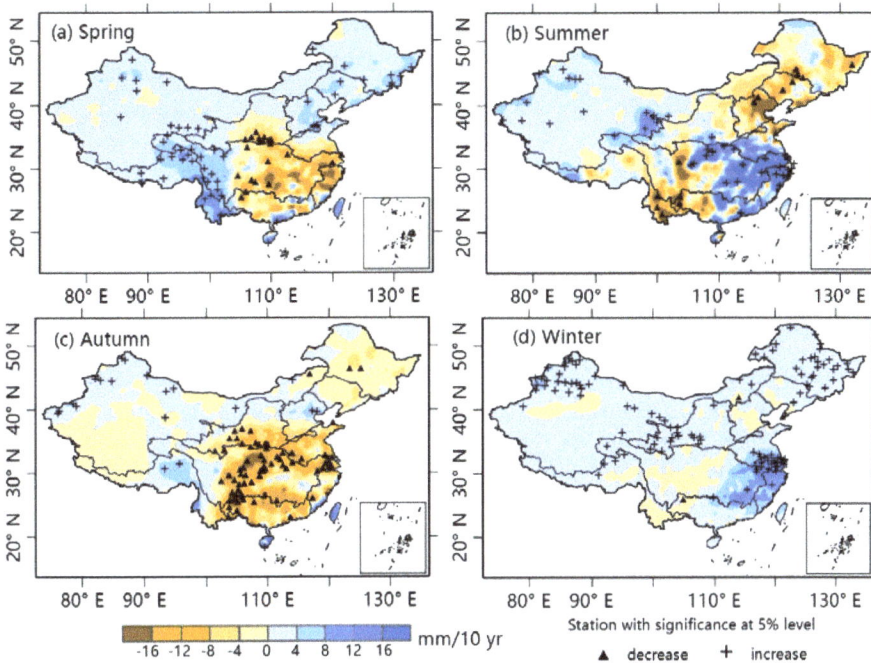

Figure 5. Trends in seasonal precipitation during 1960–2013. (**a**) Spring; (**b**) summer; (**c**) autumn; and (**d**) winter.

Table 2. Seasonal precipitation trends over China's large river basins during 1960–2013 (mm/10yr).

Basin	Spring	Summer	Autumn	Winter	Annual
Songhuajiang River	2.03	−4.90	−2.16[a]	1.33[a]	−2.89
Liaohe River	2.21	−7.75	−3.42	0.17	−8.60
Northwestern rivers	0.97[b]	1.36	0.33	0.75[a]	3.69[a]
Haihe River	4.91[a]	−12.53[a]	1.85	−0.18	−5.59
Yellow River	−0.62	−0.84	−3.38	0.34	−2.69
Yangtze River	−3.35	6.68[b]	−6.15[a]	2.56	−1.13
Huaihe River	−1.48	3.30	−7.60	1.86	−7.20
Southeastern rivers	−5.14	15.99[b]	0.55	7.37[a]	22.21
Southwestern rivers	5.72[a]	−2.70	0.59	−0.09	5.29
Pearl River	−4.27	0.94	−9.47	2.11	−5.71
All China	0.54	0.10	−2.76	1.26	1.10

[a] Significance at 95% level; [b] Significance at 90% level.

In summer, precipitation has markedly changed over the past half-century, showing large uptrends and downtrends with absolute values as large as >16 mm/10yr in most monsoon regions of China (Figure 5b). Specifically, significantly uptrends were observed in the Yangtze River and southeastern rivers, while significant downtrends were observed in Haihe River (Table 2).

Autumn precipitation showed a significant decrease from 1960 to 2013 in the whole of China, as well as in Songhuajiang River and Yangtze River (Table 2). Precipitation increases occurred only in limited areas of the Haihe River, northwestern and southwestern rivers (Figure 5c).

However, in winter, precipitation has increased significantly in China since 1960s, and nearly all basins experienced positive trends; significant in Songhuajiang River, northwestern rivers, and southeastern rivers (Table 2). The increase was more than 8 mm/10yr in the southeastern rivers, downstream areas of the Huaihe River and Yangtze River (Figure 5d).

Comparing changes in annual precipitation (Figure 2) to those in summer precipitation (Figure 5b), we can see that the large increase in summer rainfall in Southeastern China and large decrease in Northeastern and Southwestern China dominated the annual change during 1960–2013. At the basin scale, the strong impact of summer rainfall on annual precipitation change could be found in southeastern rivers and the Haihe River (Table 2). In the Yellow River basin, however, which experienced only small changes in summer rainfall, the annual decreasing trend represented precipitation decreases in the spring and autumn (Table 2). This was also observed in the Yangtze River, Huaihe River, and Pearl River, even though summer rainfall increased in these basins (Table 2). In northwestern rivers, the increase in annual precipitation was the result of increases during all seasons.

3.2.1. Seasonal Precipitation

As Figure 6a,b shows, there was less light and moderate precipitation during spring in regions from the mid-Yellow River to the Pearl River. Table 3 shows significant increasing trends in spring precipitation were observed in the Haihe River, northwestern rivers, and southwestern rivers (Table 3). In northwestern rivers, the increase in spring precipitation resulting from increase in light and moderate categories, while in southwestern rivers all categories led to the increase. The significant changes of light precipitation in basins, such as Songhuajiang River, northwestern rivers, Huaihe River, and southeastern rivers, and significant changes of moderate precipitation in basins, such as Liaohe River, Yangtze River, and southeastern rivers dominated changes in spring precipitation in these basins (Table 3). Moreover, in many basins, trends in moderate precipitation made a greater contribution to spring trends than trends in light precipitation.

Heavy and extreme precipitation were also observed in spring; these events occurred mainly in Southeastern China (Figure 6c,d). Specifically, heavy precipitation increased significantly in the Haihe River and southwestern rivers, decreased largely in the Yangtze River and Pearl River, but without significance (Table 3). Stations with significant extreme precipitation increase occurred in southeastern coastal regions (Figure 6d), and there was significant increase in Songhuajiang River, Haihe River, and southwestern rivers (Table 3).

Decreasing trends in light precipitation during the summer were observed in most basins, except for over the northwestern rivers, and moderate precipitation also decreased in many basins, such as Liaohe River, Haihe River, and Pearl River (Table 3). As Figure 6g shows, heavy precipitation increased significantly during the summer over the southeastern rivers, downstream areas of the Yangtze River and in some regions of the Huaihe River and Pearl River. Extreme precipitation occurred mainly in the summer (Figure 6h). The significant increase in extreme precipitation over the southeastern rivers and Yangtze River and significant decrease in Haihe River largely represented the changes in summer rainfall over these regions.

In autumn, both light and moderate precipitation exhibited significant decrease over China, and decreasing trends of light and moderate precipitation were observed in most basins during the study period (Table 3). Moreover Table 3 shows that in the Songhuajiang River, Yellow River, Yangtze River, Huaihe River, and Pearl River, light and moderate precipitation largely represented the overall decrease. Downward trends in heavy autumn precipitation in the Yangtze River, Huaihe River, and Pearl River also contributed to the autumn decrease. In the Haihe River; however, autumn precipitation increased substantially (1.85 mm/10yr), mainly resulting from an increase in moderate precipitation (1.03 mm/10yr).

Figure 6m–p shows winter precipitation trends. Stations that displayed significant trends in moderate and heavy precipitation were only in downstream areas of the Yangtze River and southeastern rivers (Figure 6n,o), while stations was rarely observed with changes in extreme precipitation across China during the past half-century (Figure 6p). Light precipitation increased significantly nearly in the whole country (Figure 6m) and, regionally, significant in Songhuajiang River, northwestern rivers, and Yellow River, which largely explained the increases in winter precipitation over these basins (Table 3). In the southeastern rivers, wetter winter conditions largely derived from increases in moderate and heavy precipitation, and in the Yangtze River and Pearl River, these conditions largely derived from increases in heavy precipitation (Table 3).

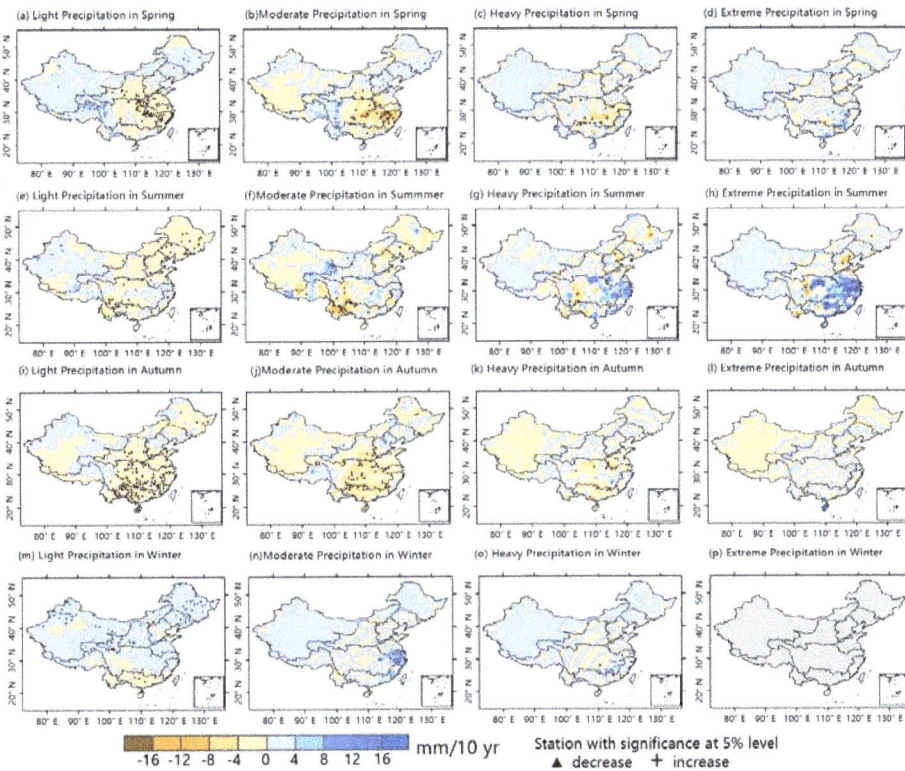

Figure 6. Seasonal trends in precipitation categories during 1960–2013. (**a–d**) Spring; (**e–h**) summer; (**i–l**) autumn; and (**m–p**) winter. For each season, the maps show light, moderate, heavy, and extreme precipitation from left to right.

Table 3. Seasonal trends of precipitation intensity over China's large river basins during 1960–2013 (mm/10yr).

Season	Intensity	Songhuajiang River	Liaohe River	Northwestern Rivers	Haihe River	Yellow River	Yangtze River	Huaihe River	Southeastern Rivers	Southwestern Rivers	Pearl River	All China
Spring	Light rain	1.02[b]	0.56	0.43[b]	0.28	−0.56	−0.69	−1.67[a]	−2.52[a]	1.21[a]	−0.82	0.11
	Moderate rain	0.84	1.29[a]	0.31[b]	1.93[a]	−0.11	−1.95[a]	−0.87	−5.89[a]	1.93[a]	−1.71	−0.15
	Heavy rain	0.03	0.35	0.13	1.15[a]	0.16	−1.53	−0.20	0.05	1.34[a]	−1.28	0.02
	Extreme rain	0.11[a]	0.08	0.15	0.28[a]	−0.02	0.40	0.95	2.66	0.89[a]	−0.66	0.36
	Total	2.03	2.21	0.97[b]	4.91[a]	−0.62	−3.35	−1.48	−5.14	5.72[a]	−4.27	0.54
Summer	Light rain	−2.21[a]	−2.28[a]	0.23	−1.58[a]	−0.56	−0.23	−0.74[b]	−0.23	−0.36	−1.22[b]	−0.57[b]
	Moderate rain	−2.12	−2.05[b]	0.52[b]	−2.71[b]	0.49	0.19	0.06	−0.27	−1.83	−2.13	−0.60
	Heavy rain	−1.22	−1.01	0.33	−2.08	−0.19	1.44	2.63	6.05[a]	−0.79	0.67	0.29
	Extreme rain	0.06	−2.49	0.36	−5.12[b]	−0.53	4.79[a]	2.95	8.88[b]	0.55	5.16	0.88[b]
	Total	−4.90	−7.75	1.36	−12.53[a]	−0.84	6.68[b]	3.30	15.99[b]	−2.70	0.94	0.10
Autumn	Light rain	−0.53	−0.48	0.20	−0.17	−1.46[b]	−1.95[a]	−1.81[b]	−1.42	−0.19	−3.55[a]	−0.86[a]
	Moderate rain	−0.97	−0.52	−0.03	1.03	−1.44	−3.04[a]	−1.88	−0.77	0.28	−3.54[a]	−1.09[a]
	Heavy rain	−0.31	−0.90	0.02	0.39	0.05	−1.01	−3.33[b]	0.58	0.23	−3.09	−0.38
	Extreme rain	−0.03	−0.19	0.05	0.37	−0.43[b]	−0.59	−0.40	1.90	0.18	0.29	−0.06
	Total	−2.16	−3.42	0.33	1.85	−3.38	−6.15[a]	−7.60	0.55	0.59	−9.47	−2.76[a]
Winter	Light rain	1.12[a]	0.36	0.53[a]	0.01	0.38[b]	0.46	1.00	0.17	−0.03	−1.21	0.38[b]
	Moderate rain	0.18[a]	0.03	0.05	−0.01	−0.01	0.63	1.11	3.75[b]	−0.10	1.00	0.27
	Heavy rain	0.05	0.00	0.03	−0.01	−0.01	1.15[a]	0.04	3.08[b]	0.24	1.93[a]	0.44[a]
	Extreme rain	0.00	0.00	0.02	−0.01	0.00	0.09	0.00	0.52[b]	0.01	0.55[b]	0.09
	Total	1.33	0.17	0.75	−0.18	0.34	2.56	1.86	7.37	−0.09	2.11	1.26

[a] Significance at 95% level; [b] Significance at 90% level.

3.2.2. Seasonal Distribution

As discussed above, the annual proportion of light precipitation decreased across nearly the whole country; the only regions where it increased were over the Songhuajiang River and Liaohe River, but without significance (Table 1). Seasonally, Table 4 shows the proportional changes in precipitation categories for each basin. For light precipitation, its decreasing proportion was observed across the country in seasons except winter, and was significant in spring and summer (Table 4), even when its amounts were observed to have increased in spring (Table 3). In winter, however, seven of the ten basins still displayed negative trends (Table 4).

In spring, the proportion of moderate, heavy and extreme precipitation all exhibited significant increase over China, as well as in northwestern rivers, Haihe River, and southwestern rivers (Table 4). However, in southeastern rivers, the proportion of moderate precipitation showed significant increase, and the distribution of precipitation shifted toward heavy and extreme categories (Table 4). Besides, the proportion of extreme precipitation increased significantly in Songhuajiang River and Huaihe River (Table 4).

Changes in proportions of summer precipitation categories are shown in Figure 7e–h; these were similar to those representing the whole year. In summer, the proportion of moderate, heavy and extreme precipitation significantly increased across China (Table 4). Light precipitation decreased in proportion nearly across all basins except Haihe River. In the Yangtze River, Huaihe River, southeastern rivers, and Pearl River, the proportion of moderate precipitation decreased significantly over 1960–2013, indicating precipitation became much more intense than basins such as northwestern rivers and Yellow River (Table 4). Contrary to other regions, the Liaohe River, Haihe River, and Yellow River experienced a negative trend in its proportion of extreme summer precipitation.

In winter, the proportion of heavy precipitation significantly increased over China, as well as in Yangtze River, southeastern rivers and Pearl River (Table 4). In Songhuajiang River, northwestern rivers, and Huaihe River, moderate precipitation exhibited significant increase in proportion (Table 4). These suggested the distribution of winter precipitation shifted toward the intense category in these basins. Changes in the distribution of autumn precipitation during the study period were inconspicuous across China. In general, however, decreasing trends in autumn precipitation were observed over nearly the whole of China, as discussed above.

Table 4. Seasonal trends in precipitation intensity proportions over China's large river basins during 1960–2013 (%/10yr).

Season	Intensity	Songhuajiang River	Liaohe River	Northwestern Rivers	Haihe River	Yellow River	Yangtze River	Huaihe River	Southeastern Rivers	Southwestern Rivers	Pearl River	All China
Spring	Light rain	−0.28	−1.93[a]	−0.51	−3.00[a]	−0.58	−0.13	−1.35[b]	−0.49[b]	−1.17[a]	0.14	−0.59[a]
	Moderate rain	0.46	1.05[b]	0.68[a]	1.49[a]	0.42	0.04	−0.05	−0.58[a]	0.91[a]	−0.29	0.40[a]
	Heavy rain	−0.24	0.34	0.11	1.14[a]	0.39	−0.03	0.40	0.45[b]	0.23[b]	0.00	0.13[b]
	Extreme rain	0.04[a]	0.08	0.02	0.21[a]	0.02	0.27	0.87[a]	0.66[b]	0.17[b]	0.17	0.16[a]
Summer	Light rain	−0.14	−0.08	−1.17[a]	0.05	−0.32	−0.35[a]	−0.32[b]	−0.32[a]	−0.14	−0.26[b]	−0.56[a]
	Moderate rain	0.03	0.17	0.76[a]	0.19	0.41[b]	−0.29[a]	−0.31[b]	−0.61[a]	0.14	−0.42[a]	0.30[a]
	Heavy rain	0.04	0.24	0.14	0.22	0.13	0.08	0.13	0.28	0.05	0.05	0.16[a]
	Extreme rain	0.08	−0.28	0.09	−0.47	−0.11	0.54[a]	0.58	0.64	0.14[b]	0.65[b]	0.15[b]
Autumn	Light rain	0.29	0.60	−0.52	−0.97	−0.09	0.06	0.33	−0.77[b]	−0.67[b]	−0.26	−0.16
	Moderate rain	−0.38	0.14	0.23	0.98[b]	−0.15	−0.27	0.44	−0.12	0.48[b]	−0.06	0.05
	Heavy rain	−0.14	−0.18	0.02	0.43	0.59[b]	0.13	−1.07[a]	0.17	0.07	−0.34	0.06
	Extreme rain	−0.03	−0.06	0.05	0.22	−0.17	0.14	−0.04	0.66	0.06	0.42	0.09
Winter	Light rain	−0.36[a]	0.44	0.55	−0.03	−0.10	−0.24	−1.07	−1.88[a]	1.10[b]	−1.95[b]	−0.16
	Moderate rain	0.29[a]	0.12	0.14[b]	0.03	0.11	−0.19	1.66[b]	0.58	0.10	0.18	0.14
	Heavy rain	0.03	0.00	0.05	−0.02	0.00	0.35[a]	0.10	1.21[b]	0.12	1.19[a]	0.20[a]
	Extreme rain	0.00	0.00	0.01	0.00	0.00	0.02	0.00	0.21	0.01	0.29[a]	0.03

[a] Significance at 95% level; [b] Significance at 90% level.

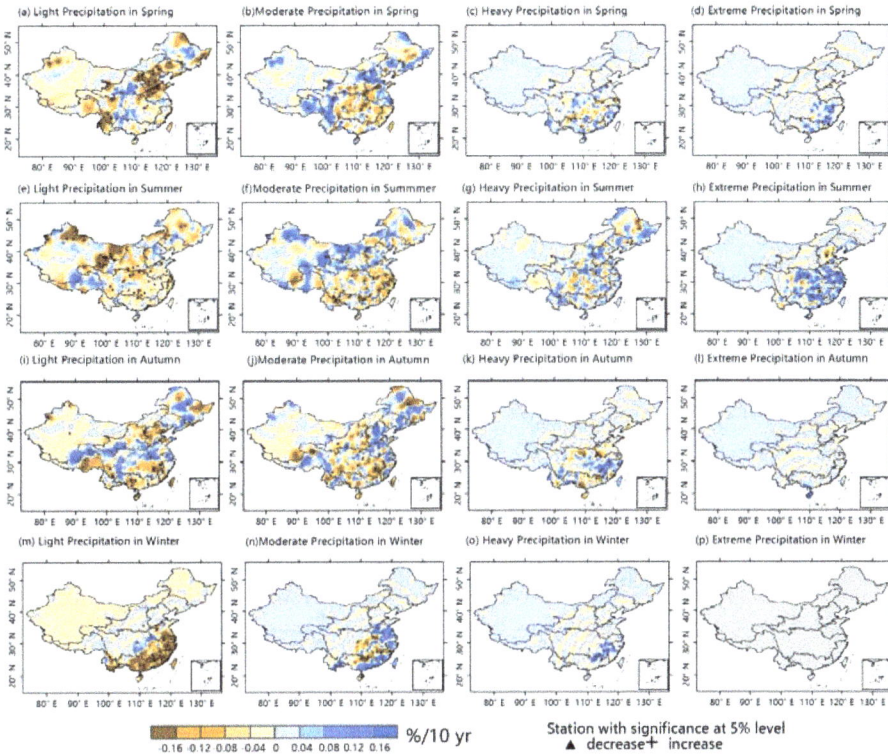

Figure 7. Seasonal trends in precipitation intensity proportions during 1960–2013. (**a–d**) Spring; (**e–h**) summer; (**i–l**) autumn; and (**m–p**) winter. For each season, the maps show light, moderate, heavy, and extreme precipitation from left to right.

4. Discussion and Conclusion

Long-term changes of precipitation are regarded as a practical subject for monitoring changes in water resources. In China, the result of our study showed that annual precipitation had increased slightly in the past half-century, which was consistent with some previous studies [7,42]. At the basin scale, significant increasing trends over the northwestern rivers was found, while annual precipitation decreased in large amplitude over the Liaohe River, Haihe River, Huaihe River, and Pearl River. Seasonally, Liu *et al.* found that precipitation had increased in winter and summer but decreased in spring and fall, over 1960–2000 [1]. Our study showed that autumn precipitation significantly decreased, while in spring, precipitation increased by 0.54 mm/10yr across China during the past half-century.

In hydrology, precipitation intensity is an important factor in rainfall-runoff processes [43]. Many studies have shown that over the past several decades, daily precipitation has become more intense in China [1,11,21,44]. In this study, changes in daily precipitation were analyzed by fixed categories. Results from trends in proportion showed the distribution of precipitation significantly shifted toward heavy and extreme precipitation over China. At the basin scale, precipitation intensity significantly increased in basins except Songhuajiang River, Liaohe River, Haihe River, and Yellow River. These increases are consistent the study of Zhang and Cong, which showed that precipitation intensity increased principally in the southern basins of China [11]. Seasonally, the shift in precipitation distribution from light to intense occurred in all seasons, significantly in spring and summer. However, this study showed, in northwestern rivers, precipitation also became more intense than before. Within the ten large basins, only the Haihe River experienced a significant decrease in extreme precipitation, while the proportions of the moderate and heavy categories increased there. In the Yellow and Songhuajiang Rivers, the proportions of the extreme category also decreased, but without significance.

Liu *et al.* showed that changes in annual precipitation could be attributed mostly to changes in the frequency and intensity of precipitation events in the top decile (90%) in China [1]. Using fixed thresholds, the results of this study showed only in the Haihe River, southeastern rivers, and southwestern rivers, extreme precipitation significantly influenced total precipitation. Furthermore, the impact of extreme precipitation was also obvious in spring, not merely in summer. The results also revealed, however, that changes in light and moderate precipitation also played an important role in the trends observed in the total precipitation in many basins, such as the northwestern rivers and Huaihe River, even in the Yangtze River and Pearl River. Seasonally, the increase in winter precipitation and the decrease in autumn precipitation also mainly derived from changes in the amount of light and moderate precipitation. In hydrology, precipitation in lower intensities also has considerable influence on basic flow, ground water, and so on [43,45]. For example, water infiltration of the soil surface during rainfall tends to be higher during lower-intensity rainfalls [1]. Therefore, significant increases in light precipitation over China during winter, and some basins during spring, could alleviate the dry conditions during the cold seasons in these regions.

China is prone to natural hazards from extreme weather events and flooding has always been a major problem [7]. Recently, numerous studies have documented precipitation extremes in China. Zhai *et al.* found significant increases in extreme precipitation events in Western China, the mid- and lower reaches of the Yangtze River, and coastal areas of China during 1951–2000 [42]. Xu *et al.* found that extreme precipitation amounts and extreme precipitation days significantly increased in the mid-and lower Yangtze River valley and Northwestern China [20].In those studies, extreme precipitation derived from percentile thresholds. As discussed above, extreme events by top 5% was just equal to moderate or light precipitation (defined in this study) in Western China. In this study, extreme precipitation is defined by fixed threshold (>50 mm/d), which could be more closely linked to floods. Results showed extreme precipitation in the Yangtze River, southeastern rivers, and southwestern rivers, increased significantly during 1960–2013, indicating a trend toward high risk of floods within these basins, as these basins are relatively developed regions of China with high population densities. What is more, the seasonal changes in extreme precipitation revealed in this study indicate that the risk of floods in some regions may be higher than suggested by previous studies; this should be highlighted in water resources management as follows. (1) Over the Yangtze River and southeastern rivers, a significant increase in extreme precipitation at the annual scale mainly derived from the summer, reflecting extreme events being concentrated in summer. (2) A significant increase in extreme precipitation was observed across the southwestern rivers and Songhuajiang River during spring, and Pearl River during winter, when vegetation cover is sparse and rainfall-runoff processes would be quick and heavy. (3) In the Haihe River, extreme precipitation during spring also significantly increased, although the yearly extreme precipitation showed a marked decrease. (4) However, in northwestern rivers, the increase in precipitation intensity was not related to an increase in the risk of floods over the past half-century, because it mainly resulted from an increase in moderate precipitation.

The amount and distribution of precipitation also play crucial roles in the occurrence of drought. As light precipitation generally occupied a fairly large proportion in rainfall events, a significant decrease in light precipitation over Yangtze, Huaihe, and Pearl Rivers might indicate an increase in dry-spells. Seasonally, autumn precipitation significantly decreased in the Songhuajiang and Yangtze Rivers, which might enhance the possibility of seasonal drought in these regions.

Acknowledgments: This work was supported by the Key Programs of the Chinese Academy of Sciences (KFZD-SW-301-5), the Important Science & Technology Specific Projects of Qinghai Province (2014-NK-A4-1).

Author Contributions: Aifeng Lv conceived the study. Bo Qu performed the data analysis and wrote the paper. Aifeng Lv, Shaofeng Jia and Wenbin Zhu read and edited the manuscript. All authors reviewed and approved the manuscript.

Water **2016**, *8*, 185

Conflicts of Interest: The authors declare no conflict of interest.

References

1. Liu, B.; Xu, M.; Henderson, M.; Qi, Y. Observed trends of precipitation amount, frequency, and intensity in China, 1960–2000. *J. Geophys. Res. Atmos.* **2005**, *110*. [CrossRef]
2. Rodrigo, F.S.; Trigo, R.M. Trends in daily rainfall in the Iberian Peninsula from 1951 to 2002. *Int. J. Climatol.* **2007**, *27*, 513–529. [CrossRef]
3. Field, C.B.; Van Aalst, M. Freshwater resource. In *Climate Change 2014: Impacts, Adaptation, and Vulnerability. Part A: Global and Sectoral Aspects*; Cambridge University Press: Cambridge, UK; New York, NY, USA, 2014; Volume 1.
4. Hulme, M.; Osborn, T.J.; Johns, T.C. Precipitation sensitivity to global warming: Comparison of observations with HadCM2 simulations. *Geophys Res. Lett.* **1998**, *25*, 3379–3382. [CrossRef]
5. Karl, T.R.; Knight, R.W. Secular Trends of Precipitation Amount, Frequency, and Intensity in the United States. *Bull. Am. Meteorol. Soc.* **1998**, *79*, 231–241. [CrossRef]
6. Brunetti, M.; Maugeri, M.; Monti, F.; Nanni, T. Changes in daily precipitation frequency and distribution in Italy over the last 120 years. *J. Geophys. Res. Atmos.* **2004**, *109*. [CrossRef]
7. Wu, Y.; Wu, S.Y.; Wen, J.; Xu, M.; Tan, J. Changing characteristics of precipitation in China during 1960–2012. *Int. J. Climatol.* **2015**, *36*, 1387–1402. [CrossRef]
8. Karl, T.R.; Knight, R.W.; Plummer, N. Trends in high-frequency climate variability in the twentieth century. *Nature* **1995**, *377*, 217–220. [CrossRef]
9. Zinke, P.J. Forest interception studies in the United States. In *International Symposium on Forest Hydrology*; Pergamon Press: New York, NY, USA, 1967; pp. 137–161.
10. Li, W.; Zhai, P.; Cai, J. Research on the Relationship of ENSO and the Frequency of Extreme Precipitation Events in China. *Adv. Clim. Chang. Res.* **2011**, *2*, 101–107. [CrossRef]
11. Zhang, X.; Cong, Z. Trends of precipitation intensity and frequency in hydrological regions of China from 1956 to 2005. *Glob. Planet Chang.* **2014**, *117*, 40–51. [CrossRef]
12. Chen, Y.; Chen, X.; Ren, G. Variation of extreme precipitation over large river basins in China. *Adv. Clim. Chang. Res.* **2011**, *2*, 108–114. [CrossRef]
13. Su, B.D.; Jiang, T.; Jin, W.B. Recent trends in observed temperature and precipitation extremes in the Yangtze River basin, China. *Theor. Appl. Climatol.* **2006**, *83*, 139–151. [CrossRef]
14. Zhang, Q.; Xu, C.; Gemmer, M.; Chen, Y.D.; Liu, C. Changing properties of precipitation concentration in the Pearl River basin, China. *Stoch. Environ. Res. Risk Assess.* **2009**, *23*, 377–385. [CrossRef]
15. Yang, T.; Shao, Q.; Hao, Z.; Chen, X.; Zhang, Z.; Xu, C.; Sun, L. Regional frequency analysis and spatio-temporal pattern characterization of rainfall extremes in the Pearl River Basin, China. *J. Hydrol.* **2010**, *380*, 386–405. [CrossRef]
16. Shi, W.; Yu, X.; Liao, W.; Wang, Y.; Jia, B. Spatial and temporal variability of daily precipitation concentration in the Lancang River basin, China. *J. Hydrol.* **2013**, *495*, 197–207. [CrossRef]
17. Hu, Y.; Maskey, S.; Uhlenbrook, S. Trends in temperature and rainfall extremes in the Yellow River source region, China. *Clim. Chang.* **2012**, *110*, 403–429. [CrossRef]
18. Vincent, L.A.; Zhang, X.; Bonsal, B.R.; Hogg, W.D. Homogenization of daily temperatures over Canada. *J. Clim.* **2002**, *15*, 1322–1334. [CrossRef]
19. Liu, W.; Fu, G.; Liu, C.; Charles, S.P. A comparison of three multi-site statistical downscaling models for daily rainfall in the North China Plain. *Theor. Appl. Climatol.* **2013**, *111*, 585–600. [CrossRef]
20. Xu, X.; Du, Y.; Tang, J.; Wang, Y. Variations of temperature and precipitation extremes in recent two decades over China. *Atmos. Res.* **2011**, *101*, 143–154. [CrossRef]
21. Gong, D.; Shi, P.; Wang, J. Daily precipitation changes in the semi-arid region over northern China. *J. Arid Environ.* **2004**, *59*, 771–784. [CrossRef]
22. Feng, S.; Hu, Q.; Qian, W. Quality control of daily meteorological data in China, 1951–2000: a new dataset. *Int. J. Climatol.* **2004**, *24*, 853–870. [CrossRef]
23. Mao-chang, C.; Hai, Z.; Shu-min, L.; Arpe, K.; Dumenil, L. Variability of daily precipitation in China (1980–1993): PCA and wavelet analysis of observation and ECMWF reanalysis data. *Chin. J. Oceanol. Limnol.* **2000**, *18*, 117–125. [CrossRef]

24. Svensson, C. Empirical orthogonal function analysis of daily rainfall in the upper reaches of the Huai River basin, China. *Theor. Appl. Climatol.* **1999**, *62*, 147–161. [CrossRef]

25. Qian, W.; Lin, X. Regional trends in recent precipitation indices in China. *Meteorol. Atmos. Phys.* **2005**, *90*, 193–207. [CrossRef]

26. Easterling, D.R.; Evans, J.L.; Groisman, P.Y.; Karl, T.R.; Kunkel, K.E.; Ambenje, P. Observed variability and trends in extreme climate events: A brief review. *Bull. Am. Meteorol. Soc.* **2000**, *81*, 417–425. [CrossRef]

27. Vicente Serrano, S.M.; Beguería, S.; López Moreno, J.I.; García Vera, M.A.; Stepanek, P. A complete daily precipitation database for northeast Spain: Reconstruction, quality control, and homogeneity. *Int. J. Climatol.* **2010**, *30*, 1146–1163. [CrossRef]

28. Brunetti, M.; Maugeri, M.; Monti, F.; Nanni, T. Temperature and precipitation variability in Italy in the last two centuries from homogenised instrumental time series. *Int. J. Climatol.* **2006**, *26*, 345–381. [CrossRef]

29. Wijngaard, J.B.; Klein Tank, A.; Können, G.P. Homogeneity of 20th century European daily temperature and precipitation series. *Int. J. Climatol.* **2003**, *23*, 679–692. [CrossRef]

30. Peterson, T.C.; Easterling, D.R.; Karl, T.R.; Groisman, P.; Nicholls, N.; Plummer, N.; Torok, S.; Auer, I.; Boehm, R.; Gullett, D. Homogeneity adjustments of *in situ* atmospheric climate data: A review. *Int. J. Climatol.* **1998**, *18*, 1493–1517. [CrossRef]

31. Vincent, L.A. A technique for the identification of inhomogeneities in Canadian temperature series. *J. Clim.* **1998**, *11*, 1094–1104. [CrossRef]

32. Alexandersson, H.; Moberg, A. Homogenization of Swedish temperature data. Part I: Homogeneity test for linear trends. *Int. J. Climatol.* **1997**, *17*, 25–34. [CrossRef]

33. Szentimrey, T. Statistical procedure for joint homogenization of climatic time series. In Proceedings of the 1st Seminar of Homogenisation of Surface Climatological Data, Budapest, Hungary, 6–12 October 1996; pp. 47–62.

34. Potter, K.W. Illustration of a new test for detecting a shift in mean in precipitation series. *Mon. Weather Rev.* **1981**, *109*, 2040–2045. [CrossRef]

35. Moberg, A.; Bergström, H. Homogenization of Swedish temperature data. Part III: The long temperature records from Uppsala and Stockholm. *Int. J. Climatol.* **1997**, *17*, 667–699. [CrossRef]

36. Vincent, L.A.; Mekis, E. Changes in daily and extreme temperature and precipitation indices for Canada over the twentieth century. *Atmosphere-Ocean* **2006**, *44*, 177–193. [CrossRef]

37. Kumar, S.; Merwade, V.; Kam, J.; Thurner, K. Streamflow trends in Indiana: Effects of long term persistence, precipitation and subsurface drains. *J. Hydrol.* **2009**, *374*, 171–183. [CrossRef]

38. Yue, S.; Pilon, P.; Phinney, B.; Cavadias, G. The influence of autocorrelation on the ability to detect trend in hydrological series. *Hydrol. Process.* **2002**, *16*, 1807–1829. [CrossRef]

39. Khattak, M.S.; Babel, M.S.; Sharif, M. Hydro-meteorological trends in the upper Indus River basin in Pakistan. *Clim. Res.* **2011**, *46*, 103–119. [CrossRef]

40. Sen, P.K. Estimates of the regression coefficient based on Kendall's tau. *J. Am. Stat. Assoc.* **1968**, *63*, 1379–1389. [CrossRef]

41. Bargaoui, Z.K.; Chebbi, A. Comparison of two kriging interpolation methods applied to spatiotemporal rainfall. *J. Hydrol.* **2009**, *365*, 56–73. [CrossRef]

42. Zhai, P.; Zhang, X.; Wan, H.; Pan, X. Trends in total precipitation and frequency of daily precipitation extremes over China. *J. Clim.* **2005**, *18*, 1096–1108. [CrossRef]

43. Wang, H.; Gao, J.E.; Zhang, M.; Li, X.; Zhang, S.; Jia, L. Effects of rainfall intensity on groundwater recharge based on simulated rainfall experiments and a groundwater flow model. *Catena* **2015**, *127*, 80–91. [CrossRef]

44. Wang, Y.; Zhou, L. Observed trends in extreme precipitation events in China during 1961–2001 and the associated changes in large-scale circulation. *Geophys. Res. Lett.* **2005**, *32*. [CrossRef]

45. Liu, H.; Lei, T.W.; Zhao, J.; Yuan, C.P.; Fan, Y.T.; Qu, L.Q. Effects of rainfall intensity and antecedent soil water content on soil infiltrability under rainfall conditions using the run off-on-out method. *J. Hydrol.* **2011**, *396*, 24–32. [CrossRef]

water

MDPI

Article

Space–Time Characterization of Rainfall Field in Tuscany

Alessandro Mazza [1,2,*]

[1] LaMMA Consortium, Via Madonna del Piano 10, Sesto Fiorentino 50019, Italy; info@lamma.rete.toscana.it
[2] National Research Council, Institute of Biometeorology, Via Gino Caproni 8, Florence 50141, Italy; info@ibimet.cnr.it
* Correspondence: mazza@lamma.rete.toscana.it; Tel.: +39-55-448-301
* Correspondence: mazza@lamma.rete.toscana.it; Tel.: +39-55-448-301

Academic Editor: Tommaso Moramarco
Received: 28 October 2016; Accepted: 23 January 2017; Published: 31 January 2017

Abstract: Precipitation during the period 2001–2016 over the northern and central part of Tuscany was studied in order to characterize the rainfall regime. The dataset consisted of hourly cumulative rainfall series recorded by a network of 801 rain gauges. The territory was divided into 30×30 km^2 square areas where the annual and seasonal Average Cumulative Rainfall (ACR) and its uncertainty were estimated using the Non-Parametric Ordinary Block Kriging (NPOBK) technique. The choice of area size was a compromise that allows a satisfactory spatial resolution and an acceptable uncertainty of ACR estimates. The daily ACR was estimated using a less computationally expensive technique, averaging the cumulative rainfall measurements in the area. The trend analysis of annual and seasonal ACR time series was performed by means of the Mann–Kendall test. Four climatic zones were identified: the north-western was the rainiest, followed by the north-eastern, north-central and south-central. An overall increase in precipitation was identified, more intense in the north-west, and determined mostly by the increase in winter precipitation. On the entire territory, the number of rainy days, mean precipitation intensity and sum of daily ACR in four intensity groups were evaluated at annual and seasonal scale. The main result was a magnitude of the ACR trend evaluated as 35 mm/year, due mainly to an increase in light and extreme precipitations. This result is in contrast with the decreasing rainfall detected in the past decades.

Keywords: precipitation distribution; extreme events; seasonality; rain gauge; kriging; stationary random function; exponential variogram model; trend detection

---

1. Introduction

In recent decades, there has been a gradual increase in the average temperature of the atmosphere. Global warming has determined a higher water vapor concentration and thus an increase in global precipitation [1]. Despite this overall scenario, a decrease in rainfall was found in the Mediterranean area; for example, experimental results for Italy and Spain indicated a reduction in precipitation [2,3]. Along with the decrease in the average rainfall amount, there has been a change in rainfall patterns, with extreme events becoming more frequent [4]; in particular, Brunetti [5] and Martinez [6] found similar results for Italy and northern Spain (Catalonia). The pattern was not homogeneous on the Italian territory, being more pronounced in the north [7].

Climate change may cause increased drought and extreme events, therefore the study of the evolution of the rainfall regime is a very important task in order to construct a hydrological model, prevent flooding and hydraulic risk [8,9]. It is thus necessary to study rainfall phenomena in different time periods. Evaluation of the rainfall regime for relatively long time periods, annual or seasonal, is essential for the study of climate change and hydrological models. Prevention of hydraulic risks requires an assessment of rainfall phenomena for relatively short time periods, daily or hourly. In this perspective, the amount of rain falling on the territory has to be estimated and its uncertainty evaluated in order to establish the reliability of the results.

In this study, the rainfall regime in the Tuscany region was analyzed during the years 2001–2016. A rain gauge network was used that provides measurements of the cumulative rainfall relative to a small area (in the order of hundreds of square centimeters). There were only a limited number of rain gauges in the territory, the typical distance between them was a few km in more populated areas and more than ten km in hilly areas.

The spatial variability of the cumulative rainfall field implies that for a distance of a few km, the values of the correlation coefficient of such an observable is very low [10–12]. Therefore, by means of a rain gauge network with such a low density, it was not possible to reconstruct the details of the rainfall field using some spatialization techniques. Given the limitation of the instrumentation, an observable was chosen that evaluates the overall rainfall amount on a selected area: the Average Cumulative Rainfall (ACR) i.e., the cumulative rainfall averaged over the area. The spatialization technique chosen to estimate the ACR was Non-Parametric Ordinary Block Kriging (NPOBK) [13–18]. It allows the best value to be estimated and the standard deviation of the ACR, which is a measure of the uncertainty.

An observable closely related to the ACR is the Rain Volume (RV) i.e., the ACR multiplied by the area; RV can be estimated during thunderstorms by means of the area time integral method, a technique based on non-quantitative rainfall detection [19,20]. Griffith et al. [21] used geosynchronous visible or infrared satellite imagery to estimate rainfall over large space and time scales. All these techniques are characterized by lower spatial and temporal resolution.

The estimation of ACR degrades the spatial resolution compared to rain gauge measurements but it allows the amount of rainfall to be assessed on a selected area. In this context, the Tuscany territory was divided into square areas where the ACR was estimated. The uncertainty of the ACR decreases by increasing the area under consideration, but the increase in the area deteriorates the spatial resolution. A compromise was therefore found for the area size that allows a satisfactory spatial resolution and an acceptable uncertainty of ACR estimates.

The annual, seasonal and daily ACR were estimated for all the areas, highlighting the difference in the precipitation amount in different parts of the territory. The trend detection of annual and seasonal ACR time series was performed by means of the Mann–Kendall test. The analysis showed higher rainfall in the north-western area, and an overall increase during the period, more pronounced on the Tyrrhenian coast. This increase occurred mainly in winter. The precipitation increase in Tuscany is contrary to that of previous decades, when a null or negative trend was recorded [22,23]. The analysis of daily ACR time series showed an increase in extreme events, therefore in this case, in agreement with the results obtained for the previous years [5,7].

2. Data and Methodology

2.1. Data

The instrumentation used to evaluate the rainfall field was a network of weather stations equipped with rain gauges and other meteorological sensors. The rain gauges measure cumulative rainfall with an hourly accumulation time. The network is managed by the Hydrological Service of Tuscany Region that provides the quality control of the data. The area covered extends over the Tuscany region and neighboring areas of Emilia-Romagna, Lazio and Umbria. Figure 1 reports the rain gauges located in Tuscany in March 2015.

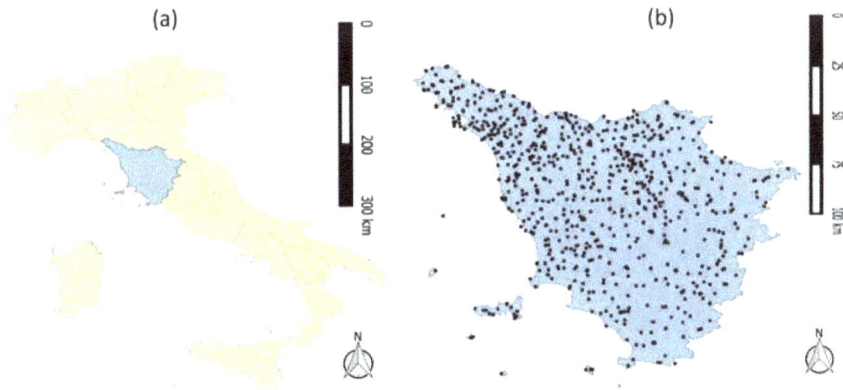

Figure 1. The Tuscany rain gauge network. (**a**) shows the position of the Tuscany region on the Italian peninsula; (**b**) the rain gauges installed in the region (March 2015).

The number of rain gauges increased during the period considered (2001–2016) as new gauges were installed: 325 in 2001 and 801 in 2016. The cumulative rainfall for different accumulation times was evaluated for each rain gauge by adding the hourly measurements. The accumulation times considered were annual (from 1 September to 31 August); seasonal (from 1 September to 30 November in autumn; from 1 December to 28 or 29 February in winter; from 1 March to 31 May in spring; and from 1 June to 31 August in summer); and daily.

Only cumulative rainfall measurements of rain gauges that had recorded more than 90 percent of the hourly measurements during the accumulation time were considered. The results obtained were multiplied by the ratio between the number of hours of the accumulation time and the number of hourly measurements recorded in order not to underestimate the result. This procedure eliminates the bias due to a loss of hourly measurements, but introduces an increase in the uncertainty of the cumulative rainfall.

2.2. The NPOBK Technique

The complete characterization of the rainfall regime on the territory can be obtained by evaluating the amount that falls on an arbitrary area for an arbitrary time period; this can be obtained by means of ACR, which can be expressed in terms of rain rate as:

$$ACR(A,T) = \frac{1}{A}\int_A \int_0^T r(x,t)dxdt = \frac{1}{A}\int_A C(x)dx, \qquad (1)$$

where t is an arbitrary instant of time period T, x an arbitrary point inside area A, r the rain rate evaluated in point x and time t, $C(x)$ the cumulative rainfall in point x.

Determination of the rain rate field for every instant and every point is a necessary and sufficient condition to determine the ACR for any space–time domain. This is obviously impossible given the limitations of the measurements provided by the available instrumentation; the rain gauges perform cumulative rainfall measurements that represent an integral of the rain rate, furthermore they are only installed at a limited number of points. The first limitation implies that 1 h is the shortest accumulation time to evaluate the ACR. The second implies that the ACR estimates should be made by means of a spatialization of the measurements. The spatialization technique used was the NPOBK method; it is based on the assumption that the cumulative rainfall field $C(x)$ is statistically describable by means of a second-order stationary random function [13–18] i.e., the first and second moments are invariant under translations; they are described by means of the following equation:

$$m = E[C(x)]$$
$$F(h) = E[(C(x) - m)(C(x+h) - m)] \quad (2)$$

where $E[]$ is the mean value operator, m the mean value of cumulative rainfall independent of point x, $F(h)$ the correlation function that depends only on the distance h of the two points as the isotropy of the field was also assumed [13]. In this perspective, ACR is described by means of a random variable (R_ACR), its mean was considered the best estimate of ACR and its variance a measure of uncertainty. The technique produces an interpolation function that gives the best unbiased linear estimate of the mean of R_ACR (Equation (3)) and its variance (σ^2_{ACR}) (Equation (4)):

$$ACR(A, T) = \sum_{\alpha=1}^{N} \lambda_\alpha C(x_\alpha), \quad (3)$$

$$\sigma^2_{ACR} = \sum_{\alpha=1}^{n} \lambda_\alpha \gamma_{\alpha V} - \gamma_{VV} - \mu \gamma_{\alpha V} = \frac{1}{|A|} \int_A \gamma(x_\alpha, x)dx \quad \gamma_{VV} = \frac{1}{|A|^2} \int_A \int_A \gamma(x, y)dxdy, \quad (4)$$

where $C(x_\alpha)$ is the rain gauge measurement value on the points x_α, A the area, T the accumulation time, $\gamma(x, y)$ the variogram relative to the pair of points x and y. The parameters λ_α and μ are the solution of the following linear equations system:

$$\sum_{\alpha=1}^{N} \lambda_\alpha \gamma(x_\alpha, x_\beta) - \mu = \gamma_{\beta V} \beta = 1, ..., N. \quad (5)$$

The uncertainty of the ACR estimate was a decreasing function of the number of rain gauges installed within and near the analyzed area, so the size of the area had to be large enough to contain a sufficient number of these in order to perform acceptably accurate estimates. It is worth noting that to estimate ACR, only rain gauge measurements placed within or in proximity to area A were used. The use of measurements away from the area decreases the value of the variance very little at the expense of a substantial increase in the calculation time required.

The variogram, for the isotropy of the model, is a function that depends only on the distance of the points and it can be evaluated by the following equation:

$$\gamma(h) = \frac{1}{2} Var[C(x) - C(x+h)] = \frac{1}{2} \left[\left\langle (C(x) - C(x+h))^2 \right\rangle - \left\langle (C(x) - C(x+h)) \right\rangle^2 \right]. \quad (6)$$

$C(x)$ is second-order stationary therefore the second term of the second member is equal to zero (Equation (2)). The sample values of the variogram were estimated nonparametrically, i.e., without supposing the shape a priori, evaluating approximately the second member of Equation (6) by means of the following:

$$\gamma(h) = \frac{1}{2n(h)} \sum_{i=0}^{n(h)} [C(x_i) - C(x_i + h)]^2, \quad (7)$$

where $C(x_i)$ is the cumulative rainfall measurement of the rain gauge installed in point x_i. The sample values of the variogram were determined accurately by considering large values of n(h), therefore using the measurements of rain gauges installed in a large area N. To determine the largest area N where the field was stationary, the variogram was evaluated using measurements of the rain gauges located within increasingly larger areas centered on the same point. When increasing the size of the area, a significant increase was observed in the sill evaluation of the variogram, so the field cannot be considered stationary. Indeed, the apparent growth of the sill is due to the non-stationarity of the field, so in this case, the second term of the second member of Equation (6) is not negligible and therefore

Equation (7) does not accurately determine the sample values of the variogram. In all the cases studied (Section 3.1), the largest area N where the cumulative rainfall field was stationary contains the area A where the ACR was estimated (a 30×30 km^2 square area), therefore the cumulative rainfall is also stationary in area A and the NPOBK method can be applied to estimate the ACR in this area. To avoid negative values of the R_ACR variance, a conditionally negative defined variogram [15–18] was assumed. For this reason, the variogram was determined by fitting the sample values with the following exponential function [18]:

$$\gamma(h) = Q\left(1 - \exp\left(-\frac{|h|}{r}\right)\right), \tag{8}$$

where Q and r are the fitting parameters.

The rain rate field was always different for each area and in each time as every rainfall event has its own peculiarities, so the characteristics of the accumulation rainfall field differed for each analyzed case. For this reason, for each ACR estimate, the procedure described above was repeated; first the maximum area was determined where the accumulation rainfall field was stationary, then the variogram was calculated using Equations (7) and (8).

2.3. ACR Estimation

The territory was divided into square areas, the annual, seasonal and daily ACR(A_i,T_j) were evaluated for each area during the period 1 March 2001–31 May 2016. A_i denotes an arbitrary area in Figure 2, T_j an arbitrary year, season or day. Each area was identified by means of its center, latitude and longitude, expressed in decimal degrees. The annual and seasonal ACR were estimated by means of the NPOBK technique described in Section 2.2. The relative error, defined by $3\sigma_{ACR}/ACR$, was evaluated. Only ACR estimates with relative error lower than 90 percent were considered sufficiently accurate.

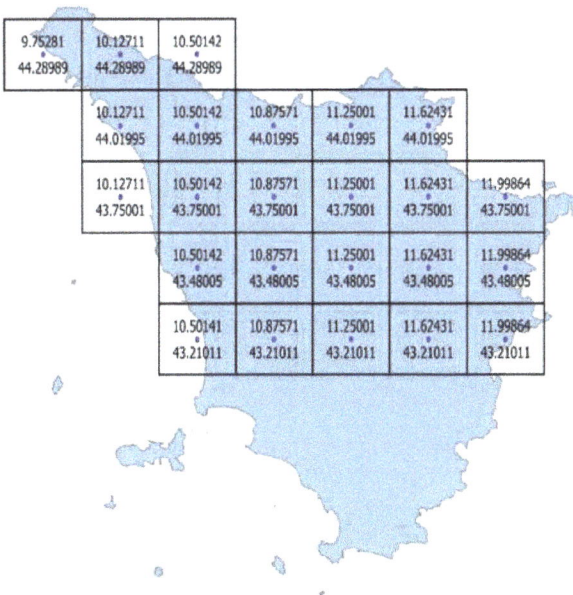

Figure 2. Areas where the ACR was estimated. Latitude and longitude of the centers of the areas expressed in decimal degrees are indicated below and above them.

The size of the areas had to be large enough to contain a sufficient number of rain gauges in order to perform an acceptably accurate ACR estimate. Conversely, the spatial resolution of the accumulation rainfall field is better for small areas. Based on these statements the choice of the size of square areas was 30×30 km^2, which is the best compromise between a good resolution and an acceptable uncertainty of the ACR estimates. Only the northern and central part of Tuscany was examined, as the low density of rain gauges installed in the southern part of the region does not allow the ACR to be estimated with sufficient accuracy (Figure 2).

Over a hundred thousand daily ACR had to be estimated, which is an enormous number. The algorithm to perform the NPOBK is computationally expensive so it was not possible to estimate the daily ACR in a reasonable amount of time. For this reason, the NPOBK technique was not used and another computationally less expensive technique had to be considered. The daily ACR were therefore estimated averaging the cumulative rainfall measurements recorded by rain gauges within the study area. This technique, identified as AM, provides less accurate estimates than the NPOBK technique, and also does not allow the uncertainty to be evaluated.

To establish whether the estimates obtained by means of the AM technique were sufficiently accurate for the aims of this study, the daily ACR estimated by AM and NPOBK techniques was compared for some special cases. The discrepancy of the values in all cases was in the order of a few tens of percentage points, a quantity considered small enough.

2.4. Local Rainfall Regime

2.4.1. Time Average of Local Rainfall

For each area (Figure 2), in order to establish the difference of the typical rainfall amount falling in different zones of Tuscany, the means of the annual and seasonal ACR over the entire period (ACR_AV) were considered:

$$
\begin{aligned}
ACR\_AV(A_i) = \frac{1}{N} \int\limits_{-\infty}^{+\infty} \int\limits_{-\infty}^{+\infty} \cdots \int\limits_{-\infty}^{+\infty} f(R\_ACR(A_i, T_1), \ldots, R\_ACR(A_i, T_N)) \cdot \\
\left(\sum_{n=1}^{N} (R\_ACR(A_i, T_n)) \right) dR\_ACR(A_i, T_1) \ldots dR\_ACR(A_i, T_N)
\end{aligned}
\tag{9}
$$

where A_i is an arbitrary area (Figure 2), $T_1, \ldots, Tn$, all the years or all autumn, winter, spring and summer seasons of the considered period, f the joint probability density function of the random variables $R\_ACR(A_i, T_1), \ldots, R\_ACR(A_i, T_N)$.

The exact value of ACR_AV could not be established as the joint probability density function was unknown. ACR_AV was therefore evaluated approximately averaging the values of the ACR estimated for each year or season:

$$
ACR\_AV(A_i) = \frac{\sum\limits_{n=1}^{N} (ACR(A_i, T_n))}{N}.
\tag{10}
$$

2.4.2. Trend Analysis of Local Rainfall

To establish the rainfall regime evolution in each area, in particular if an increase or a decrease in ACR occurred, the trend analysis of annual and seasonal $ACR(A_i, T_j)$ time series was conducted together with an evaluation of the magnitude of the trend.

The trend analysis of the time series was performed by means of the non-parametric Mann–Kendall test for autocorrelated data [24,25]. The Z statistic used to perform the trend analysis is a random variable whose probability density function is known and can be approximated to a Gaussian with zero mean and standard deviation equal to 1 for time series with a number of elements greater than or equal to 10. The significance level chosen to reject the null hypothesis, i.e., the absence of a trend, was 0.05, corresponding to $Z > 1.96$ (positive trend) or $Z < -1.96$ (negative trend).

The magnitude of the trends was computed using the Theil–Sen estimator (TSA), which is less sensitive to outliers than the least square linear fitting [26,27].

2.5. Global Rainfall Regime

2.5.1. Global Rainfall

Annual, seasonal and daily ACR evaluated over the entire territory (ACR_TOT) was analyzed in order to study the overall evolution of the rainfall regime. The cumulative rainfall field was not second-order stationary as the area was too large, so the NPOBK technique was not applicable. It was therefore chosen to evaluate ACR_TOT by means of the mean of ACR in all areas (Figure 2) corresponding to the same accumulation time:

$$
ACR\_TOT(T_j) = \frac{1}{M} \int_{-\infty}^{+\infty} \int_{-\infty}^{+\infty} \cdots \int_{-\infty}^{+\infty} f\left(R\_ACR(A_1, T_j), \ldots, R\_ACR(A_M, T_j)\right) \cdot \\
\left(\sum_{m=1}^{M} \left(R\_ACR(A_m, T_j)\right)\right) dR\_ACR(A_1, T_j) \ldots \ldots dR\_ACR(A_M, T_j)
\tag{11}
$$

where T_j is an arbitrary year for annual ACR_TOT, or an arbitrary autumn, winter, spring, summer season for seasonal ACR_TOT, or an arbitrary day for daily ACR_TOT; $A_1 \ldots A_m$ are all the areas in Figure 2, f the joint probability density function of random variables $R\_ACR(A_1,T_j), \ldots, R\_ACR(A_M,T_j)$.

Once again, the joint probability density function of the R_ACR was unknown so the ACR_TOT was evaluated, approximately averaging the ACR estimates;

$$
ACR\_TOT(T_j) = \frac{\sum\limits_{m=1}^{M} \left(ACR(A_m, T_j)\right)}{M}.
\tag{12}
$$

2.5.2. Observables for the Study of the Global Rainfall Regime

The following observables were analyzed on annual and seasonal time scales, [27,28]:
ACR_TOT.

Number of rainy days (NRD): the number of days during which daily ACR was greater than 1 mm. If the ACR was less than 1 mm, the day was considered not rainy.

Mean intensity of precipitation (MIP): the ACR_TOT divided by NRD.

The sum of daily ACR_TOT falling into these intensity groups: ACR_TOT <10 mm (SD0), 10 mm < ACR_TOT < 25 mm (SD10), 25 mm < ACR_TOT < 40 mm (SD25), 40 mm < ACR_TOT (SD40).

The time scale will be indicated in subscript to the name of the observable (ann, aut, win, spr, sum indicate the annual and seasonal time scales respectively).

It should be noted that the daily ACR evaluated over relatively large areas never presents the peak of daily cumulative rainfall values observed by some rain gauges, which take measurements on a much smaller area. The correlation coefficient of daily cumulative rainfall presents values significantly lower than 1, even for distances of a few km [10–12]. This implies that the peak values may not remain on the whole area. For this reason, the threshold value (40 mm) chosen for the last intensity group is significantly lower than the cumulative rainfall measurements of some rain gauges during an extreme event, which may exceed 100 mm.

The trend analysis was performed for all time series and the magnitude of the trends was computed.

3. Results

3.1. Evaluation of the Variogram

The sample values of variograms for all the years and seasons were derived by means of Equation (7), using the cumulative rainfall measurements inside 30×30, 40×40, 50×50, 60×60, 70×70 km^2 square areas. The variograms were evaluated by fitting the sample values obtained by the measurements within the largest areas where the condition of stationarity occurred. In almost every case, the stationarity of the field is checked inside areas of 60×60 km^2 or smaller. In all cases, the value of the SILL (Q parameter of Equation (8)) was between 40,000 and 600,000 mm^2 for the annual ACR and between 3000 and 100,000 mm^2 for the seasonal ACR. The r parameter (Equation (8)) was between 7000 and 30,000 m for the annual ACR and 15,000 and 70,000 for the seasonal ACR. As an example, Figure 3 reports the variograms of annual and seasonal ACR for the year 2009–2010 (areas centered at lat. 43.75°, lon. 11.25°). In all cases, the 60×60 km^2 area was the largest where the stationarity condition was verified, except in spring 2010, where this condition was also verified for the 70×70 km^2 area.

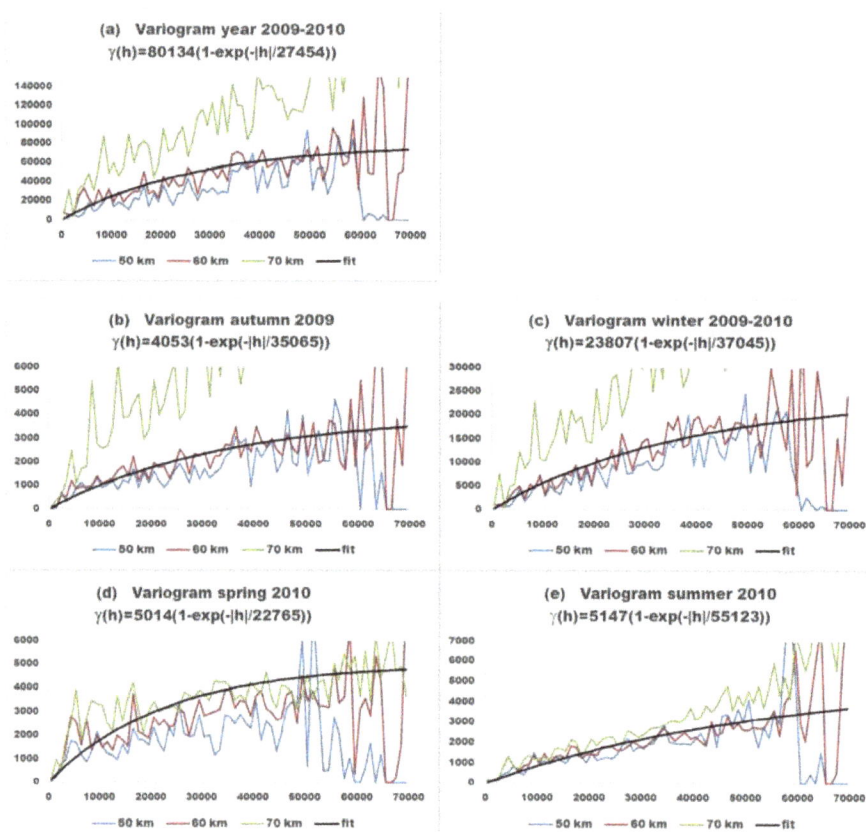

Figure 3. Variograms of annual (**a**) and seasonal (**b–e**) ACR (year: 2009–2010, areas centered at lat. 43.75°, lon. 11.25°). The sample values obtained with the measurements inside 50×50, 60×60, and 70×70 km^2 areas are reported. The distances of the points (meters) are reported in the abscissa, the values of the variogram (mm^2) in the ordinate. The fitting function is also reported.

3.2. Estimation of Rainfall Amount

The annual and seasonal ACR were estimated for each area (Figure 2) by means of the NPOBK technique during the considered period. Examples are given in Table 1 (for some areas and all years and seasons) and in Figures 4 and 5 (for all areas and some years and seasons). The typical values of the relative error were some tens of percentage points; in all cases, it was lower than 75 percent. The density of rain gauges was not homogeneous, it was greater near the main towns and in the most populated areas in general (Figure 1), therefore the relative error decreases accordingly. New rain gauges were installed during the considered period, so in some areas the uncertainty was lower when evaluated more recently. For some seasons and areas, it was not possible to estimate the ACR since the measurements acquired by rain gauges were less than 90 percent of the total.

Table 1. Examples of annual and seasonal ACR (mm) and their standard deviations (σ) and relative errors (ERR). M denotes missing values. Latitude and longitude are expressed in decimal degrees.

	lat.: 43.480053 lon.: 11.250000														
YEAR	ANNUAL			AUTUMN			WINTER			SPRING			SUMMER		
	ACR	σ	ERR	ACR	σ	ERR	ACR	σ	ERR	ACR	σ	ERR	ACR	σ	ERR
2001–2002	795	99	0.38	229	29	0.38	117	17	0.43	197	25	0.38	172	43	0.75
2002–2003	780	80	0.31	290	30	0.31	238	24	0.31	158	16	0.31	104	26	0.75
2003–2004	870	89	0.31	278	28	0.31	238	24	0.31	227	23	0.31	208	52	0.75
2004–2005	798	89	0.34	143	36	0.75	222	25	0.34	158	18	0.34	226	57	0.75
2005–2006	983	110	0.34	513	57	0.34	220	25	0.34	M	M	M	100	11	0.34
2006–2007	842	86	0.31	240	18	0.23	189	14	0.22	200	13	0.19	78	9	0.34
2007–2008	681	51	0.23	135	11	0.25	188	14	0.23	224	16	0.22	109	11	0.31
2008–2009	931	95	0.31	324	23	0.22	272	17	0.19	221	14	0.19	139	14	0.31
2009–2010	982	61	0.19	200	13	0.19	365	23	0.19	265	15	0.17	96	10	0.31
2010–2011	937	60	0.19	443	30	0.20	230	15	0.19	123	8	0.19	138	14	0.31
2011–2012	541	36	0.20	127	8	0.20	114	8	0.21	224	15	0.20	64	6	0.28
2012–2013	1155	83	0.22	446	31	0.21	261	19	0.22	309	21	0.20	136	34	0.75
2013–2014	1266	95	0.23	418	29	0.21	330	24	0.22	170	13	0.23	233	41	0.53
2014–2015	824	51	0.19	316	19	0.18	203	12	0.18	176	10	0.18	246	27	0.34

	lat.: 44.019947 lon.: 10.875702														
YEAR	ANNUAL			AUTUMN			WINTER			SPRING			SUMMER		
	ACR	σ	ERR	ACR	σ	ERR	ACR	σ	ERR	ACR	σ	ERR	ACR	S	ERR
2001–2002	1375	243	0.53	386	97	0.75	284	35	0.38	324	21	0.19	359	24	0.20
2002–2003	1434	96	0.20	750	48	0.19	408	27	0.20	194	13	0.21	63	5	0.22
2003–2004	1940	130	0.20	701	47	0.20	617	45	0.22	464	31	0.20	171	11	0.19
2004–2005	1036	116	0.34	M	M	M	251	17	0.20	291	19	0.19	194	20	0.31
2005–2006	1459	94	0.19	509	33	0.19	490	33	0.20	M	M	M	199	11	0.17
2006–2007	1328	83	0.19	364	21	0.17	504	29	0.17	251	14	0.16	159	10	0.18
2007–2008	1231	71	0.17	341	20	0.17	353	20	0.17	427	23	0.16	105	6	0.16
2008–2009	1843	109	0.18	677	37	0.16	700	37	0.16	392	21	0.16	111	6	0.15
2009–2010	1906	97	0.15	453	23	0.15	786	39	0.15	361	18	0.15	317	16	0.15
2010–2011	1744	89	0.15	740	38	0.15	575	29	0.15	233	11	0.15	210	10	0.15
2011–2012	1162	59	0.15	384	20	0.15	287	15	0.16	424	21	0.15	80	4	0.15
2012–2013	2193	112	0.15	744	37	0.15	570	29	0.15	815	40	0.15	140	7	0.15
2013–2014	2241	117	0.16	599	31	0.15	1057	54	0.15	290	15	0.16	300	15	0.15
2014–2015	1478	75	0.15	643	33	0.15	374	19	0.15	325	17	0.15	138	7	0.15

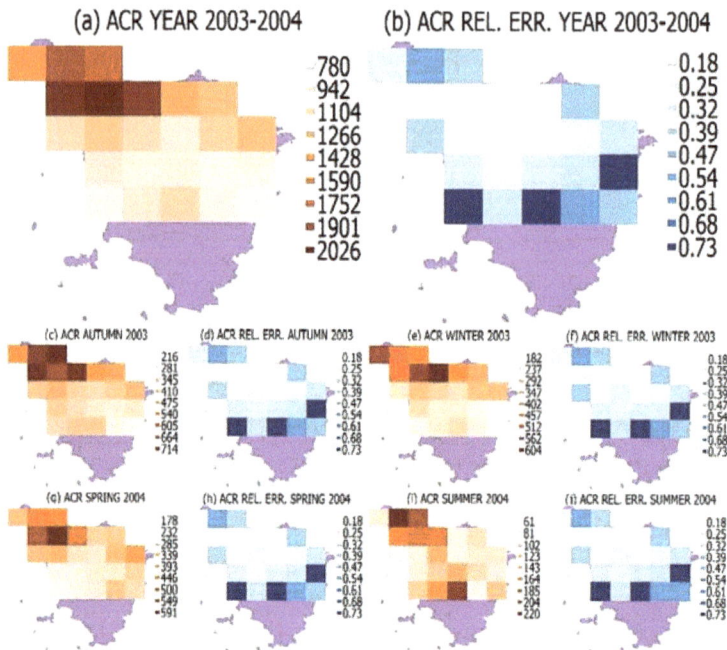

Figure 4. Annual (**a**) and seasonal (**c,e,g,i**) ACR (mm) and their relative errors (**b,d,f,h,j**), for all areas, year: 2003–2004.

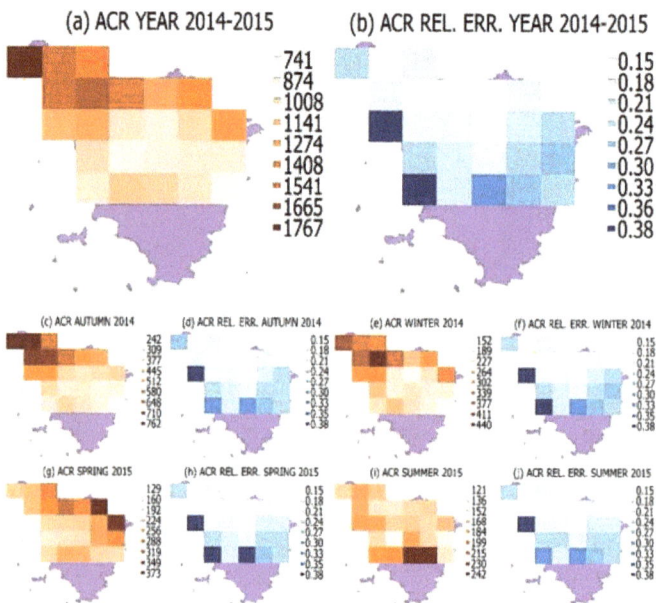

Figure 5. Annual (**a**) and seasonal (**c,e,g,i**) ACR (mm) and their relative errors (**b,d,f,h,j**), for all areas, year: 2014–2015.

It should be noted that, for each year, the sum of seasonal ACR estimates is not exactly equal to the annual one. This discrepancy is due to two distinct causes. First, the data quality control sometimes excluded the measurements of some rain gauges; the rain gauge measurements excluded by the annual estimates may differ from the seasonal ones. Therefore, the set of data in the interpolation equations (Equation (3)) to estimate annual and seasonal ACR were different. Second, the annual and seasonal cumulative rainfall fields and thus the shape of the related variograms did not coincide, so the coefficients of the NPOBK linear systems (Equation (5)) related to the same rain gauges were different.

The daily ACR were estimated by the AM technique. Table 2 reports examples of estimates obtained by means of the NPOBK and AM techniques. The results show a discrepancy in the order of some tens of percentage points, which is sufficiently accurate.

Table 2. Comparison of daily ACR estimates obtained by means of AM and NPOBK techniques.

Lat.	Lon.	Day_Month_Year	ACR AM	ACR NPOBK
43.75	10.13	11_11_2001	4.4	5.2
44.02	10.88	7_10_2003	0.8	0.5
44.02	10.13	29_10_2004	44.1	37.0
44.29	9.75	22_3_2006	12.2	8.9
44.20	10.13	29_2_2008	1.2	1.9
43.75	12.00	24_7_2011	7.5	11.0
43.21	11.25	5_10_2013	17.1	18.7

3.3. Characterization of the Local Rainfall Regime

3.3.1. Estimation of the Time Average of the Local Rainfall

The annual and seasonal ACR_AV of each area was calculated (Equation (10)); the results are reported in Table 3 and Figure 6. The annual values of ACR_AV highlight different rainfall regimes for different areas. In central-southern areas (centered at lat. $43.21°$ and $43.48°$) ACR_AV were between 800 and 900 mm, in central-northern areas (centered at lat. $43.75°$) they were about 1000 mm. In north-western areas (centered at lat. $44.01°$, $44.28°$ and lon. $10.12°$, $10.87°$) ACR_AV were about 1600 mm, in north-eastern areas (lat. $44.01°$, lon. $11.25°$, $11.62°$) 1100 mm.

In spring, the ACR_AV values in central-southern areas were about 210 mm. In central-northern areas a gradient in the east–west direction was observed; in the east, the ACR values were about 200 mm, while in the west 280 mm. In north-western areas, they were about 350 mm, while in north-eastern areas 300 mm. In the summer, the ACR_AV values in central-southern and central-northern areas were about 130 mm, and in northern areas 200 mm. In autumn, in central-southern areas, a gradient in the east–west direction was observed; in the west, the ACR_AV values were about 350 mm, in the east 250 mm. In central-northern and northern areas, values were about 350 mm and 500 mm respectively. In the winter, the ACR_AV values in central-southern areas were about 250 mm, in central-northern areas about 300 mm and 500 mm in northern areas.

Based on these results, it may be noted that the annual volume of rainfall decreases moving in the north–south direction; furthermore, the north-western part of the territory was rainier than the north-east. The rainfall regime, on a seasonal basis, reproduces the same pattern; if a region is rainier annually, the same pattern occurs for all seasons.

Table 3. Annual and seasonal ACR_AV estimates, values of Z statistic and magnitudes of the trend. The Z values >1.96 are reported in bold.

Lat	Lon	ANNUAL			AUTUMN			WINTER			SPRING			SUMMER		
		ACR AV	Z	T	ACR AV	Z	T	ACR AV	Z	T	ACR AV	Z	T	ACR AV	Z	T
43.21	10.50	914	1.17	23.1	335	−0.48	−3.6	269	1.89	7.9	202	−0.21	−1.8	104	0.62	5.2
43.21	10.88	955	**3.09**	34.3	350	0.06	−0.1	284	**3.03**	9.5	210	0.99	3.5	136	0.44	2.6
43.21	11.25	870	0.95	−0.7	329	−0.22	−1.8	249	1.31	7.0	199	0.43	0.8	146	0.33	2.5
43.21	11.62	827	**5.28**	20.9	273	0.44	1.1	229	0.74	3.0	196	0.99	4.7	130	1.31	7.6
43.21	12.00	730	**2.01**	13.3	235	0.22	0.6	200	0.90	3.5	180	1.48	4.3	129	−0.33	−1.4
43.48	10.50	910	**3.02**	28.1	343	−0.52	−2.5	268	**4.70**	10.7	205	0.59	1.8	109	0.55	1.6
43.48	10.88	824	1.64	23.3	289	0.11	1.5	235	**2.24**	8.4	203	0.59	2.4	113	0.55	2.6
43.48	11.25	885	1.42	19.6	290	1.12	2.2	234	0.89	6.3	213	0.40	1.6	139	0.66	3.4
43.48	11.62	838	**2.31**	15.8	293	−0.44	−1.6	236	1.90	6.0	215	1.09	6.6	123	0.99	1.4
43.48	12.00	852	0.31	3.4	288	−0.11	−2.0	239	1.82	6.8	222	0.43	2.2	124	1.40	3.1
43.75	10.13	956	1.75	30.2	352	−0.32	0.3	279	**2.67**	16.6	203	−0.44	−1.9	123	0.00	1.3
43.75	10.50	1075	1.75	40.6	380	0.32	2.0	339	**4.22**	19.6	242	0.69	3.3	135	1.31	3.2
43.75	10.88	896	1.86	37.8	292	0.55	3.2	270	**2.35**	9.3	210	0.69	2.9	133	0.55	3.2
43.75	11.25	841	**2.08**	19.1	286	0.10	1.3	242	1.39	9.5	201	0.30	2.6	132	1.73	5.4
43.75	11.62	1079	**2.01**	11.2	351	0.10	1.0	318	1.66	7.0	268	0.20	1.1	148	0.89	4.3
43.75	12.00	1264	**2.87**	35.3	402	0.44	3.6	371	**8.50**	9.3	317	0.30	2.1	166	0.66	3.2
44.02	10.13	1652	1.40	60.4	565	0.33	7.1	523	**3.39**	24.3	354	0.58	1.9	192	0.88	3.3
44.02	10.50	1777	1.65	52.1	614	0.18	3.0	580	**1.98**	28.4	383	0.66	4.8	202	0.00	0.2
44.02	10.88	1598	1.31	44.2	557	0.11	2.0	533	1.68	24.7	378	0.00	0.1	182	−0.22	−3.0
44.02	11.25	1156	1.64	31.8	383	0.10	0.4	371	1.58	14.8	282	0.10	0.4	133	0.79	5.4
44.02	11.62	1113	1.31	29.6	378	0.22	2.8	313	1.85	5.0	301	1.68	6.5	146	0.95	3.7
44.29	9.75	1588	**3.50**	104.6	554	**2.47**	22.2	521	1.78	26.5	355	0.30	2.3	199	0.88	5.0
44.29	10.13	1792	**1.99**	70.8	608	0.43	8.7	522	1.31	19.7	369	0.33	4.2	205	0.34	3.9
44.29	10.50	1517	**2.12**	40.0	525	−0.06	−0.4	434	1.64	18.9	345	0.55	3.4	195	−0.18	−2.0

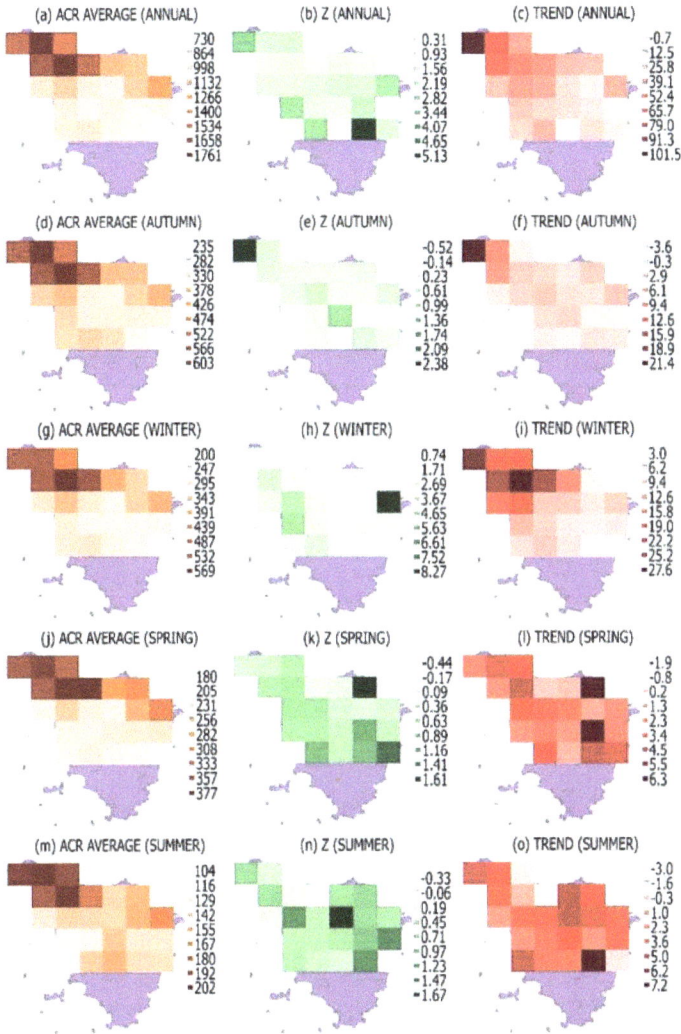

Figure 6. Annual and seasonal ACR_AV estimates (**a,d,g,j,m**); values of Z statistic (**b,e,h,k,n**); and magnitudes of the trend (**c,f,i,l,o**).

3.3.2. Evolution of the Local Rainfall Regime

The trend analysis was performed for the time series of annual and seasonal ACR of each area (Figures 2 and 6, Table 3). For almost all areas, the value of the Z statistic of the Mann–Kendall test was positive, which is a clue to a positive trend. The Z value of many time series was lower than 1.96, the threshold chosen to reject the null hypothesis, therefore it is not possible to state with sufficient confidence the presence of a positive trend in many of the areas considered.

The magnitude of trends of annual ACR was higher in the north-western coastal areas; in general the gradient of the trend magnitude was directed towards the north-west; in other words, there is a trend increase by moving in the western and northern directions. The typical trend magnitude was some tens of mm/year. In autumn, winter and summer, the trend analysis highlights a similar pattern to the annual one; in spring, the magnitude of the trend was overall homogeneous.

3.4. Characterization of the Global Rainfall Regime

The annual and seasonal values of ACR_TOT, NRD, MIP, SD0, SD10, SD25, SD40 were evaluated (Figures 7 and 8). The Mann–Kendall test was performed for all time series (Table 4).

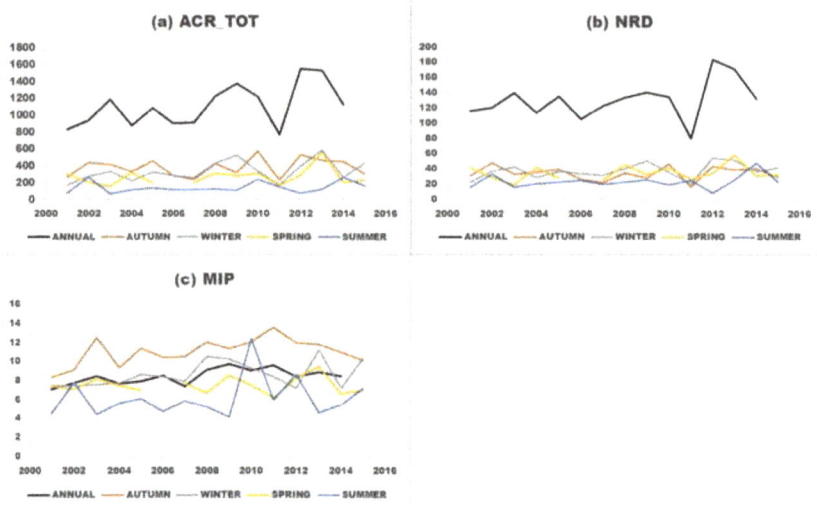

Figure 7. Annual and seasonal time series of ACR_TOT (**a**); NRD (**b**) and MIP (**c**).

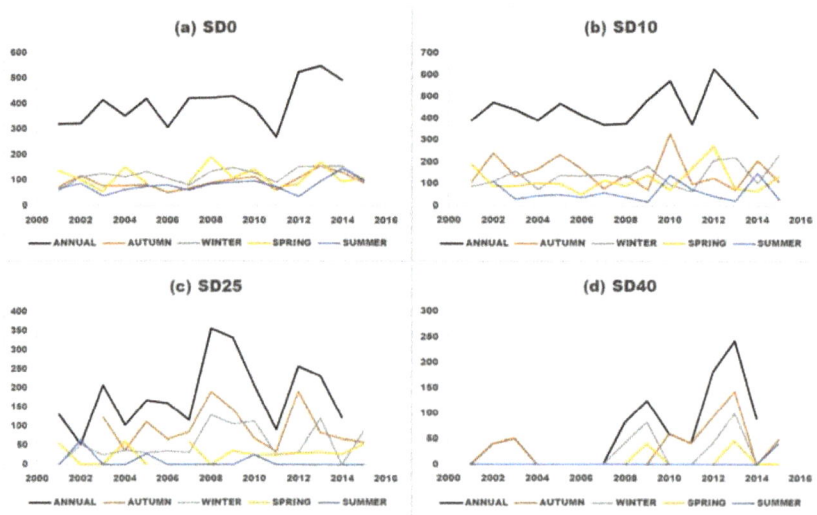

Figure 8. Annual and seasonal time series of SD0 (**a**); SD10 (**b**); SD25 (**c**) and SD40 (**d**).

ACR_TOT$_{ann}$ highlights an increase in precipitation; the Z statistic value of 1.86 was very close to the threshold value, suggesting the presence of a growing trend. The trend magnitude of 35.4 mm/year indicates a significant increase in the rainfall amount. The Z values of ACR_TOT$_{aut}$, ACR_TOT$_{win}$, ACR_TOT$_{spr}$, ACR_TOT$_{sum}$ show an increase in precipitation only in winter.

Table 4. Values of Z statistic and trend magnitudes of annual and seasonal observables. The Z values >
1.96 are reported in bold.

	ANNUAL		AUTUMN		WINTER		SPRING		SUMMER	
	Z	T	Z	T	Z	T	Z	T	Z	T
ACR_TOT	1.86	35.4	0.49	2.9	**3.57**	14.8	0.20	1.1	1.48	3.9
NRD	1.20	2.2	0.00	−0.1	**3.37**	1.1	0.69	0.4	1.14	0.5
MIP	1.86	0.1	1.29	0.2	1.58	0.1	0.00	0.0	0.79	0.1
SD0	**2.36**	13.3	1.68	3.3	1.78	3.1	0.40	0.6	**4.16**	2.7
SD10	0.71	4.3	−0.79	−1.6	1.48	6.5	0.10	0.5	0.40	1.7
SD25	0.88	7.6	0.30	0.5	0.94	2.8	0.10	0.0	−1.14	0.0
SD40	**2.58**	10.5	1.51	0.2	**3.96**	2.689	0.68	0.0	1.50	0.0

The trend analysis of the NRD_{ann} time series does not clearly highlight an increase in this
observable. However, the Z values of seasonal time series show that a positive trend was detected
only in winter.

The Z value of MIP_{ann} was 1.86; also in this case, the presence of a positive trend is very likely.
The analysis of the Z statistic of the seasonal time series did not show an increase in this observable;
although the Z value of MIP_{win} suggests the possible presence of a positive trend in winter.

The study of these observables highlights that one of the causes of the increase in annual
precipitation was the growth of MIP_{ann}. The Mann–Kendall test of NRD_{ann} does not clearly show
a positive trend, although the Z value might suggest it; therefore, it cannot be excluded that this
phenomenon could also have caused the increase in annual rainfall. For all the considered time series,
the positive trend was always attributable to a change of the rainfall regime in winter; a clear growth
of the observables was not highlighted in the other seasons.

The trend analysis of $SD0_{ann}$ shows a positive trend; the value of the Z statistic was 2.36.
The magnitude of trend equal to 13.3 mm/year was very high, so the rainfall amount caused by
low-intensity events has increased a lot during the considered period. $SD10_{ann}$ and $SD25_{ann}$ did not
show any positive trend, so the rainfall amount due to medium intensity events can be considered
stationary. The value of the Z statistic of $SD40_{ann}$ was 2.58, so the presence of a positive trend was
established. The magnitude of the trend equal to 10.5 mm/year was very high; it shows a significant
increase in the contribution of extreme events to the rainfall amount.

The analysis of seasonal time series shows that the rainfall regime in autumn, winter and summer
was similar to the annual one; there was an increase in rainfall amount for the light and extreme
precipitations and there was substantial stability for the intermediate ones. In the spring, no trend was
observed in any of the intensity groups.

4. Discussion and Conclusions

The estimation of rainfall amount is fundamental to evaluate the rainfall regime on the territory.
The observable introduced to perform this task was the ACR [13,14], evaluated on some square
areas in the studied territory. In addition to the best value of the observable, it is very important
to quantitatively evaluate the uncertainty to establish the validity of the results. The annual and
seasonal ACR and its uncertainty was therefore estimated, spatializing the measurements of a rain
gauge network by means of the NPOBK technique [13–18]. The daily measurements were evaluated
without an assessment of the error.

The uncertainty was due mainly to the limited density of rain gauges on the territory; indeed,
the standard deviation of the ACR estimate depends essentially on the number of rain gauges in the
area. In this context, a compromise was reached about the size of the square areas (30×30 km^2) that
allows a satisfactory spatial resolution and an acceptable uncertainty. Only the northern and central
part of Tuscany was studied as the low density of rain gauges in the south does not allow acceptable
estimates to be obtained. The period considered was 1 March 2001–31 May 2016.

The results can be summarized as follows:

the territory can be divided into four areas distinguished by the ACR_AV values during the considered period. The north-western area is the rainiest (ACR_AV about 1600 mm) and experienced the highest increase in precipitation. The north-eastern area (ACR_AV about 1200 mm), central area (ACR_AV about 1000 mm) and southern-central area (ACR_AV about 800 mm) had the most limited increase in precipitation. For all areas, the precipitation in autumn and winter was more abundant than in spring; summer precipitation was the least abundant.

The ACR_TOT time series showed a very strong increase in precipitation; this was assessed as 35 mm/year and was mainly due to winter rainfall. It was caused by the positive trend in MIP and probably also in NRD. The ACR_TOT positive trend was not homogenous on the territory but the magnitude of the trend was higher in the north-western area. Analysis of the distribution of daily precipitation shows an increase in the contribution of low and extreme intensity events, while the contribution of medium intensity events can be considered stationary.

The increase in rainfall that occurred during the considered period is in contrast with the overall trend of the last decades. Bartolini et al. [23] showed a tendency to a decrease in rainfall during the period 1948–2009 in two distinct sites in Tuscany, while Fatichi et al. [22], during the period 1916–2003, showed an absence of any trends in precipitation amount or intensity of extreme events.

The results of this paper are only apparently surprising. Within a generally decreasing rainfall trend, a period of precipitation increase may occur without affecting the overall trend. For example, Romano et al. [28] showed an overall decreasing trend in annual precipitation in the Tiber river basin, an area close to Tuscany. The details of the pattern of the rainfall time series showed a succession of dry and wet periods when the rainfall amount increased.

The values of SD40 have increased, so extreme rainfall events have become more frequent. This result is in agreement with Crisci et al. [29] who reported an increase of extreme events in Tuscany during the period 1973–1994, while Fatichi et al. [22] did not detect any trend on the basis of a longer time series. The increase in extreme precipitation aggravates the risk of flooding; in this context, the monitoring of these events, also with the technique described in this paper, becomes essential.

In the future, the ACR estimates will be improved by means of the measurements from radar operating on the territory. The radar can measure the average rain rate on 1×1 km square pixels, so can improve the spatial resolution of ACR estimates. The radar sweeps all Tuscany, so the measurements will allow the rainfall regime to be studied over the whole region.

Acknowledgments: The Tuscany region meteorological network managed by the Hydrological Service of Tuscany provided rain gauge data.

Conflicts of Interest: The author declares no conflict of interest.

References

1. Trenberth, K.E.; Dai, A.; Rasmussen, R.M.; Parsons, D.B. The changing character of precipitation. *Bull. Am. Meteorol. Soc.* **2003**, *84*, 1205–1217. [CrossRef]
2. Brunetti, M.; Maugeri, M.; Monti, F.; Nanni, T. Temperature and precipitation variability in Italy in the last two centuries from homogenised instrumental time series. *Int. J. Climatol.* **2006**, *26*, 345–381. [CrossRef]
3. Romero, R.; Guijarro, J.A.; Ramis, C.; Alonso, S. A 30-year (1964–1993) daily rainfall data base for the Spanish Mediterranean regions: First exploratory study. *Int. J. Climatol.* **1998**, *18*, 541–560. [CrossRef]
4. Alpert, P. The paradoxical increase of Mediterranean extreme daily rainfall in spite of decrease in total values. *Geophys. Res. Lett.* **2002**, *29*, 31.1–31.4. [CrossRef]
5. Brunetti, M.; Maugeri, M.; Nanni, T. Changes in total precipitation, rainy days and extreme events in northeastern Italy. *Int. J. Climatol.* **2001**, *21*, 861–871. [CrossRef]
6. Martínez, M.D.; Lana, X.; Burgueño, A.; Serra, C. Spatial and temporal daily rainfall regime in Catalonia (NE Spain) derived from four precipitation indices, years 1950–2000. *Int. J. Climatol.* **2007**, *27*, 123–138. [CrossRef]
7. Brunetti, M.; Colacino, M.; Maugeri, M.; Nanni, T. Trends in the daily intensity of precipitation in Italy from 1951 to 1996. *Int. J. Climatol.* **2001**, *21*, 299–316. [CrossRef]

8. Jang, J.H. An advanced method to apply multiple rainfall thresholds for urban flood warnings. *Water* **2015**, *7*, 6056–6078. [CrossRef]

9. Notaro, V.; Liuzzo, L.; Freni, G.; La Loggia, G. Uncertainty analysis in the evaluation of extreme rainfall trends and its implications on urban drainage system design. *Water* **2015**, *7*, 6931–6945. [CrossRef]

10. Pedersen, L.; Jensen, N.E.; Christensen, L.E.; Madsen, H. Quantification of the spatial variability of rainfall based on a dense network of rain gauges. *Atmos. Res.* **2010**, *95*, 441–454. [CrossRef]

11. Mandapaka, P.V.; Krajewski, W.F.; Ciach, G.J.; Villarini, G.; Smith, J.A. Estimation of radar-rainfall error spatial correlation. *Adv. Water Resour.* **2009**, *32*, 1020–1030. [CrossRef]

12. Ciach, G.J.; Krajewski, W.F. On the estimation of radar rainfall error variance. *Adv. Water Resour.* **1999**, *22*, 585–595. [CrossRef]

13. Mazza, A.; Antonini, A.; Melani, S.; Ortolani, A. Recalibration of cumulative rainfall estimates by weather radar over a large area. *J. Appl. Remote Sens.* **2015**, *9*, 095993. [CrossRef]

14. Mazza, A.; Antonini, A.; Melani, S.; Ortolani, A. Estimates of cumulative rainfall over a large area by weather radar. In Proceedings of the SPIE Remote Sensing and Security, Remote Sensing of Clouds and the Atmosphere XIX and Optics in Atmospheric Propagation and Adaptive Systems XVII, Amsterdam, The Netherlands, 22–25 September 2014.

15. Chiles, J.P.; Delfiner, P.B. Structural analysis. In *Geostatistics: Modeling Spatial Uncertainty*, 2nd ed.; Balding, D.J., Cressie, N.A., Eds.; John Wiley & Sons: Hoboken, NJ, USA, 2012; pp. 28–146.

16. Chiles, J.P.; Delfiner, P.B. Kriging. In *Geostatistics: Modeling Spatial Uncertainty*, 2nd ed.; Balding, D.J., Cressie, N.A., Eds.; John Wiley & Sons: Hoboken, NJ, USA, 2012; pp. 147–237.

17. Wackernagel, H. Geostatistics. In *Multivariate. Geostatistics an Introduction with Applications*, 3rd ed.; Springer: Berlin, Germany, 2003; pp. 35–120.

18. Armstrong, M. The theory of kriging. In *Basic Linear Geostatistics*, 1st ed.; Springer: Berlin, Germany, 1998; pp. 83–102.

19. Doneaud, A.; Ionescu-Niscov, S.; Priegnitz, D.L.; Smith, P.L. The area-time integral as an indicator for convective rain volumes. *J. Clim. Appl. Meteorol.* **1984**, *23*, 555–561. [CrossRef]

20. Doneaud, A.A.; Smith, P.L.; Dennis, A.S.; Sengupta, S. A simple method for estimating convective rain volume over an area. *Water Resour. Res.* **1981**, *17*, 1676–1682. [CrossRef]

21. Griffith, C.G.; Woodley, W.L.; Grube, P.G.; Martin, D.W.; Stout, J.; Sikdar, D.N. Rain estimation from geosynchronous satellite imagery—Visible and infrared studies. *Mon. Weather Rev.* **1978**, *106*, 1153–1171. [CrossRef]

22. Fatichi, S.; Caporali, E. A comprehensive analysis of changes in precipitation regime in Tuscany. *Int. J. Climatol.* **2009**, *29*, 1883–1893. [CrossRef]

23. Bartolini, G.; Grifoni, D.; Torrigiani, T.; Vallorani, R.; Meneguzzo, F.; Gozzini, B. Precipitation changes from two long-term hourly datasets in Tuscany, Italy. *Int. J. Climatol.* **2014**, *34*, 3977–3985. [CrossRef]

24. Hamed, K.H.; Ramachandra Rao, A. A modified Mann-Kendall trend test for autocorrelated data. *J. Hydrol.* **1998**, *204*, 182–196. [CrossRef]

25. Hipel, K.W.; McLeod, A.I. Non parametric test for trend detection. In *Time Series Modelling of Water Resources and Environmental Systems*, 1st ed.; Elsevier: Amsterdam, The Netherlands, 1994; pp. 853–938.

26. Sen, P.K. Estimates of the regression coefficient based on Kendall's Tau. *J. Am. Stat. Assoc.* **1968**, *63*, 1379–1389. [CrossRef]

27. Qu, B.; Lv, A.; Jia, S.; Zhu, W. Daily precipitation changes over large river basins in China. 1960–2013. *Water* **2016**, *8*, 185. [CrossRef]

28. Romano, E.; Preziosi, E. Precipitation pattern analysis in the Tiber River basin (central Italy) using standardized indices. *Int. J. Climatol.* **2013**, *33*, 1781–1792. [CrossRef]

29. Crisci, A.; Gozzini, B.; Meneguzzo, F.; Pagliara, S.; Maracchi, G. Extreme rainfall in a changing climate: Regional analysis and hydrological implications in Tuscany. *Hydrol. Process.* **2002**, *16*, 1261–1274. [CrossRef]

water

MDPI

Article

Rainfall Characteristics and Regionalization in Peninsular Malaysia Based on a High Resolution Gridded Data Set

Chee Loong Wong [1,2,*], Juneng Liew [3], Zulkifli Yusop [1,4,*], Tarmizi Ismail [1,4], Raymond Venneker [5] and Stefan Uhlenbrook [6]

[1] Faculty of Civil Engineering, Universiti Teknologi Malaysia, Johor Bahru 81310, Malaysia; tarmiziismail@utm.my

[2] Department of Irrigation and Drainage, Jalan Sultan Salahuddin, Kuala Lumpur 50626, Malaysia

[3] School of Environment and Natural Resource Sciences, Universiti Kebangsaan Malaysia, Bangi 43600, Malaysia; juneng@ukm.edu.my

[4] Centre for Environmental Sustainability and Water Security, Research Institute for Sustainable Environment, Universiti Teknologi Malaysia, Johor Bahru 81310, Malaysia

[5] UNESCO-IHE Institute for Water Education, Westvest 7, Delft 2611 AX, The Netherlands; r.venneker@unesco-ihe.org

[6] World Water Assessment Programme (WWAP), UNESCO Villa La Colombella—Località di Colombella Alta, Perugia 06134, Italy; s.uhlenbrook@unesco.org

* Correspondence: wongcl_my@yahoo.com (C.L.W.); zulyusop@utm.my (Z.Y.); Tel.: +60-19-573-8018 (C.L.W.)

Academic Editor: Tommaso Moramarco

Received: 14 September 2016; Accepted: 25 October 2016; Published: 2 November 2016

Abstract: Daily gridded rainfall data over Peninsular Malaysia are delineated using an objective clustering algorithm, with the objective of classifying rainfall grids into groups of homogeneous regions based on the similarity of the rainfall annual cycles. It has been demonstrated that Peninsular Malaysia can be statistically delineated into eight distinct rainfall regions. This delineation is closely associated with the topographic and geographic characteristics. The variation of rainfall over the Peninsula is generally characterized by bimodal variations with two peaks, i.e., a primary peak occurring during the autumn transitional period and a secondary peak during the spring transitional period. The east coast zones, however, showed a single peak during the northeast monsoon (NEM). The influence of NEM is stronger compared to the southwest monsoon (SWM). Significantly increasing rainfall trends at 95% confidence level are not observed in all regions during the NEM, with exception of northwest zone (R1) and coastal band of west coast interior region (R3). During SWM, most areas have become drier over the last three decades. The study identifies higher variation of mean monthly rainfall over the east coast regions, but spatially, the rainfall is uniformly distributed. For the southwestern coast and west coast regions, a larger range of coefficients of variation is mostly obtained during the NEM, and to a smaller extent during the SWM. The inland region received least rainfall in February, but showed the largest spatial variation. The relationship between rainfall and the El Niño Southern Oscillation (ENSO) was examined based on the Multivariate ENSO Index (MEI). Although the concurrent relationships between rainfall in the different regions and ENSO are generally weak with negative correlations, the rainfall shows stronger positive correlation with preceding ENSO signals with a time lag of four to eight months.

Keywords: rainfall zonation; rainfall trends; rainfall variability; rainfall regionalization; Peninsular Malaysia

1. Introduction

Delineating climatic zones facilitates the investigation of the general climate characteristics of the region, and helps improve understanding of climate variability across a range of spatial and temporal scales. It plays an important role in gaining knowledge of water balance dynamics on various scales for water resources planning and management [1–3]. At low latitudes, where temperatures vary little within and over years, climate differences are largely determined by differences in rainfall patterns. Climatic boundaries are often plotted according to mean monthly rainfall, varying gradually except where the gradients steepen on mountain slopes or along seacoasts [4]. Distinction between zones is always arbitrary and relies on the cartographical convenience. For example, the climatic regions are grouped according to the same category of isohyet range, the climatic zones boundaries are not clear and are always subjectively determined [4]. Thus, an objective quantitative climatic classification is essential for discovering definite and distinctive climate zones boundaries using the climatic data.

The relationship between monsoon seasons and rainfall of Peninsular Malaysia was studied by Lim [5]. He delineated rainfall regions based on 29 rainfall stations and assessed pentad rainfall patterns that could be grouped roughly into five rainfall regions. These results became the reference for the researchers in investigating rainfall distribution across the country [6–9]. Bishop [10] who researched on integration of land use and climatic data, believed that this could produce good region classification. However, his efforts generated some doubts over certain regions, which were mainly attributed to the reliability of climatic data used in his research. He recommended that a more sophisticated statistical analysis procedure and better data sets be applied to overcome the uncertainty in defining the climatic regions.

With the defined climatic regions, the knowledge of past rainfall distribution and patterns can be assessed in order to qualify the nature of changes in time and space. In the past, most of the hydroclimatic studies focused on rainfall distribution and patterns of change over time. For example, Nieuwolt [11] adopted an agricultural rainfall index to quantify rainfall variability over time. Researchers further investigated the temporal and spatial characteristics of rainfall during 1990s, but often restricted their analyses to small catchments, e.g., an urbanized area [12] or a forested catchment [13].

Peninsular Malaysia has undergone development at a rapid pace over last decades. Thus, the overall picture of country water resources distribution has become important for future water resources planning and management. However, a comprehensive regional study of rainfall patterns is still very limited. The mean annual rainfall maps, derived from long-term monthly records (1950–1990) by the Economic Planning Unit [14,15], are only able to show the spatial distribution of average rainfall in the country instead of high temporal and spatial variability of rainfall patterns. Regional trends of rainfall distribution also have not received much attention from researchers.

The climate over Peninsular Malaysia is subjected to pronounced interannual variability which modulates hydrological variability, including floods and droughts [16–18]. The relation between Malaysian rainfall anomalies, sea surface temperature and El Niño-Southern Oscillation (ENSO) were studied by Tangang and Juneng [17] and Juneng and Tangang [19,20]. The global warming/temperature trends and variations were investigated by Tangang et al. [21] and Ng et al. [22]. However, these studies were based solely on station observations which have insufficient spatial coverage [23], biases associated with the gauge measurement process and homogeneity of rainfall time series [24], and problems of missing data [23]. The need for data with better accuracy and space-time resolution has been emphasized by many researchers [12,25–27].

The development of gridded rainfall data sets [27] can provide better spatio-temporal coverage across a region. It is also timely to reassess the climatic regions as defined by Lim [5] and Bishop [10], and investigate regional rainfall trends and distribution to update the understanding of the spatial and temporal variability in the country and support further water resource planning and management.

The aim of this study is two-fold. First, to delineate climatic regions using clustering algorithm based on the daily gridded rainfall data set for Peninsular Malaysia. Second, to explore the regional characteristics of rainfall distribution through trend analysis and analysis of spatial rainfall variability on annual and monsoon-seasonal basis for different climatic regions.

2. Materials and Methods

2.1. Study Area

Peninsular Malaysia is located in the tropics between 1° and 7° north and between 99° and 105° east. The total area of 132,000 km^2 is composed of highlands, floodplains and coastal zones. The Titiwangsa range forms the backbone of the Peninsula, running approximately south-southeast from southern Thailand over a distance of 480 km and separating the eastern part from the western part. Surrounding the central high regions are the coastal lowlands. In general, the Peninsula experiences a warm and humid tropical climate all year round, with uniform temperatures ranging from 25 °C to 32 °C. Rainfall is characterized by two rainy seasons associated with the southwest monsoon (SWM) from May to September and the northeast monsoon (NEM) from November to March [7,16,28].

2.2. Data Sources

The rainfall data used in this study is the gridded data set described by Wong et al. [27] for a period of 31 years from 1976 to 2006. The daily rainfall data originates from four data sources, i.e., the automatic station database of the Department of Irrigation and Drainage (DID), the Malaysian Meteorological Department (MMD), the Global Summary of the Day (GSOD) that archived at the National Climatic Data Center of the National Oceanic and Atmospheric Administration (NOAA/NCDC) was transmitted over the Global Telecommunication System of the World Meteorological Organisation (WMO-GTS), and additional Global Energy and Water Balance Experiment (GEWEX) research programme Asian Monsoon Experiment (GAME) data which also originate from MMD. The combination of all data sources brings an average of 123 daily point observations with maxima of 167 in years 2001 and 2003.

2.3. Methodology

Rainfall regions are delineated from annual rainfall cycles based on monthly climatological values using a clustering algorithm. The delineated regions were then analyzed statistically to assess the spatio-temporal rainfall distribution. Rainfall trends were investigated using Mann-Kendall nonparametric trend detection method. The spatial variability of annual and monsoon rainfall of all regions is calculated. The relationship and influence of El Niño-Southern Oscillation (ENSO) over the delineated regions is assessed by using the Multivariate ENSO Index (MEI) indicator.

2.3.1. Data Interpolation

The rainfall is interpolated directly from point observation values. The input data stations are mostly located at elevations below 300 m above sea level (m.a.s.l) with relatively few stations at higher altitudes between 1000 and 2000 m.a.s.l. Analysis of the available data did not reveal a clearly identifiable orographic relationship in daily rainfall between the station time series. The available rainfall data have been interpolated into daily grids with 0.05 degree resolution (approximately 5.5 km). The interpolation scheme is based on an adaption of Shepard's [29] angular distance weighting (ADW) procedure.

2.3.2. Climatic Regions Delineation

Zones of different annual rainfall cycles based on monthly climatological values were identified using a clustering algorithm. The identification of groups or clusters is based on the similarity between data properties. There are several algorithms developed for this purpose. Here, a non-hierarchical K-means clustering algorithm [30] coupled to an empirical orthogonal function (EOF) analysis was applied.

There are 4289 rainfall grid cells in the study areas. Hence, the size of monthly rainfall climatological data matrix is 4289 × 12. In order to reduce the dimensionality of the data and to filter the random noise, the EOF analysis [31] was applied to the monthly rainfall climatological matrix. The EOF analysis is also commonly known as principal component analysis (PCA) in statistical literature [32]. A Monte-Carlo randomization test suggests a cut-off point at the sixth principal component which describes a total of ~98% of the cumulative variance in the data matrix. The resulting data matrix (4289 × 6) was subjected to the non-hierarchical K-means clustering algorithm.

The K-means clustering algorithm minimizes the Euclidean distance between the data items and the cluster centroid by partitioning the data into k non-overlapping regions. Each of the data items is represented by six values corresponding to the six principal components selected earlier. The algorithm iteratively minimizes the total intra-cluster variance (V) described by the following:

$$V = \sum_{i=1}^{k} \sum_{x_j \in C_i} (x_j - \mu_i)^2 \tag{1}$$

where C_i shows the ith cluster ($i = 1, 2, \ldots, k$) and μ_i is the centroid of all the cluster points $x_j \in C_i$. The algorithm however requires the number of clusters, k, to be set a priori. Here, the algorithm was applied to the 4289 data items, setting the value of k from 3 to 10. The validity of the optimum number of cluster was then determined by the silhouette width [33] for each data items is calculated as:

$$S(i) = \frac{b(i) - a(i)}{\max\{a(i), b(i)\}} \tag{2}$$

where $a(i)$ is the averaged Euclidean distance of the i data item to all other data items in the same cluster, $b(i)$ is the minimum of the averaged distance of i data item to all data items in the other clusters. The averaged silhouette width, $\overline{S}$ evaluates the quality of the clustering solution by considering both compactness (distance between data items within the same cluster) and separation (distance between data items in two neighboring clusters). The closer $\overline{S}$ value is to 1, the better the grouping.

2.3.3. Trend Detection

The annual and monsoon seasonal averaged time series of rainfall of the climatic regions were analyzed using the Mann-Kendall nonparametric trend detection test [34,35]. The Mann-Kendall method has been widely used and tested to be an effective method to evaluate the presence of a statistically significant trend in climatological and hydrological time series [2,36–38]. In the trend test, the null hypothesis H_0 is that there is no trend in the series (the data are independent and identically distributed). The alternative hypothesis H_1 is that a trend exists in the data. The Kendall S-statistic is obtained from comparison between all possible (x, y) pairs of data and is given by:

$$S = \sum_{i=1}^{n-1} \sum_{j=i+1}^{n} \text{sgn}(x_i - x_j)\text{sgn}(y_i - y_j) \tag{3}$$

where, the sign function is defined (for any variable u) as

$$\text{sgn}(u) = \begin{cases} -1 \text{ if } u < 0 \\ 0 \text{ if } u = 0 \\ +1 \text{ if } u > 0 \end{cases} \tag{4}$$

Under the null hypothesis, the statistic S is approximately normally distributed with zero mean and variance given by

$$\text{Var}(S) = \frac{n(n-1)(2n+5) - \sum_{i=1}^{n} t_i i(i-1)(2i+5)}{18} \tag{5}$$

where t_i is the number of ties of extent i in either the x or y data. The summation term in the numerator is used only if the data series contains tied values. The standardized test statistic Z is obtained as

$$Z = \begin{cases} \frac{S+1}{\sqrt{\text{Var}(S)}} & \text{if } S < 0 \\ 0 & \text{if } S = 0 \\ \frac{S-1}{\sqrt{\text{Var}(S)}} & \text{if } S > 0 \end{cases} \tag{6}$$

The test statistic Z is used to measure the significance of a trend. In a two sided test, H_0 should be accepted if $|Z|$ is greater than $Z_{\alpha/2}$, where α represents the chosen significance level (e.g., 5% with $Z_{0.025} = 1.96$) then the null hypothesis is rejected implying that the trend is significant. A positive Z value indicates an upward trend, whereas a negative Z value indicates a downward trend. If a significant trend is present, the average rate of increase or decrease can be obtained from the slope of a simple linear regression.

2.3.4. Spatial Variation of Rainfall

Assessment of spatial variation is based on computation of the spatial variance within a particular delineated gridded region, as

$$\sigma^2 = \frac{1}{n} \sum_{i=1}^{n} (x_i - x)^2 \tag{7}$$

where $\bar{x}$ is the average areal rainfall and, x_i is the rainfall depth in each grid cell. The average areal rainfall and the spatial variance are computed for each region. The coefficient of variation of rainfall which is used to characterize the monthly spatial variability of rainfall is defined as the standard deviation divided by the average areal rainfall.

2.3.5. Rainfalls Correlation Analysis with El Ninõ-Southern Oscillation (ENSO)

The influence of ENSO over the delineated regions in Peninsular Malaysia is assessed by correlating the rainfall time series with the Multivariate ENSO Index (MEI). The MEI is based on six observed variables over the tropical Pacific Ocean, i.e., sea-level pressure, zonal and meridional components of the surface wind, sea surface temperature, surface air temperature, and total cloudiness fraction of the sky [39]. Negative values of the MEI represent the cold ENSO phase, i.e., La Niña, while positive MEI values represent the warm ENSO phase (El Niño). This index was chosen because it is less vulnerable to errors compared to station-based or single variable field based indices (e.g., Southern Oscillation Index, SOI) and has more robust key features in spatial variations and seasonal cycle [39].

To examine the relationship between the interannual rainfall and ENSO, each regional monthly rainfall index is correlated with the MEI time series concurrently. To further explore potential lagged influence of ENSO on NEM and SWM rainfall, the monsoon seasonal rainfall index is created to allow for comparison with the monthly MEI values preceding the seasons. The lag is defined by the time distance between the first month of monsoon season and monthly MEI value. For example, the correlation between the NEM rainfall and MEI at one month lag is the correlation coefficient value between the NEM, which is aggregated from November(0) to March(1)) rainfall and MEI value in October(0). The lagged correlation coefficient values were calculated with a lag time of 0–8 months (MEI leads the rainfall) as well as each individual monsoon month considered.

3. Results and Discussion

3.1. Characteristics of Delineated Climatic Regions

Table 1 shows the averaged silhouette width, $\bar{S}$ suggesting a best grouping of eight different regions for a maximum $\bar{S}$ value of 0.53. The eight delineated rainfall zones are shown in Figure 1. The classification patterns are generally consistent with Lim [5] that classified subjectively the Peninsular Malaysia rainfall into five different zones based on pentad data. The general similarity appears to be the existence of the northwest zone and the central interior rainfall zone over the central mountainous valley areas of Peninsular Malaysia. However, there exist some differences over the southern part of Peninsular Malaysia where current clustering algorithm further delineates the southwest zone of Lim [5] into two different rainfall zones. This rainfall characteristic is also observed by Dale [40] who reported the existence of a distinctive Port Dickson-Muar coastal rain belt in conjunction with other four larger rainfall zones. In addition, the east coast zone is further divided north-south into the northeastern zone and southeastern zone. The latter covers a large part of Johor state. Another difference between the current analysis and that of Lim [5] is the east-west separation of the west coast zone which is represented by region R6 (refer to Figure 1).

Table 1. Averaged silhouette width, $\bar{S}$ indicates better grouping if closer to one.

K, Non-Overlapping Regions	3	4	5	6	7	8	9	10
Averaged silhouette width, $\bar{S}$	0.35	0.41	0.43	0.46	0.43	0.53	0.50	0.48

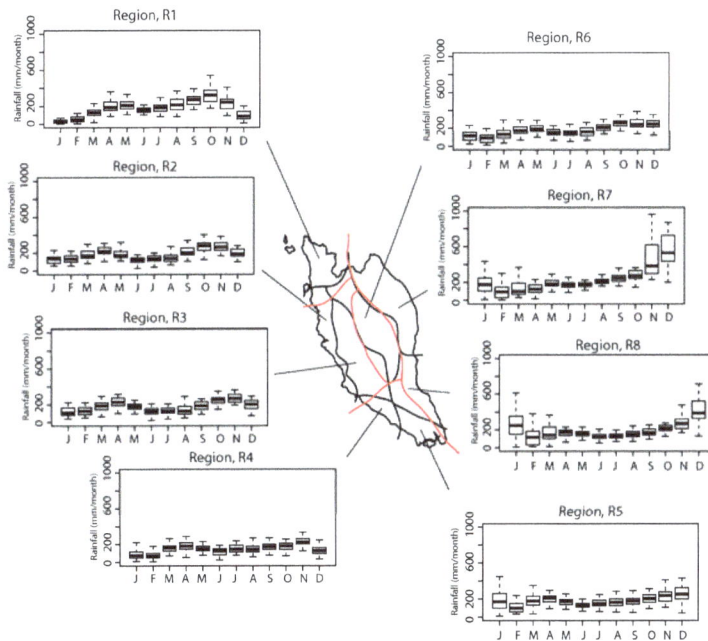

Figure 1. Box and whisker plot of areal average mean monthly rainfall (1976–2006) of the eight delineated rainfall regions in Malaysia. The solid line is the median, the height of the box is the difference between the third and first quartiles (IQR). The zones delineated by Lim [5] are shown in red lines.

An examination of the rainfall zonation patterns and the regional topographic structure suggests that the delineation is closely associated with the topographic characteristics. The temporal variation of rainfall over Peninsular Malaysia is generally characterized by a bimodal annual cycle, i.e., a primary peak occurring during the autumn transitional period and a secondary peak during the spring transitional period as shown in Figure 1. However, over the east coast regions (R7), the rainfall appears to have a single peak during the NEM. This region received 2940 mm/year of mean annual rainfall, 52% and 33% of which occurred during NEM and SWM periods, respectively, as shown in Table 2. The monsoons contribute 85% of the total annual rainfall in this region. During the NEM, the dry northeasterly cold surge winds become moist during the passage over the South China Sea. The interaction with the land along the east coast area creates deep convection clouds and rainfall during the NEM [41–43]. The difference between the northeast zone (R7) and southeast zone (R8) appears to be the higher monsoon rainfall during November–December over the northeastern coast region (see Figure 1). Based on numerical experiments, Juneng et al. [43] argued that the higher elevation in the interior parts of the north-eastern coast of Peninsular Malaysia (in R7) plays a crucial role in providing additional lifting and enhances convection and rainfall while interacting with the mesoscale circulations. Over the south of the eastern region (R8), the rather flat terrain allows the storm systems to reach further inland [43], hence resulting in higher inland rainfall.

Table 2. Rainfall contributions in Peninsular Malaysia for the period of 1976–2006.

Region	Mean Elevation (m) (min, max)	Annual Rainfall (mm/Year)	Standard Deviation	Northeast Monsoon (Nov.–Mar.) Rainfall		Southwest Monsoon (May–Sep.) Rainfall		Total Monsoon Rainfall	
				mm/Year	%	mm/Year	%	mm/Year	%
R1	272 (0, 860)	2118	440	567	26	1035	48	1602	75
R2	116 (0, 773)	2229	314	922	41	806	36	1728	77
R3	326 (16, 1475)	2154	347	921	42	765	35	1686	78
R4	74 (0, 189)	1807	464	705	39	743	41	1448	80
R5	50 (0, 183)	2206	604	1004	45	794	35	1798	81
R6	457 (68, 1280)	2174	329	868	39	868	39	1736	79
R7	266 (0, 1044)	2940	321	1530	52	993	33	2523	85
R8	128 (0, 352)	2383	550	1268	53	735	30	2004	84

Over the southwestern coast of Peninsular Malaysia, the bi-modal seasonal variation of the rainfall is significantly weaker. The terrain is relatively flat to undulating compared to the northern regions. The seasonal rainfall cycle shows that the southern area (R5) of the southwest zones has a rainfall peak in December and maintains a high amount of rainfall in January. Over the northern parts of the southwest zones (R4), rainfall peaks in November and decreases sharply in January the following year. The monsoon rainfall contributes 80% to the mean annual of 1807 mm/year (see Table 2). Note that this zone lies at the edge of the central mountain range (Titiwangsa mountain range) and the local meteorology may interact with the land-sea breeze to nurture storm intensity system over this part of Peninsular Malaysia [44,45].

The west coast region of Peninsular Malaysia, which is characterized by strong bi-modality, is delineated into the coastal band (R2) and the interior zone (R3). The seasonal variations of these two zones show similarity with two rainfall peaks during the monsoon transitional periods. The meteorology during this time of the year is dominated by strong diurnal variations [46] with high frequency of rainfall occurrence in the evening hours. However, the interior zone, which is characterized by rough terrain of the Titiwangsa mountain range, has a sharper drop in monthly rainfall after the annual peaks. The diurnal rainfall cycle over the Straits of Malacca is largely dominated by land-sea breeze forcing [44,45]. The blockage by the high north-south oriented terrain structures (Titiwangsa range) limits the in-land penetration of the sea breezes. This results in minor differences of rainfall climatology over this lower coastal band (R2) of the west coast region. The contribution of monsoon rainfall for both zones (R2 and R3) is relatively equal (see Table 2), but region R2 has higher annual rainfall of 2229 mm/year, compared to R3 with 2154 mm/year.

The northwest zone (R1) also shows a bimodal annual rainfall distribution. However, the primary peak during the autumn transitional period in October is higher compared to the secondary peak during the spring transitional period in April/May. The primary monthly rainfall peak drops dramatically and reaches at minimum level in January. Another significant difference that delineates this zone from the rest of the west coast region is the relatively high amount of rainfall received during the May–July months (>200 mm/month). This is likely due to the exposure of the northern region to the southwest monsoon wind influence. On the other hand, the rest of the west coast areas are shielded by the mountain ranges in Sumatra during this time of the year. Geographically, the northwest zone has a relatively flat topography separated from the rest of the regions by high and rough terrains in the northern parts of Peninsular Malaysia. The R6 region in the central interior part of the country received 2174 mm of mean annual rainfall, 79% of which occurred during the NEM and SWM (see Table 2). The reduction of rainfall amount in this region compared to the west coastal regions during NEM is due to the Titiwangsa mountain range, that appears to block the westward progression of the climatic system and therefore inhibits excessive rainfall over the central interior zone [43]. According to Nieuwolt [47], the rainfall produced in this region is mainly due to local convection caused by intense heating of land surface.

3.2. Rainfall Trends

Most climatic regions do not show any significant trends at the 95% confidence level as indicated in Table 3. The west coast interior region (R3) and south of eastern region (R8) were the only two regions with significant (at 90% confidence level) positive trends in the total annual rainfall. During SWM monsoon, most of the areas are drier over the years, in particular the northwest region (R1) which shows significant trends at the 95% confidence level, i.e., a −5.1 mm/season/year of rainfall declining trend. In contrast, significant increasing trends at the 95% confidence level have been found in the NEM rainfall for the northwest (R1) and west coast interior region (R3) of the Peninsula. During the NEM, the northwest (R1) and west coast interior regions (R3) show large increasing rainfall trends, i.e., 7.5 mm/season/year and 6.7 mm/season/year, respectively.

Table 3. Annual and monsoons rainfall trends (mm/year) of eight distinct climatic regions for the period of 1976–2006.

Region	Annual	NEM	SWM
R1	6.3	7.5	−5.1
R2	3.7	4.3	−1.5
R3	7.7	6.7	0.0
R4	2.1	3.1	−1.8
R5	7.1	4.9	−0.8
R6	5.8	5.0	−1.0
R7	13.3	11.4	−1.1
R8	9.1	5.8	−0.3

Notes: Light grey indicates significant rainfall (mm) trend at 90% confidence level; Dark grey indicates significant rainfall (mm) trend at 95% confidence level; NEM: northeast monsoon; SWM: southwest monsoon.

Focusing on monthly rainfall trend, Table 4 shows that all regions, with exception of the northern parts of the southwest zones (R4), have a significant increasing trend with at least 90% confidence level in January. It is noted that although most of the monthly rainfall throughout the NEM do not exhibit a significant trend at 95% confidence level, the monthly rainfall is at an increasing trend over years (with exception of regions R4 to R8 in November, R4 and R5 in February). On the other hand, most of the monthly rainfall for SWM (May to September) no trend was found, but the monthly rainfall is in decreasing condition over the years. It is interesting to note that although the monthly rainfall does not necessary show significant trends, the consistent sign of changes throughout the seasons resulted in significant trends (at 95% confidence level) when the seasonal means were analyzed.

Table 4. Monthly rainfall trends (mm/month) of eight distinct climatic regions.

Region	Jan.	Feb.	Mar.	Apr.	May	Jun.	Jul.	Aug.	Sep.	Oct.	Nov.	Dec.
R1	1.3	1.3	2.5	0.7	−3.5	0.8	−0.8	0.5	−2.0	2.6	0.2	2.8
R2	1.6	0.3	−0.2	0.1	−1.7	0.3	−0.3	0.5	−0.2	0.3	1.7	1.5
R3	2.2	0.2	1.2	0.8	−1.0	0.3	−0.1	0.5	0.4	−0.7	1.6	2.5
R4	2.0	−1.3	1.1	−0.2	−0.3	−0.2	−0.6	1.1	−2.0	0.1	−0.9	3.1
R5	5.6	−1.1	1.5	−0.2	−0.0	−0.9	0.0	0.8	−0.7	1.1	−0.4	1.5
R6	2.2	0.2	1.7	0.3	−1.4	0.9	−0.9	0.9	−0.6	0.1	−0.5	2.8
R7	5.0	2.4	3.5	−0.4	−0.6	1.1	−1.4	0.0	−0.2	1.0	−1.2	4.2
R8	5.8	0.1	1.1	−0.1	−0.5	−0.1	−0.2	0.9	−0.4	0.8	−0.3	2.0

Notes: Light grey indicates significant monthly rainfall (mm/month) trend at 90% confidence level; Dark grey indicates significant monthly rainfall (mm/month) trend at 95% confidence level.

3.3. Spatial Variability

Figure 2 shows the mean monthly spatial variability for the eight climatic regions. It is noted that the spatial rainfall variation is more uniform throughout the year in the east coast regions (R7 and R8) as shown in Figure 2g,f, although both regions have the largest variation in mean monthly rainfall as indicated in Figure 1. The southwestern coast (R4 and R5) and west coast (R2 and R3) regions receive smaller amounts of monthly rainfall, but the spatial variation differs, particularly during monsoons. For the inland region (R6), the largest spatial variability occurs in February, the month with the smallest rainfall amount of 94 mm/month (refer to Figure 1). The larger range of coefficients of variation is most visible for all regions during the NEM, when the largest amount of rainfall is received over the east coast region (see Table 1), and to a smaller extent during the SWM.

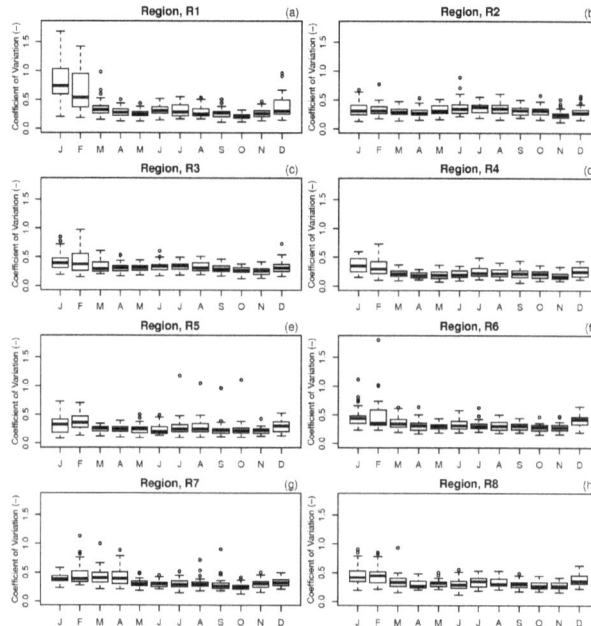

Figure 2. Box and whisker plot of mean monthly coefficients of spatial variations (1976–2006) of the eight delineated rainfall regions (Panel **a** to **h**). The solid line is the median, the height of the box is the difference between the third and first quartiles (IQR). Any data observation which lies 1.5 IQR lower than the first quartile or 1.5 IQR higher than the third quartile is considered an outlier in the statistical sense, indicated by open circles.

The northern region (R1) shows the largest interquartile range (0.19) and mean monthly coefficients of variation (0.38) compared to the other regions. In contrast, the northern part of the southwest region (R4) has a smaller interquartile range (0.13) and lowest mean monthly coefficients of variation (0.25), indicating that the areal rainfall is more uniformly distributed. The calculated median values of the coefficients of variation are close to the mean values for all regions, which suggest that the spatial rainfall variation is symmetrically distributed.

Figure 3 shows the coefficients of variation as a function of the spatial mean monthly rainfall for each climatic region. The relative coefficients of variation generally decreases exponentially with increasing mean monthly rainfall before reaching about 150 mm/month. The coefficients of variation remain fairly constant at 0.30 for mean monthly rainfall exceeding 150 mm/month. This implies the rainfall is more uniformly distributed across the region when the average rainfall amount is high. It is usually associated with heavy monsoon rainfall over the east coast of Peninsular Malaysia [43].

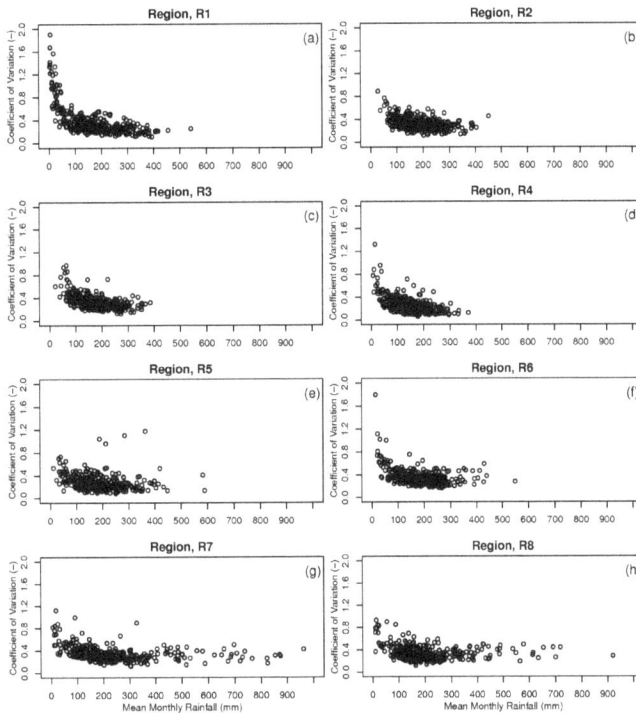

Figure 3. Mean monthly rainfall coefficients of variation over the eight delineated rainfall regions (Panel **a** to **h**) for the period of 1976–2006.

The topography and monsoon winds are likely the main factors controlling the magnitude of the spatial rainfall variation in the country. The Titiwangsa Range is a mountain range that forms the backbone of the Peninsula. During the northeast monsoon (NEM), stronger winds blow to the exposed areas, e.g., the east coast of Peninsular Malaysia [28,43,48], thus these areas receive substantial high amount of rainfall. Higher wind speeds promote more evaporation, which destabilizes the boundary layer and triggers deep convection, and hence, increases rainfall [49]. The inland and west coast areas, which are sheltered by mountain ranges, are relatively free from its influence. On the other hand, the southwest monsoon (SWM) wind speed is generally lighter. It tends to be wetter at the west coast of Peninsular Malaysia compared to the inland and east coast areas. The presence of mountain ranges separating the eastern and western parts of the Peninsula could be the best reason explaining the differences between the rainfall distributions of each region. The relatively flat landscape of the southern region (R5) and east coast regions (R7 and R8) result in reduced spatial rainfall variability.

3.4. Influence of ENSO

The evolution of the ENSO [19,50] may have implications on the rainfall distribution over the Peninsula. The 31-year long-term rainfall anomaly for Peninsular Malaysia was, as a whole, found to be rather weakly and insignificantly (at 95% level) correlated (−0.19) with MEI. In addition, the monthly rainfall of all eight regions generally showed weak concurrent relationships with MEI. Figure 4 shows the correlation coefficients between the rainfalls over the Peninsula and MEI on monthly basis. The most obvious monthly rainfall-MEI correlation (−0.58) was seen in April during the inter-monsoon season. The relationships between monthly rainfalls and MEI of each region were also relatively weak (see Figure 5). These weak relationships are found to be consistent with findings of Juneng and Tangang [19]. The impact of ENSO on rainfall anomalies over the maritime continent has shown considerable spatial variation throughout the different phases of ENSO. During the SWM months, ENSO signature is largely confined to the south of the equator, induced by the interaction between the regional anomalous circulation and the background flow. This anomalous circulation is maintained by the regional feedback processes between the regional seas and the atmosphere. Hence the impact does not extend far northward to the Peninsular Malaysia. During the NEM months, the ENSO signature is shifted northeastward due to the establishment of the northwestern Pacific anticyclone/cyclone anomalies as the Rossby wave response to the surface heating/cooling. Due to the location of this anomalous circulation in the northwestern Pacific, the impact is largely confined to the northern Borneo and southern Philippines, but does not extend far westward to the Malay Peninsula. Thus, the concurrent impact of ENSO is generally weaker over the Peninsular Malaysia.

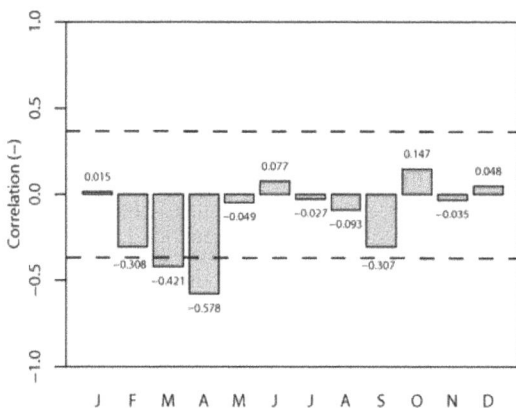

Figure 4. The correlation of concurrent monthly rainfall (mm/month) and Multivariate El Niño-Southern Oscillation (ENSO) Index (MEI) for Peninsular Malaysia. The dotted line indicates significance at 95% level.

A lagged correlation analysis to determine potential lag/lead relationship between monsoon rainfall and MEI is presented in Table 5, showing the lagged correlation between monthly NEM rainfall and MEI for the eight climatic regions. The months of NEM rainfall are mostly correlated negatively with MEI during the NEM months, i.e., from November to March, suggesting modulation of drier conditions during the warm phase of ENSO events during ENSO mature months. The northwest zone (R1), west coast interior zone (R3) and central interior part (R6) of Peninsular Malaysia are correlated negatively with MEI at 95% level of significance. It is interesting to note that the NEM rainfalls posed positive correlation at 95% level of significance with MEI, with a delay of between five and eight months. The correlation magnitude is generally larger compared to the concurrent coefficient values. This means any ENSO (La Niña or El Niño) events could affect Peninsular Malaysia with a time lag of between five and eight months. The lagged correlation between monthly rainfalls of SWM and MEI was found to have similar correlation results as shown in Table 6 where monthly rainfalls are mostly correlated negatively with the MEI indicator during SWM monsoon months, with few regions (as highlighted in dark grey colour in Table 6) posting negative correlation at 95% level of significance in August and September. Generally, the monthly rainfalls of SWM monsoon correlated positively with MEI at 95% level of significance, with a delay of 4–8 months. The analysis clearly shows that the influence of ENSO on rainfall in Peninsular Malaysia is delayed by several months before posing implications to rainfall distributions, especially during SWM. The stronger lagged impact of ENSO on Peninsular rainfall may be explained by the capacitor effect [51] of ENSO which sees the tropical Indian Ocean acting as a heat capacitor during the decaying of a warm ENSO event. It is noted that after the dissipation of El Nino in the spring over the equatorial Pacific, the sea surface temperature over the tropical Indian Ocean remains anomalously warm. This causes the tropospheric temperature to increase by a moist adiabatic adjustment in deep convection, emanating a baroclinic Kelvin wave into the Pacific [51]. Numerical experiment by Xie et al. [51] suggests that this generates low pressure on the equator and anomalous anticyclone over the western north Pacific. The sea surface temperature over the South China Sea is anomalously warm and this promotes deep convective activities over the region. Hence, together with the lower pressure, the increases the rainfall anomalously over the region as shown by the positive lagged correlations in Table 6. Conventionally, the warm ENSO phases are associated with dry rainfall anomalies over Malaysia [17] during the development of ENSO events and this relationship was used to forecast the seasonal rainfall anomalies in the country [20]. The lagged positive relationship found in this study is an important consideration in predicting seasonal rainfall anomalies over Peninsular Malaysia with a longer lead time.

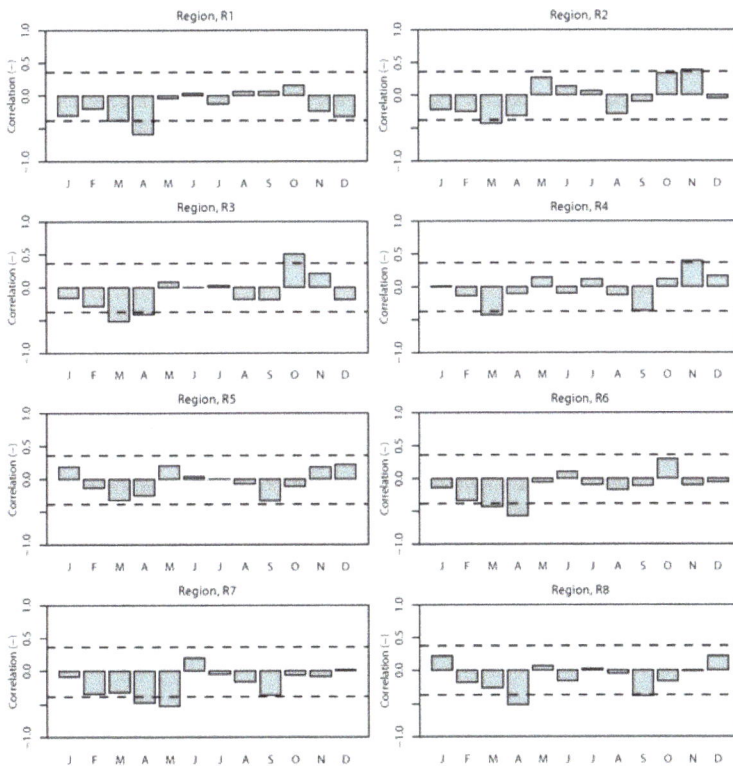

Figure 5. The correlation of concurrent monthly rainfall (mm/month) and Multivariate ENSO Index (MEI) of eight distinct climatic regions. The dotted line indicates significance at 95% level.

Table 5. Correlation coefficient values between the NEM rainfall and the MEI at various concurrent and leading months.

Region	Mar.	Apr.	May	Jun.	Jul.	Aug.	Sep.	Oct.	Nov.	Dec.	Jan.	Feb.	Mar.
R1	0.232	0.215	0.269	0.053	−0.118	−0.222	−0.267	−0.342	−0.435	−0.478	−0.489	−0.537	−0.536
R2	0.489	0.563	0.585	0.453	0.230	0.045	−0.051	−0.057	−0.153	−0.223	−0.243	−0.284	−0.287
R3	0.308	0.344	0.348	0.203	0.002	−0.143	−0.244	−0.236	−0.319	−0.393	−0.394	−0.444	−0.470
R4	0.293	0.392	0.408	0.414	0.280	0.146	0.030	0.055	−0.023	−0.087	−0.071	−0.114	−0.128
R5	0.396	0.439	0.395	0.363	0.259	0.166	0.100	0.147	0.059	0.045	0.056	0.012	−0.011
R6	0.542	0.527	0.559	0.300	0.048	−0.106	−0.212	−0.265	−0.383	−0.394	−0.401	−0.486	−0.478
R7	0.426	0.451	0.523	0.335	0.153	0.046	−0.036	−0.112	−0.247	−0.235	0.201	−0.258	−0.264
R8	0.538	0.582	0.589	0.529	0.341	0.222	0.145	0.136	0.022	0.040	0.052	−0.007	−0.016

Notes: Light grey indicates significant correlation coefficient at 90% confidence level; Dark grey indicates significant correlation coefficient at 95% confidence level.

Table 6. Correlation coefficient values between the SWM rainfall and the MEI at various concurrent and leading months.

Region	Sep.	Oct.	Nov.	Dec.	Jan.	Feb.	Mar.	Apr.	May	Jun.	Jul.	Aug.	Sep.
R1	0.395	0.411	0.381	0.398	0.405	0.355	0.330	0.280	0.255	0.218	0.086	−0.024	−0.090
R2	0.462	0.423	0.425	0.451	0.448	0.419	0.401	0.380	0.382	0.213	0.040	−0.117	−0.219
R3	0.373	0.323	0.285	0.341	0.332	0.265	0.250	0.253	0.249	0.061	−0.128	−0.249	−0.308
R4	0.527	0.450	0.371	0.389	0.371	0.328	0.322	0.288	0.201	−0.060	−0.205	−0.375	−0.464
R5	0.576	0.563	0.539	0.590	0.637	0.589	0.565	0.558	0.465	0.201	−0.085	−0.270	−0.321
R6	0.292	0.279	0.212	0.252	0.254	0.223	0.242	0.209	0.205	0.040	−0.156	−0.280	−0.326
R7	0.356	0.346	0.263	0.258	0.225	0.217	0.202	0.054	−0.035	−0.189	−0.335	−0.472	−0.499
R8	0.505	0.446	0.340	0.381	0.390	0.329	0.294	0.239	0.130	−0.094	−0.254	−0.378	−0.423

Notes: Light grey indicates significant correlation coefficient at 90% confidence level; Dark grey indicates significant correlation coefficient at 95% confidence level.

4. Conclusions

Statistically, Peninsular Malaysia can be differentiated into eight distinct rainfall regions. The results are generally in line with Lim [5] who classified subjectively the regional rainfall into five different zones based on pentad rainfall data. The differences between the current analysis and that of Lim [5] are (i) the east-west separation of the west coast zone; (ii) two different rainfall zones over south-west zone; and (ii) the east coast zone is divided into the northeastern zone covering parts of the states of Kelantan, Terengganu and Pahang, and southeastern zone covering large parts of Johor state.

The delineated rainfall zones are closely associated with the geography and topographic characteristics. The temporal variation of rainfall over the Peninsula is generally characterized by a bimodal distribution, i.e., a primary peak occurring during autumn transitional period and a secondary peak during the spring transitional period, except in the east coast zones with a single pronounced rainfall peak during the NEM. Generally, the NEM influence is relatively stronger and contributes more rainfall over all regions than the SWM. Significant increasing trends at the 95% confidence level have been found in the NEM rainfall for the northwest and west coast interior regions. The opposite is true during the SWM where most areas get drier due to decreases of rainfall over the year.

The east coast regions receive high monthly rainfall compared to other regions, but the rainfall is uniformly distributed spatially. For the southwestern coast and west coast regions, the spatial variation is more pronounced during monsoon periods, especially for NEM rainfalls. The larger range of the coefficients of variation is most visible during the NEM, and to a smaller extent during the SWM. The NEM also brings the largest amount of rainfall to the east coast region. It is noted that the inland regions, which received less rainfall in February, showed the largest spatial variation. The topography and monsoon winds are the main factors controlling the spatial rainfall variation in the country.

The relationship between monthly time series and MEI throughout the study period (31 years) showed a weakly negative correlation in the concurrent monthly rainfall and MEI for all regions. The influence of ENSO showed a correlation lag in the range of a 4–8 month period. Such delays may have implications for the rainfall distribution over the country, especially during NEM and SWM.

The results presented in this study demonstrate the advantages of using high quality gridded data sets in understanding hydrological regimes over Peninsular Malaysia. With a consistent spatio-temporal pattern of the data set, various precipitation indices can be properly defined, offering new insights to different hydrological regimes over the regions in the future. In addition, the enhanced understanding of the spatial and temporal rainfall variability is useful for more effective planning and management of water resources.

Acknowledgments: The authors thank the Department of Irrigation and Drainage Malaysia (DID), Malaysia Meteorological Department (MMD), and NOAA/NCDC for providing the data. Special thanks are also dedicated to Universiti Teknologi Malaysia for the financial support from Vote number RJ130000.7809.4F910.

Author Contributions: Chee Loong Wong and Juneng Liew designed the analytical framework and analyzed the data; Chee Loong Wong wrote the paper; Zulkifli Yusop, Tarmizi Ismail, Raymond Venneker and Stefan Uhlenbrook provided valuable comments on the papers.

References

1. Sato, Y.; Ma, X.; Xu, J.; Matsuoka, M.; Zheng, H. Analysis of long-term water balance in the source area of the Yellow river basin. *Hydrol. Process.* **2008**, *22*, 1618–1629. [CrossRef]
2. Yang, D.; Li, C.; Hu, H.; Lei, Z.; Yang, S.; Kususa, T.; Koike, T.; Musiake, K. Analysis of water resources variability in the Yellow river of china during the last half century using historical data. *Water Resour. Res.* **2004**, *40*, W06502. [CrossRef]
3. Abbaspour, K.C.; Rouholahnejad, E.; Vaghefi, S.; Srinivasan, R.; Yang, H.; Kløve, B. A continental-scale hydrology and water quality model for europe: Calibration and uncertainty of a high-resolution large-scale SWAT model. *J. Hydrol.* **2015**, *524*, 733–752. [CrossRef]

4. Thornthwaite, C.W. An approach toward a rational classification of climate. *Geogr. Rev.* **1948**, *38*, 55–94. [CrossRef]

5. Lim, J.T. Rainfall minimum in peninsular Malaysia during the northeast monsoon. *Mon. Weather Rev.* **1976**, *104*, 96–99. [CrossRef]

6. Deni, S.M.; Jemain, A.A.; Ibrahim, K. Fitting optimum order of markov chain models for daily rainfall occurrences in peninsular Malaysia. *Theor. Appl. Climatol.* **2009**, *97*, 109–121. [CrossRef]

7. Suhaila, J.; Jemain, A.A. Investigating the impacts of adjoining wet days on the distribution of daily rainfall amounts in peninsular Malaysia. *J. Hydrol.* **2009**, *368*, 17–25. [CrossRef]

8. Suhaila, J.; Yusop, Z. Spatial and temporal variabilities of rainfall data using functional data analysis. *Theor. Appl. Climatol.* **2016**. [CrossRef]

9. Deni, S.M.; Suhaila, J.; Wan, Z.W.Z.; Jemain, A.A. Spatial trends of dry spells over peninsular Malaysia during monsoon seasons. *Theor. Appl. Climatol.* **2009**, *99*, 357–371. [CrossRef]

10. Bishop, I.D. Provisional climatic regions in peninsular Malaysia. *Pertanika* **1984**, *7*, 19–24.

11. Nieuwolt, S. Tropical rainfall variability—The agroclimatic impact. *Agric. Environ.* **1982**, *7*, 135–148. [CrossRef]

12. Desa, M.N.; Niemczynowicz, J. Temporal and spatial characteristics in Kuala Lumpur, Malaysia. *Atmos. Res.* **1996**, *42*, 263–277. [CrossRef]

13. Noguchi, S.; Nik, A.R. Rainfall characteristics of tropical rain forest and temperate forest: Comparison between bukit tarek in peninsular Malaysia and hitachi ohta in Japan. *J. Trop. For. Sci.* **1996**, *9*, 206–220.

14. Economic Planning Unit. *Masterplan for the Development of Water Resources in Peninsular Malaysia 2000–2050*; Economic Planning Unit, Prime Minister's Department: Kuala Lumpur, Malaysia, 1999.

15. Department of Irrigation and Drainage. *Review of the National Water Resources Study (2000–2050) and Formulation of National Water Resources Policy*; Department of Irrigation and Drainage Malaysia: Kuala Lumpur, Malaysia, 2011.

16. Tangang, F.T. Low frequency and quasi-biennial oscillations in the Malaysian precipitation anomaly. *Int. J. Climatol.* **2001**, *21*, 1199–1210. [CrossRef]

17. Tangang, F.T.; Juneng, L. Mechanism of Malaysian rainfall anomalies. *J. Clim.* **2004**, *17*, 3616–3622. [CrossRef]

18. Chan, N.W. Impacts of disasters and disaster risk management in Malaysia: The case of floods. In *Resilience and Recovery in Asian Disasters: Community Ties, Market Mechanisms, and Governance*; Aldrich, D.P., Oum, S., Sawada, Y., Eds.; Springer: Tokyo, Japan, 2015; pp. 239–265.

19. Juneng, L.; Tangang, F.T. Evolution of enso-related rainfall anomalies in southeast Asia region and its relationship with atmosphere-ocean variations in indo-pacific sector. *Clim. Dyn.* **2005**, *25*, 337–350. [CrossRef]

20. Juneng, L.; Tangang, F.T. Level and source of predictability of seasonal rainfall anomalies in Malaysia using canonical correlation analysis. *Int. J. Climatol.* **2008**, *28*, 1255–1267. [CrossRef]

21. Tangang, F.T.; Juneng, L.; Ahmad, S. Trend and interannual variability of temperature in Malaysia: 1961–2002. *Theor. Appl. Climatol.* **2006**, *89*, 127–141. [CrossRef]

22. Ng, M.W.; Camerlengo, A.; Khairi, A.A.W. A study of global warming in Malaysia. *Jurnal Teknol.* **2005**, *42*, 1–10.

23. Moten, S. Multiple time scales in rainfall variability. *J. Earth Syst. Sci.* **1993**, *102*, 249–263.

24. Groisman, P.Y.; Legates, D.R. Documenting and detecting long-term precipitation trends: Where are we and what should be done. *Clim. Chang.* **1995**, *31*, 601–622. [CrossRef]

25. Yatagai, A.; Alpert, P.; Xie, P. Development of a daily gridded precipitation data set for the middle east. *Adv. Geosci.* **2008**, *12*, 1–6. [CrossRef]

26. New, M.; Todd, M.; Hulme, M.; Jones, P. Precipitation measurements and trends in the twentieth century. *Int. J. Climatol.* **2001**, *21*, 1899–1922. [CrossRef]

27. Wong, C.L.; Venneker, R.; Jamil, A.B.M.; Uhlenbrook, S. Development of a gridded daily hydrometeorological data set for peninsular Malaysia. *Hydrol. Process.* **2011**, *25*, 1009–1020. [CrossRef]

28. Camerlengo, A.; Demmler, M.I. Wind-driven circulation of peninsular Malaysia's eastern continental shelf. *Sci. Mar.* **1997**, *61*, 203–211.

29. Shepard, D. A Two-dimensional Interpolation Function for Irregularly-spaced Data. In Proceedings of the 1968 23rd ACM National Conference, New York, NY, USA, 27–29 August 1968; ACM: New York, NY, USA, 1968; pp. 517–524.

30. MacQueen, J.B. Some Methods for Classification and Analysis of Multivariate Observations. In Proceedings of the Fifth Berkeley Symposium on Mathematical Statistics and Probability, Oakland, CA, USA, 21 June–18 July 1965; University of California Press: Berkeley, CA, USA, 1967; Volume 1, pp. 281–297.

31. Von Storch, H.; Zwiers, F.W. *Statistical Analysis in Climate Research*; Cambridge University Press: Cambridge, UK, 1999.

32. Joliffe, I.T. *Principal Component Analysis*; Springer: New York, NY, USA, 1986; p. 271.

33. Rousseeuw, P.J. Silhouettes: A graphical aid to the interpretation and validation of cluster analysis. *J. Comput. Appl. Math.* **1987**, *20*, 53–65. [CrossRef]

34. Kendall, M.G. *Rank Correlation Methods*; Hafner: New York, NY, USA, 1948.

35. Mann, H.B. Non-parametric test against trend. *Econometrika* **1945**, *13*, 245–259. [CrossRef]

36. Zhang, X.; Harvey, K.D.; Hogg, W.D.; Yuzyk, T.R. Trends in canadian streamflow. *Water Resour. Res.* **2001**, *37*, 987–998. [CrossRef]

37. Liu, Q.; Yang, Z.; Cui, B. Spatial and temporal variability of annual precipitation during 1961–2006 in yellow river basin, China. *J. Hydrol.* **2006**, *361*, 330–338. [CrossRef]

38. Burn, D.H.; Elnur, M.A.H. Detection of hydrologic trends and variability. *J. Hydrol.* **2002**, *255*, 107–122. [CrossRef]

39. Wolter, K.; Timlin, M.S. El niño/southern oscillation behaviour since 1871 as diagnosed in an extended multivariate enso index (mei.Ext). *Int. J. Climatol.* **2011**, *31*, 1074–1087. [CrossRef]

40. Dale, W.L. The rainfall of Malaya, Part I. *J. Trop. Geogr.* **1959**, *13*, 23–37.

41. Chang, C.P.; Harr, P.A.; Chen, H.J. Sypnotic disturbancess over the equatorial south china sea and western maritime continent during boreal winter. *Mon. Weather Rev.* **2005**, *113*, 489–503. [CrossRef]

42. Salimum, E.; Tangang, F.T.; Juneng, L. Simulation of heavy precipitation episode over eastern peninsular Malaysia using mm5: Sensitivity to cumulus parameterization schemes. *Meteorol. Atmos. Phys.* **2010**, *107*, 33–49. [CrossRef]

43. Juneng, L.; Tangang, F.T.; Reason, C.J.C. Numerical case study of an extreme rainfall event during 9–11 December 2004 over the east coast of peninsular Malaysia. *Meteorol. Atmos. Phys.* **2007**, *98*, 81–98. [CrossRef]

44. Joseph, B.; Bhatt, B.C.; Koh, T.Y.; Chen, S. Sea breeze simulation over the malay peninsular in an intermonsoon period. *J. Geophys. Res.* **2008**, *113*, D20122. [CrossRef]

45. Sow, K.S.; Juneng, L.; Tangang, F.T.; Hussin, A.G.; Mahmud, M. Numerical simulation of a severe late afternoon thunderstorm over peninsular Malaysia. *Asmos. Res.* **2011**, *99*, 248–262. [CrossRef]

46. Varikoden, H.; Samah, A.A.; Babu, C.A. Spatial and temporal characteristics of rain intensity in the peninsular malaysia using trmm rain rate. *J. Hydrol.* **2010**, *387*, 312–319. [CrossRef]

47. Nieuwolt, S. Diurnal rainfall variation in Malaya. *Ann. Assoc. Am. Geogr.* **1968**, *58*, 313–326. [CrossRef]

48. Lim, E.S.; Das, U.; Pan, C.J.; Abdullah, K.; Wong, C.J. Investigating variability of outgoing longwave radiation over peninsular Malaysia using wavelet transform. *J. Clim.* **2013**, *26*, 3415–3428. [CrossRef]

49. Back, L.E.; Bretherton, C.S. The relationship between wind speed and precipitation in the pacific itcz. *J. Clim.* **2005**, *18*, 4317–4328. [CrossRef]

50. Cheang, B.K. Interannual variability of monsoons in malaysia and its relationship with ENSO. *J. Earth Syst. Sci.* **1993**, *102*, 219–239.

51. Xie, S.-P.; Hu, K.; Hafner, J.; Tokinaga, H.; Du, Y.; Huang, G.; Sampe, T. Indian Ocean capacitor effect on Indo-western Pacific climate during the summer following El Niño. *J. Clim.* **2009**, *22*, 730–747. [CrossRef]

water

MDPI

Article

Snow Precipitation Measured by Gauges: Systematic Error Estimation and Data Series Correction in the Central Italian Alps

Giovanna Grossi [1],* [ID], Amerigo Lendvai [1], Giovanni Peretti [2] and Roberto Ranzi [1]

[1] Department of Civil, Environmental, Architectural Engineering and Mathematics, Università degli Studi di Brescia-DICATAM, Via Branze, 42, 25123 Brescia BS, Italy; a.lendvai@studenti.unibs.it (A.L.); roberto.ranzi@unibs.it (R.R.)

[2] ARPA Lombardia–Centro Nivometeorologico di Bormio (SO), 23032 Bormio SO, Italy; g.peretti@arpalombardia.it

* Correspondence: giovanna.grossi@unibs.it; Tel.: +39-030-3711294

Received: 28 December 2016; Accepted: 20 June 2017; Published: 25 June 2017

Abstract: Precipitation measurements by rain gauges are usually affected by a systematic underestimation, which can be larger in case of snowfall. The wind, disturbing the trajectory of the falling water droplets or snowflakes above the rain gauge, is the major source of error, but when tipping-bucket recording gauges are used, the induced evaporation due to the heating device must also be taken into account. Manual measurements of fresh snow water equivalent (SWE) were taken in Alpine areas of Valtellina and Vallecamonica, in Northern Italy, and compared with daily precipitation and melted snow measured by manual precipitation gauges and by mechanical and electronic heated tipping-bucket recording gauges without any wind-shield: all of these gauges underestimated the SWE in a range between 15% and 66%. In some experimental monitoring sites, instead, electronic weighing storage gauges with Alter-type wind-shields are coupled with snow pillows data: daily SWE measurements from these instruments are in good agreement. In order to correct the historical data series of precipitation affected by systematic errors in snowfall measurements, a simple 'at-site' and instrument-dependent model was first developed that applies a correction factor as a function of daily air temperature, which is an index of the solid/liquid precipitation type. The threshold air temperatures were estimated through a statistical analysis of snow field observations. The correction model applied to daily observations led to 5–37% total annual precipitation increments, growing with altitude (1740 ÷ 2190 m above sea level, a.s.l.) and wind exposure. A second 'climatological' correction model based on daily air temperature and wind speed was proposed, leading to errors only slightly higher than those obtained for the at-site corrections.

Keywords: precipitation measurement; precipitation correction; snow water equivalent

---

1. Introduction

Snow precipitation is one of the key hydrometeorological variables, besides snow monitoring activities connected to hydrological and meteorological processes. Planning and managing water resources, as well as preventing floods and landslides, need precise and accurate precipitation data. However, precipitation measurements by rain gauges are normally affected by systematic errors that lead to an underestimation of the real value [1–3]. The wind, disturbing the trajectory of the falling water droplets or snowflakes above the rain gauge, is the major source of error [4–6]. When precipitation occurs in its liquid phase, the maximum underestimation, even in strong wind conditions, is generally below 15%. Errors are much higher when precipitation turns into snow: even 100% underestimation is possible in this case [7,8].

Evaporation induced by heating devices (necessary for tipping bucket rain gauges) adds to the wind effect [9]. Tipping-bucket rain gauges are the most used around the world [10], but they are very often installed in those areas where snowfall exceeds rainfall throughout the year. This happens despite their well-known low reliability in such conditions [11–13].

The errors clearly show up in the Alps, a densely-monitored area, where precipitation measurements are commonly underestimated, more and more as altitude increases. It is at higher elevations that precipitation is greater, the wind is stronger and more continuous, and snowfall is the predominant form of precipitation: these are the most critical issues in correctly measuring the water volume that falls onto the ground. The increasing automatization in meteorological monitoring systems poses new challenges in this field of work, but also pushes to develop new, and maybe crucial, technologies, including advanced methods for monitoring the snow water equivalent and snow properties with remote sensing [14].

In this study a dataset from snow-meteorological monitoring stations located in Northern Italy is analysed. The focus is on the Valtellina and Vallecamonica upper areas, belonging to the Lombardy Region, with the purpose of estimating systematic errors in solid precipitation measurements collected by rain gauges (1740 ÷ 2190 m a.s.l.). To do so, daily rain gauge observations are compared to fresh Snow Water Equivalent (SWE) data from monitoring sites close to the study area. The dataset was collected thanks to cooperation between Università degli Studi di Brescia and the Snow and Avalanches ARPA centre located in Bormio, which is the official snow, avalanche, and weather monitoring institution for mountain areas in Lombardy. Field surveys supported the choice of the most suitable sites, according to the study aim. This paper not only draws the reader's attention to the issues of solid precipitation-measuring methods, but it also suggests a practical way to reconstruct the time series of precipitation observations affected by underestimation errors. Therefore, a simple statistical one-parameter model is proposed here, useful to correct daily precipitation data and based on two meteorological variables, temperature, and precipitation.

2. Case Study and the Error Estimate

Lombardy Region owns a dense meteorological monitoring network, part of which is a legacy of the former National Hydrological Service SIMN (Servizio Idrografico e Mareografico Nazionale, decommissioned in the nineties) and it is still composed of the original manual and mechanical devices, such as the mechanical tipping-bucket rain gauges with paper ink recorders manufactured by SIAP (Villanova (BO), Italy) (Figure 1a,b). With the transfer to regional administrations of the responsibility of environmental monitoring services from the national level to the regional level (ARPA Lombardia, the regional Environment Protection Agency), electronic devices and digital recording systems were installed in addition to the old ones, sometimes adjacent to the old manual stations. In many cases rain gauges manufactured by CAE (San Lazzaro di Savena (BO), Italy) (PMB2 model, Figure 1c) were chosen, so these are now the newest instruments in Lombardy.

For each one of the sites shown in Figure 2 and listed in Table 1, selected from those that are still active in the area, freshly-fallen snow water equivalent (SWE) daily measurements were compared to daily precipitation data retrieved from rain gauges nearby (Figure 3). SWE data by depth and density measurements of freshly-fallen snow collected on snowboards are assumed to be reliable because they provide a direct estimate of the real precipitation fallen on the ground [15]. In fact SWE was previously used by Sevruk [11,16] to reconstruct solid precipitation data from rain gauges and also by [17] as a representative reference measure for quality control of snow pillow measurements by the Snow Telemetry (SNOTEL) Data Collection Network by the U.S Department of Agriculture (USDA).

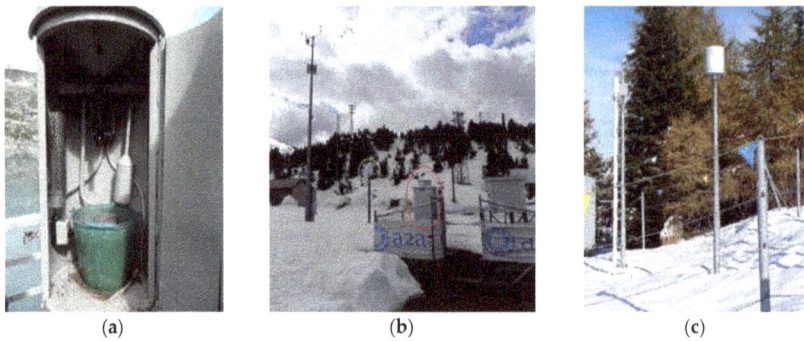

(a) (b) (c)

Figure 1. (**a**) Precipitation gauge at Pantano d'Avio (Brescia province, see also Figure 7b) monitoring site. The tank is inside a heated housing made by SIAP (Italian weather instruments manufacturer). On the left, the heating device, activated by a thermostat, here set to 4 °C; (**b**) Cancano monitoring site, where a CAE PMB2 electronic tipping-bucket heated gauge (yellow circle) is installed besidea heated mechanical SIAP one (red circle); (**c**) Electronic tipping-bucket heated gauge model CAE PMB2 at the Aprica Magnolta (SO) monitoring site.

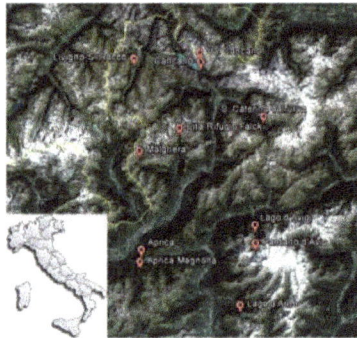

Figure 2. Map of the sites selected for the error estimate for solid precipitation measurements through precipitation gauges in the upper Valtellina and upper Valcamonica, in Northern Italy.

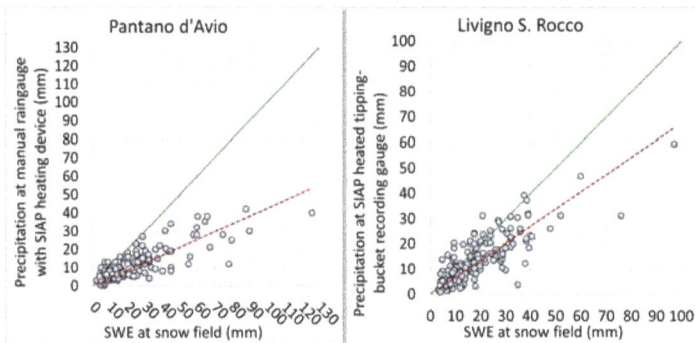

Figure 3. Comparison between melted snow measured daily by rain gauges and SWE of freshly-fallen snow manually measured at snow field sites. Plots refer to Pantano d'Avio (2390 m a.s.l., period 1987–2014) and Livigno S. Rocco (1865 m a.s.l., period 1989–2014).

Table 1. Comparison between melted snow measured by precipitation gauges and SWE of freshly-fallen snow manually measured at snow field sites nearby. Instruments abbreviation: SWE = manual measurement of fresh fallen snow with snow board and cylindrical coring device; P1 = manual rain gauge with 1000 cm^2 top hole and melting operation made at measuring time; P2 = manual rain gauge with SIAP heating device; P3 = mechanical SIAP heated tipping-bucket recording gauge; P4 = electronic heated tipping-bucket recording gauge CAE PMB2.

Sites	Position WGS84 (Lat; Long)	Elevation (m a.s.l.)	Period	Wind Exposure	Wind Data Availability	Site Context	Instruments Compared	Distance SWE/SP-Px (m)	No. of Compared Measurements	Accumulated SWE (mm)	Accumulated P (mm)	Bias (%)
S. Caterina Valfurva	46.410; 10.500 46.413; 10.493	1740	2001–2014 2003–2010	Low	2005–2010	Valley, open	SWE-P3 SWE-P4	590	75 28	866 359	649 177	−25 −51
Livigno S. Rocco	46.522; 10.125	1865	1989–2014	Low	none	Valley, open	SWE-P3	0	241	3641	2662	−27
Cancano	46.514; 10.317 46.513; 10.318	1950	1989–2013 2003–2010	High	2005–2010	Valley, open	SWE-P3 SWE-P4	120	301 73	5723 1579	2353 535	−59 −66
Aprica	46.154; 10.147	1180	1993–2014	Low	none	Valley, garden	SWE-P1	0	66	1452	912	−37
Aprica Magnolta	46.132; 10.140 46.129; 10.148	1865–1950	2009–2014	Low	2012–2014	Slope, forest	SWE-P4	740	38	744	631	−15
Lago d'Arno	46.048; 10.430	1830	1984–2014	Medium	none	Slope, open	SWE-P2	0	163	3429	2597	−24
Pantano d'Avio	46.167; 10.474	2390	1987–2014	High	none	Slope, open	SWE-P2	20	176	3956	1909	−52
Lago d'Avio	46.202; 10.474	1902	2003–2012	Low	none	Valley, open	SWE-P1	0	12	220	176	−20
Eita	46.388; 10.258	1950	2008–2014	Medium	none	Valley, open	SP-PT	0	195	2405	2554	6
Malghera	46.344; 10.142	1995	2008–2014	Medium	none	Valley, open	SP-PT	0	164	1835	1806	−2
Val Cancano	46.532; 10.313	2190	2008–2014	High	none	Slope, open	SP-PT	0	254	2113	2294	9

Starting with seasonal manual monitoring observations, only the precipitation events with the fresh snow density and depth correctly recorded by the observers were selected, in order to evaluate the snow water equivalent (SWE) with snow course data according to the following formula:

$$SWE = \frac{\rho \cdot H_N}{100} \left(mm = \frac{kg}{m^2} \right), \tag{1}$$

where the snow density ρ is expressed in kg/m^3 and the depth of fresh snow H_N is in cm and SWE in mm or kg m^{-2}.

To assess the underestimation of fresh fallen snow, no snow-mixed-to-rain events were considered (specific forms following the international standards are adopted in Italy [18–20]), nor were those with missing, incomplete, or inconsistent SWE measurements or rain gauge precipitation values. SWE manual measurements were taken between 8:00 and 10:00 am with a white wood snow board, a graduated scale and a snow sampler attached to a dynamometer (Figure 4 shows an example snow monitoring field). SWE measurements given by Equation (1) were assumed as the 'true' precipitation reference value. These data are affected by uncertainties in both depth of fresh snow and snow density measurements. In order to assess such uncertainty, results of a recent intercomparison of SWE measurement methods adopted in Italy were analysed [21]. They showed a 7% coefficient of variation of 92 SWE estimates conducted with combined depth and density measurements. This is a result of a 1 cm standard uncertainty of snow depth measurements and a typical 5% standard relative uncertainty of gravimetric snow density measurements with density cutters and portable dynamometers in the field [22]. Rain gauge types are: manual rain gauge with 1000 cm^2 hole and melting snow operation made at the measurement time (P1); manual rain gauge (1000 cm^2) in a heated housing SIAP model (P2); mechanical tipping-bucket recording gauge (1000 cm^2) SIAP or Salmoiraghi in a heated housing SIAP model (P3); CAE PMB2 electronic tipping-bucket recording heated gauge (1000 cm^2) (P4).

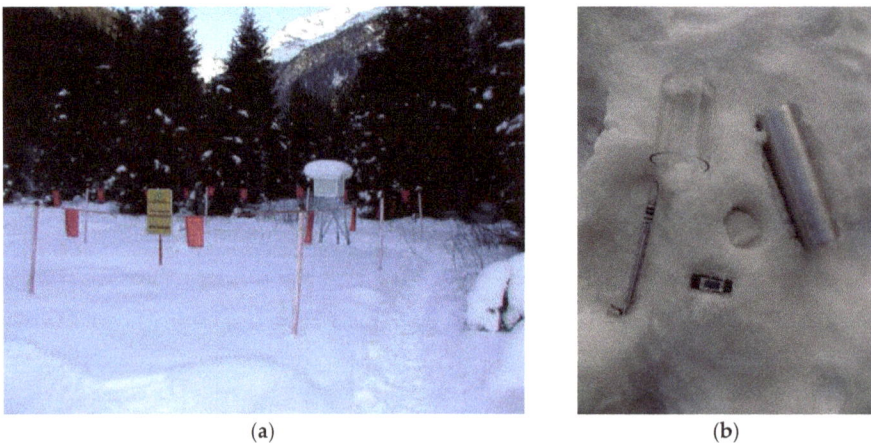

(a) (b)

Figure 4. (**a**) S. Caterina Valfurva manual monitoring site; and (**b**) instruments for the direct manual measurement of snow water equivalent: snow sampler and dynamometer.

In addition to the ARPA Lombardia monitoring network, A2A (an Italian energy generation and trading company) also manages several hydrometeorological stations, used in their management activity of hydroelectric power plants located in this area. Some of these stations are equipped with snow pillows and electronic weighing storage precipitation gauges with Alter-type wind-shields (Figure 5). Table 2 shows the comparison between daily precipitation measurements from weighing storage gauges (model Geonor T200B, with 200 cm^2 hole) and the respective snow pillows (model STS ATM/N; see also [23]). In both cases the total amount in 24 h and a 0.5 mm lower bound were considered to avoid high relative differences [24,25]. A 0 °C higher bound was also set for the daily average temperature, to leave out most of the snow-mixed-to-rain events. The obtained daily dataset was finally purged of all records with missing or incoherent data from one of the instruments. The literature [17] indicates that snow pillow SWE data are within 15% of the snow course reference value in 68% of the SNOTEL network sites and within 5% in 28% of the sites.

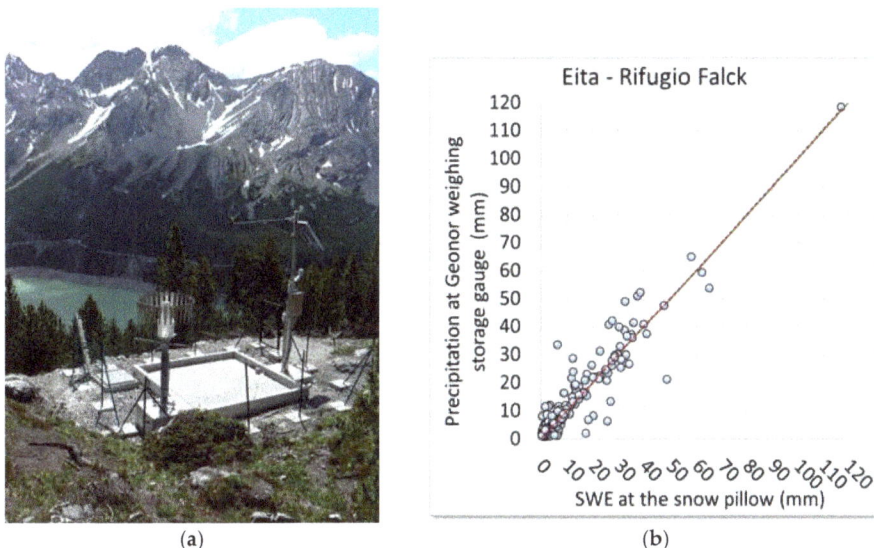

Figure 5. (**a**) Val Cancano monitoring site managed by A2A equipped with Geonor T200B weighing storage gauge with an Alter-type wind-shield and snow pillow; and (**b**) a comparison of daily precipitation measurements: melted snow measured by the Geonor weighing storage gauge with an Alter-type wind-shield versus SWE of fresh fallen snow measured by snow pillows, for the period 2008–2014.

Table 2. Comparison between melted snow measured by Geonor weighing storage gauge with Alter windshield (PT) and SWE of fresh fallen snow measured by snow pillows (SP) installed in new A2A monitoring sites.

Sites	Elevation (m a.s.l.)	Period	Instruments	No. of Measurements	Accumulated SWE (mm)	Accumulated P (mm)	Bias (%)
Eita	1950	2008–2014	SP-PT	195	2405.4	2554.3	6
Malghera	1995	2008–2014	SP-PT	164	1835.0	1805.7	−2
Val Cancano	2190	2008–2014	SP-PT	254	2112.7	2294.3	9

Regarding the manual rain gauges and tipping-bucket heated recording gauges, the underestimation is systematic and varies from 15% to 66%; windy sites (Cancano and Pantano d'Avio) are those with the highest errors, above 50%, while in low wind conditions or wind-covered stations the average error is around 25%. The wind effect is responsible for about half of the total underestimation encountered for the Cancano mechanical tipping-bucket gauge SIAP model (Figure 1b) and Pantano d'Avio manual gauge in a SIAP heated housing (Figure 1a), since these instruments have no wind shielding device; in fact, the same instrument installed in low wind sites shows about halved errors. For the majority of the sites, the lack of wind data makes it difficult to comment more precisely on the influence of the wind. For this reason our basic correction procedures of rain gauge precipitation measurements are based on air temperature only, as presented in the next section.

However, at Cancano and S. Caterina, with two types of tipping bucket gauges, and Aprica Magnolta wind data are also available, although for a limited period and with some gaps. The results of the analysis of those data confirm the influence of the wind on the SIAP mechanical tipping-bucket heated gauge systematic errors, which in S. Caterina are about halved compared to Cancano, where the station is significantly more exposed to the wind. Using the SWE, precipitation, air temperature, and wind speed data available for these three sites a correction procedure of precipitation measurements based on both daily air temperature and wind speed data will be discussed in the fourth section.

Less significant, and apparently contradictory, are the results from the CAE PMB2 electronic tipping-bucket heated gauge. In fact, at S. Caterina, the errors are large despite the fact that the wind speed is even lower when compared to the Aprica Magnolta ones, where the same instrument shows only 15% underestimation, the lowest of all the examined ones, as shown in Table 1. This discrepancy may be justified by the fact that the Aprica Magnolta precipitation gauge is equipped with a heating device only from the end of 2012. Measurements considered for this comparison are mostly from the 2013–2014 season (only one winter season), in which exceptional temperature anomalies were recorded in Northern Italy (up to 4 °C above the average; see also [26]). At Aprica Magnolta the average temperature during the days taken into account for the 2013–2014 season was −1.7 °C; the remaining measurements are from the 2012–2013 season, in which days the average temperature was −3.7 °C. Considering only the data from the 2012–2013 season the gap in accumulated precipitation is −29.3%. At S. Caterina, where the same instrument shows an underestimate near 50%, the data refer to the 2005–2010 period, with an average temperature equal to −4.0 °C. This fact, added to an on-average lower intensity precipitation occurring at this site compared to Aprica, because of the interior Alps precipitation regime, could justify this different instrument performance, less accurate at lower temperatures and lower rainfall intensities. In fact, at S. Caterina, the SIAP mechanical tipping-bucket heated gauge shows fewer underestimations than the CAE PMB2 electronic one, probably due to the different heating devices of the two instruments, given that all other conditions are nearly identical.

Regarding the automated monitoring stations equipped with snow pillows and weighing storage gauges, there is a good correspondence between the increments of precipitation recorded by the gauge and daily increases of snow water equivalent on the snow pillow (Table 2). However there might still be a non-negligible underestimation of the measurements from the snow pillow because the considered sites are not fully wind protected and the Geonor T200B weighing precipitation gauge, equipped with an Alter-type wind shield, is known to underestimate snow precipitations by about 5–10% already for 1–2 m/s wind speed [27–30]. Therefore, it is reasonable to assume that there is an underestimation in the measurements from the snow pillow, because of the well-known problem caused by the internal cohesion of the snowpack. Unfortunately, none of these stations is equipped with wind sensors, so it was not possible to resolve these uncertainties. Results are consistent with those obtained in other similar studies referring to the Alpine region [11,31–34].

3. Correction of Precipitation Data

In order to reconstruct the precipitation data series, a simple model was developed and calibrated. It is based on the application of a correction factor to the daily measurement as a function of daily average temperature [31,35–41]. It appears thusly:

$$P_c = (1 + \alpha_s C_s)P, \tag{2}$$

$$\alpha_s = \begin{cases} 1 \text{ if } T_m < T_l \\ (T_h - T_m)/(T_h - T_l) \text{ if } T_l < T_m < T_h \\ 0 \text{ if } T_m > T_h \end{cases} \qquad (3)$$

where the daily recorded precipitation P is corrected using a factor C_s as a function of the daily average temperature T_m through the coefficient α_s, which varies linearly between 0 and 1 as the temperature increases from the minimum threshold value for liquid precipitation T_h to the maximum threshold value for solid precipitation T_l (Figure 6a). In this way, theoretically, the snow-only precipitation measurements are increased, rain-only precipitation measurements remain the same and the snow-mixed-to-rain ones are partially corrected in a linear way according to the position of the daily mean temperature value between the two thresholds. This model is widely applicable because daily observations of air temperature are easy to find even in very old datasets. Wind speed data would be very useful for a more accurate correction, but they are seldom available and representative data samples are too hard to find, especially regarding the time extension. The nature of the area concerned does not even allow easy connections or interpolations with the data of other stations nearby, as the wind is influenced by the extremely complex orography. In order to determine the transition temperature between liquid and solid precipitation, a statistical analysis was carried out on the data collected by the observers at manual monitoring sites.

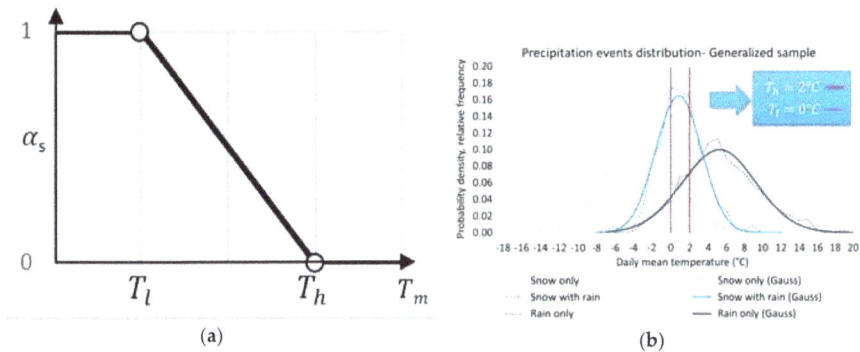

Figure 6. (a) Variability of the coefficient α_s as a function of mean daily temperature; (b) Probability density distribution of the precipitation events as a function of the daily mean temperature for the generalized sample, obtained joining many single station samples (S. Caterina Valfurva, Livigno S. Rocco, Cancano, Aprica, Lago d'Arno, Pantano d'Avio, for the time period shown in Table 1).

AINEVA (Associazione Interregionale di coordinamento e documentazione per i problemi inerenti la NEve e le VAlanghe), the Italian association for snow and avalanche studies, has set a procedure useful to clearly classify the precipitation events as: rain-only, snow-only, and sleet events. Given the availability of samples with a high abundance for each type of event and, very often, also for each individual station, it was possible to carry out a statistical analysis which can be assumed to be representative. Subdividing the samples of the three types of events mentioned above in temperature classes of 1 °C, it was possible to observe that they are well approximated by the normal distribution (Gauss). First, the samples of the three types of meteorological events selected for each station were taken into account; then, to obtain a generalized result, the samples of individual stations were also added.

The generalized sample analysis (Figure 6b) led to the setting of the following thresholds:

- 90% of snow-only events occur in days with average temperature below 1.5 °C
- 90% of rain-only events occur in days with average temperature above 0.1 °C
- 90% of sleet events occur in days with average temperature between −3.1 °C and 4.8 °C

The number of snow-mixed-to-rain events is significantly less than the number of rain-only or snow-only events (264 against 1284 and 2780, respectively). Therefore, it is logical to give more importance to the temperature thresholds identified by the rain-only and snow-only events. On the other hand the fact that the sleet events have a distribution centred at 1 °C was accounted for by choosing symmetrical temperature thresholds with respect to this value. In particular, the selected threshold temperatures are $T_h = 2$ °C and $T_l = 0$ °C. That is coherent with those specified by [31] and [35] for the application of the same model (based on a similar precipitation event analysis) to Malga Bissina and Pinzolo-Prà Rodont sites, in the Trentino-Alto Adige/Südtirol Region.

The correction factor C_s was calibrated for each monitoring site by minimizing the square bias of the individual measurements or the bias between the cumulative values (Table 3). The C_s value appeared to always be higher in the latter case. Subsequently, an intermediate value was chosen, fulfilling also the condition that the slope of the best-fit line of each measurement has to be lower than 1. In this way the resulting values of C_s were close to those obtained by minimizing the RMSE.

Table 3. Estimate of the correction factor C_s for each monitoring site and statistics of the resulting errors by considering the freshly-fallen snow water equivalent as the 'true' reference measurement.

Sites	Elevation (m a.s.l.)	Distance (m)	Instruments	No. of Measurements	Correction Factor C_s	New Bias (%)	Errors St. Dev (mm)
S. Caterina Valfurva	1740	590	SWE-P3	75	0.30	−3.6	4.9
			SWE-P4	28	0.80	−11.8	4.3
Livigno-S. Rocco	1865	0	SWE-P3	241	0.35	−2.6	6.6
Cancano	1950	120	SWE-P3	301	1.15	−17.0	13.1
			SWE-P4	73	1.50	−17.7	10.9
Aprica	1180	0	SWE-P1	66	0.60	−4.7	13.4
Aprica-Magnolta	1865/1950	740	SWE-P4	38	0.11	−6.1	5.8
Lago d'Arno	1830	0	SWE-P2	163	0.25	−6.9	10.7
Pantano d'Avio	2390	20	SWE-P2	176	1.05	−2.3	11.5
Lago d'Avio	1902	0	SWE-P1	12	0.20	−3.6	13.9

The model was applied to four precipitation data series: Pantano d'Avio, Lago d'Avio, S. Caterina and Cancano. In particular for the following years: Pantano d'Avio: 1990, 1994, 1996; Lago d'Avio: 1987–1994, 1996, 2002, 2003, 2005; S. Caterina Valfurva: 1968–1970, 1972–1976, 1978–1984, 1986, 1988–1992, 1994–1997, 1999–2007, 2009, 2012–2013; Cancano: 1979–1986; 1988–1989; 1997; 2002–1913.

An example of the correction effects on monthly precipitation averages and annual total value is reported in Figure 7a and Table 4. Corrected values for Cancano were compared to data from the nearby Val Cancano station, about 2 km apart and 250 metres higher than the first, but equipped with a Geonor T200B weighing precipitation gauge with a Alter-type wind-shield, considered a more reliable instrument in measuring solid precipitation. Regarding the 2009–2013 period, the average precipitation shows significant differences both in the annual values, and in the monthly ones, that cannot be explained only by the different exposure and the altitude gap. The corrected data show excellent consistency on the mean yearly total; at the monthly scale the variability is greater, probably due to an over-correction in the winter months but, overall, the precipitation regime (with the maximum in November) is properly reproduced.

Table 4. Mean yearly precipitation in some monitoring stations before and after the correction model application.

Sites	No. of Years	C_s	P_m (mm)	$P_{m,c}$ (mm)	Error ΔP (%)
S. Caterina Valfurva	40	0.3	860.3	923.6	7%
Cancano	23	1.15	788.3	1055.5	34%
Pantano d'Avio	3	1.05	1417.3	1897.0	34%
Lago d'Avio	12	0.2	1263.3	1324.8	5%

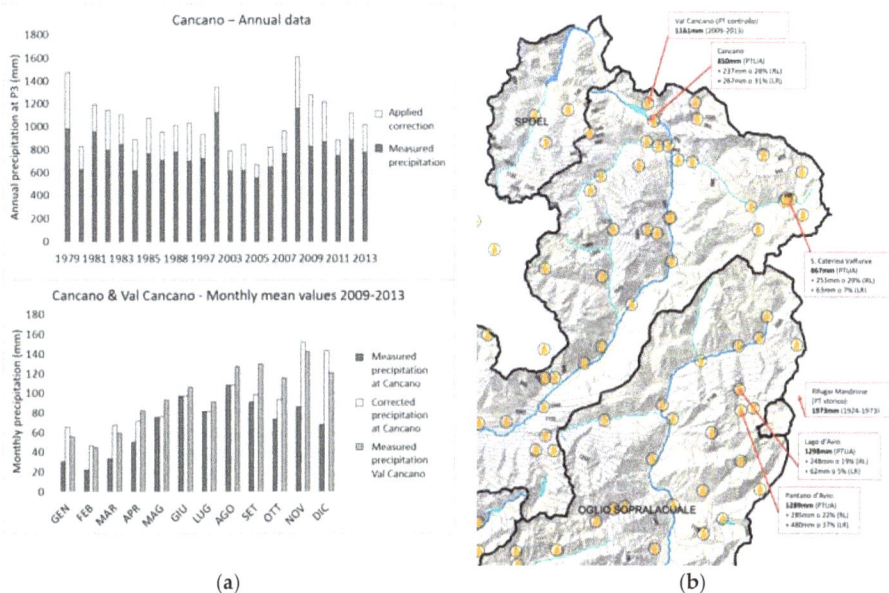

Figure 7. (**a**) The effects of the application of the correction model to the precipitation data gathered by mechanical SIAP heated tipping-bucket recording gauges in Cancano. In the upper part, the annual values and, beneath, the monthly mean values for the 2009–2013 period also compared with the precipitation measured by the close weighing storage gauge with wind-shields at the Val Cancano monitoring site; (**b**) Probability part of the official Lombardy mean annual precipitation map, made in 2003 for the General Water Management and Protection Plan (PTUA). For the highlighted stations a more representative value using the correction model was estimated: the panels show a comparison between the official values with no correction (PTUA), the correction suggested by Regione Lombardia (RL), from a mean estimated SWE, and the correction suggested in this paper using the Lendvai-Ranzi model (LR). For a better comparison, the mean annual precipitation are also measured at two weighing storage gauges with wind-shields (PT) are shown.

4. Discussion

The results obtained for this and other stations can be compared in Figure 7b with those reported on the current official Lombardy Region precipitation distribution map (2003), here limited to the area of this study. To compensate the well-known underestimation of precipitation measurements at the high elevation sites, the Lombardy Region in its General Water Management and Protection Plan-PTUA (2006) uses, and proposes to use for other hydrological studies, a procedure in order to compensate the solid precipitation missing values. This consists in adding, to the yearly precipitation average value, a variable SWE value according to the different river basins and only to the basin area above the winter mean freezing level.

Figure 7b also shows the average precipitation measured by the manual storage gauge at Rifugio Mandrone (TN). Data published on Hydrological Yearbooks by SIMN show a mean yearly precipitation of 1973 mm calculated on the hydrological year and according to the available data for the period between 1924 and 1973. In the same period, starting from the daily measurements, the manual rain gauge at Lago d'Avio shows a yearly average of 1240 mm. Although the comparison is not as rigorous as the one made for Cancano site, since there is no coincidence in time period, nor the availability of such evolved instruments, the reconstruction performed on the data of Pantano d'Avio can still be considered consistent: it gives an yearly average of about 1900 mm for the three years 1990, 1994, and 1996.

A complete dataset of wind data was available for only three stations, and for an extensive and systematic solid precipitation data correction for the past, the simple temperature-based correction procedure was implemented for all of the stations. However, for the three stations equipped with five instruments for which wind data were available some variability with respect to wind speed in the systematic underestimation of precipitation was found.

To test the possibility to extend the information collected for these stations to other environments a temperature and wind-based data correction was tested. The value of the C_s (now $C_{s,w}$) factor in Equation (2) was assigned by minimising the RMSE between the corrected value P_c and the reference 'true' value of the fresh fallen SWE with the following criteria:

- if daily wind speed V is < 1.5 m/s, then $C_{s,w} = 0.4$ (S. Caterina, Aprica Magnolta)
- if $1.5 < V < 2.5$ m/s, then $C_{s,w} = 0.7$ (Cancano P3)
- if $V > 2.5$, then $C_{s,w} = 1.5$ (Cancano P4)

Since this second procedure is no more an at-site correction through a best fit value of $C_{s,w}$ for each station, but has a more general climatological value, a slight loss of accuracy (bias) and precision (RMSE) was observed, as expected, although it has to be kept in mind that the dataset used for the climatological correction is smaller than the one used for the at-site correction, because wind data are available only for a subset of measurements. The results are still acceptable, though, as it can be seen by comparing the error standard deviation in Tables 3 and 5 and, for this reason, this second method is suggested in case both temperature and wind data are available.

Table 5. Data used for the verification of the correction procedure based on both temperature and wind speed data. The coefficient $C_{s,w}$ is calibrated by minimising the RMSE of the corrected precipitation P_c vs. the 'true' value of fresh fallen snow SWE. Uncorrected rain gauge precipitation measurements are indicated by P.

Sites	Instrument	Number of Measurements	Accumulated SWE (mm)	Accumulated Measured Precipitation P (mm)	Corrected Precipitation P_c (mm)	Error St. Dev (mm)	Event Average Wind Speed (m/s)	Annual Average Wind Speed (m/s)
SCP/S.Caterina Valfurva	SWE-P3	26	337.8	256.2	387.0	4.5	0.7	1.1
	SWE-P4	26	337.8	170.2	264.6	3.3		(2005–2009)
CAN/Cancano	SWE-P3	82	1758.3	650.4	1354.4	13.4	2.3	2.6
	SWE-P4	71	1566.8	530.0	1106.1	9.7		(2005–2009)
AMA/Aprica Magnolta	SWE-P4	37	694.6	595.4	991.9	13.4	1.1	2.2 (2012–2014)

5. Conclusions

This study concerned Valtellina and Vallecamonica upper areas, located in the Lombardy Region, where ARPA Lombardia is officially in charge for the snow monitoring and avalanche risk warning service. Regarding the precipitation measurement in high-altitude areas, the current monitoring network is mostly equipped with mechanical and electronic tipping-bucket recording gauges, and only some of them have a heating device useful to melt the snow; moreover, there are still some manual precipitation gauges in use. During the snow season, there are several sites where human observers take fresh fallen snow water equivalent (SWE) measurements that are assumed as a reference 'true value' to correct systematic errors of precipitation measured by rain gauges nearby. Heated tipping-bucket gauges and manual gauges underestimate the real precipitation depth by between 15% and 66%; windy monitoring sites are those with the highest errors, above 50%, while in low wind or wind-covered stations the average error is around 25%.

In addition the A2A hydropower generation company had some hydrometeorological stations installed in 2008, equipped with snow pillows and electronic weighing storage precipitation gauges with Alter windshield. For each station, a comparison between measurements taken by these instruments showed to be in good agreement. This fact confirms the high reliability of electronic weighing gauges in solid precipitation measurement. However, it is likely that the snow pillows also underestimate the snowpack SWE when this is subject to significant internal cohesion, so that it creates a bridge effect that threatens the correct sensor reading.

The results obtained are, overall, consistent with those obtained from other similar studies conducted in the Alpine region and confirm the great variability of the underestimation of solid precipitation observations through rain gauges, depending on both site conditions and instrument type and model. In order to correct the precipitation datasets affected by systematic errors, first, a simple model was developed, which takes into account only daily average temperature and a site and instrument-dependent correction factor as parameters to set the correction. The model is then widely applicable since the air temperature is often the only meteorological information available, especially in old monitoring sites with similar characteristics. The model evaluates the constant correction factor C_s, ranging between 0.11 and 1.50, to daily precipitation measurements when they refer to a snow-only event: it does not modify the measurements of rain-only events. It also considers a linear variation within the temperature range in which snow-mixed-to-rain event occurred. The proposed model was applied to the precipitation datasets of four stations, with different results in terms of the impact on the monthly and yearly averages, which are increased between 5% and 37% as altitude above the sea level and the site's wind exposure increase. These results were compared to those reported on the Lombardy map of yearly precipitation distribution, which is the current official reference in this field of work, and with some interesting old data from the former National Hydrological Service (SIMN). A second correction procedure with more general climatological value, based on both daily temperature and wind speed data is then considered. The verification of this procedure provides only slightly worse performances than those obtained with the at-site and instrument-dependent correction and, therefore, can be used when complete meteorological data are available.

Acknowledgments: Special thanks for their cooperation in this study: ARPA Lombardia, especially the Centro Nivometeorologico di Bormio (SO) and the Servizio Idrografico di Milano; A2A S.p.A., especially the Uffici di Grosio (SO) for providing data from their stations.

Author Contributions: Amerigo Lendvai collected and analysed the data with the help of Giovanni Peretti and his collaborators, who did most of the fieldwork. Giovanna Grossi contributed in framing the study in the international research on snow monitoring. Roberto Ranzi indicated the methodologies and the statistical analyses to be performed.

Conflicts of Interest: The authors declare no conflict of interest.

References

1.   Sevruk, B. Point precipitation measurement: Why are they not corrected? In *Water for the Future: Hydrology in Perspective*; IAHS Publication: Rome, Italy, 1987.

2.   Immerzeel, W.; Wanders, N.; Lutz, A.; Shea, J.; Bierkens, M. Reconciling high altitude precipitation in the upper Indus basin with glacier mass balances and runoff. *Hydrol. Erath Syst. Sci.* **2015**, *19*, 4673–4687. [CrossRef]

3.   Groot Zwaaftink, C.D.; Cagnati, A.; Crepaz, A.; Fierz, C.; Macelloni, G.; Valt, M.; Lehning, M. Event-driven deposition of snow on the Antarctic Plateau: Analyzing field measurements with SNOWPACK. *Cryosphere* **2013**, *7*, 333–347. [CrossRef]

4.   Sevruk, B.; Hertig, J.; Spiess, R. Wind field deformation above precipitation gauge orifices. *IAHS* **1989**, *179*, 65–70.

5.   Chvila, B.; Ondras, M.; Sevruk, B. The wind-induced loss of precipitation measurement of small time intervals as recorded in the field. In Proceedings of the WMO/CIMO Technical Conference, Bratislava, Slovaki, 25 September–3 October 2002.

6.   Rasmussen, R.; Baker, B.; Kochendorfer, J.; Meyers, T.; Landolt, S.; Fischer, A.; Black, J.; Theriault, J.; Kucera, P.; Gochis, D.; et al. How well are we measuring snow? The NOAA/FAA/NCAR winter precipitation test bed. *BAMS* **2012**, *6*, 811–829. [CrossRef]

7.   Duchon, C.; Cole, J.; Rasmussen, R. Measuring heavy snowfall using five different windshields and vibrating-wire precipitation gauges. In Proceedings of the 65th Eastern Snow Conference, Lake Morey, Fairlee, VT, USA, 28–30 May 2008.

8.   Pan, X.; Yang, D.; Li, Y.; Barr, A.; Helgason, W.; Hayashi, M.; Marsh, P.; Janowicz, J.P.R. Bias Corrections of Precipitation Measurements across Experimental Sites in Different Ecoclimatic Regions of Western Canada. *Cryosphere Discuss.* **2016**, *10*, 2347. [CrossRef]

9.   Sevruk, B. Distribution of precipitation in mountainous areas. In *Evaporation Losses from Storage Gauges*; World Meteorological Organization (WMO): Geneva, Switzerland, 1972; pp. 96–102.

10. Nitu, R.; Wong, K. *CIMO Survey on National Summaries of Methods and Instruments for Solid Precipitation Measurement at Automatic Weather Stations*; World Meteorological Organization (WMO): Geneva, Switzerland, 2010.

11. Sevruk, B. Correction of measured precipitation in the Alps using the water equivalent of new snow. *Nord. Hydrol.* **1983**, *1*, 49–58.

12. Savina, M.; Schaeppi, B.; Molnar, P.; Burlando, P.; Sevruk, B. Comparison of a tipping-bucket and electronic weighing precipitation gage for sbnowfall. *Atmos. Res.* **2012**, *103*, 45–51. [CrossRef]

13. Nitu, R.; Rasmussen, R.; Baker, B.; Lanzinger, E.; Joe, P.; Yang, D.; Smith, C.; Roulet, Y.; Goodison, B.; Liang, H.; et al. *WMO Intercomparison of Instruments and Methods for the Measurement of Solid Precipitation and Snow on the Ground: Organization of the Experiment*; World Meteorological Organization (WMO): Geneva, Switzerland, 2012.

14. Cagnati, A.; Crepaz, A.; Macelloni, G.; Pampaloni, P.; Ranzi, R.; Tedesco, M.; Tomirotti, M.; Valt, M. Study of the snow meltfreeze cycle using multi-sensor data and snow modelling. *J. Glaciol.* **2004**, *50*, 419–426. [CrossRef]

15. Bocchiola, D.; Rosso, R. The distribution of daily snow water equivalent in the central Italian Alps. *Adv. Water Resour.* **2007**, *30*, 135–147. [CrossRef]

16. Sevruk, B. Conversion of Snowfall Depths to Water Equivalents in the Swiss Alps. *Zürcher Geogr. Schr.* **1986**, *23*, 13–23.

17. Serreze, M.C.; Clark, M.P.; Armstrong, R.L.; McGinnis, D.A.; Pulwarty, R.S. Characteristics of the western United States snowpack from snowpack telemetry (SNOTEL) data. *Water Resour. Res.* **1999**, *35*, 2145–2160. [CrossRef]

18. Cagnati, A. *Strumenti di Misura e Metodi di Osservazione Nivometeorologici*; AINEVA: Trento, Italy, 2003. (In Italian)

19. Valt, M.; Cagnati, A.; Corso, T. Stima dell'equivalente in acqua della neve. *Neve e Valanghe* **2006**, *59*, 24–33. (In Italian)

20. Valt, M.; Moro, M. Average snowcover density values in eastern Alps mountain. In Proceedings of the EGU General Assembly 2009, Vienna, Austria, 19–24 April 2009.

21. D'Aosta, A.V. *Secondo Interconfronto SWE*; ARPA: Valle d'Aosta, Aosta, Italy, 2016.

22. Proksch, M.; Rutter, N.; Fierz, C.; Schneebeli, M. Intercomparison of snow density measurements: Bias, precision, and vertical resolution. *Cryosphere* **2016**, *10*, 371–384. [CrossRef]

23. Valgoi, P. Cuscinetto per la misura della densità della neve (snow pillow). *Neve e Valanghe* **2011**, *72*, 48–53. (In Italian)

24. Goodison, B.; Louie, P.; Yang, D. *WMO Solid Precipitation Measurement Intercomparison-Final Report*; World Meteorological Organization (WMO): Geneva, Switzerland, 1998.

25. Sevruk, B. Adjustment of tipping-bucket precipitation gauge measurements. *Atmos. Res.* **1996**, *42*, 237–246. [CrossRef]

26. ArCIS-Gruppo di lavoro Archivio climatologico dell'Italia centro-settentrionale. Il clima nell'inverno 2013–2014: Le eccezionali anomalie climatiche del Centro-Nord Italia. *Neve e Valanghe* **2014**, *81*, 4–9. (In Italian)

27. Colli, M.; Lanza, L.; Rasmussen, R.; Thériault, J. The collection efficiency of shielded and unshielded precipitation gauges. Part I: CFD airflow modeling. *J. Hydrometeorol.* **2016**, *17*, 231–243. [CrossRef]

28. Colli, M.; Lanza, L.; Rasmussen, R.; Thériault, J. The collection efficiency of shielded and unshielded precipitation gauges. Part II: Modeling particle trajectories. *J. Hydrometeorol.* **2016**, *17*, 245–255. [CrossRef]

29. Smith, C. Correcting the wind bias in snowfall measurements made with a Geonor T-200B precipitation gauge and alter wind shield. In Proceedings of the 15th Symposium on Meteorological Observation and Instrumentation, Atlanta, GA, USA, 17–21 January 2010.

30. Macdonald, J.; Pomeroy, J. Gauge undercatch of two common snowfall gauges in a prairie environment. In Proceedings of the 64th Eastern Snow Conference, St. John's, NL, Canada, 29 May–1 June 2007.

31. Ranzi, R.; Grossi, G.; Bacchi, B. Ten years of monitoring areal snowpack in the Southern Alps using NOAA-AVHRR imagery, ground measurements and hydrological data. *Hydrol. Process.* **1999**, *13*, 2079–2095. [CrossRef]

32. Zweifel, A.; Sevruk, B. Comparative accuracy of solid precipitation measurement using heated recording gauges in the Alps. In *Workshop on Determination of Solid Precipitation in Cold Climate Regions*; Fairbanks: Zürich, Switzerland, 2002.

33. Savina, M.; Schaeppi, B.; Molnar, P.; Burlando, P.; Sevruk, B. Comparison of a tipping-bucket and electronic weighing precipitation gage for snowfall. *Atmos. Res.* **2012**, *103*, 45–51. [CrossRef]

34. Cugerone, K.; Allamano, P.; Salandin, A.; Barbero, S. Stima della precipitazione in siti di alta quota. *Neve e Valanghe* **2012**, *77*, 36–43. (In Italian)

35. Eccel, E.; Cau, P.; Ranzi, R. Data reconstruction and homogenization for reducing uncertainties in high-resolution climate analysis in Alpine regions. *Theor. Appl. Climatol.* **2012**, *110*, 345–358. [CrossRef]

36. Ranzi, R.; Grossi, G.; Gitti, A.; Taschner, S. Energy and mass balance of the Mandrone Glacier. *Geogr. Fis. Din. Quat.* **2010**, *33*, 45–60.

37. Grossi, G.; Caronna, P.; Ranzi, R. Hydrologic vulnerability to climate change of the Mandrone glacier (Adamello-Presanella group, Italian Alps). *Adv. Water Resour.* **2013**, *55*, 190–203. [CrossRef]

38. World Meteorlolgical Organization. *Intercomparison of Models of Snowmelt Runoff*; WMO Operational Hydrology Report No. 23, WMO Publ. No. 646; World Meteorological Organization (WMO): Geneva, Switzerland, 1986.

39. Kienzle, S. A new temperature based method to separate rain and snow. *Hydrol. Process.* **2008**, *22*, 5067–5085. [CrossRef]

40. Forland, E.; Allerup, P.; Dahlström, B.; Elomaa, E.T.; Perälä, J.; Rissanen, P.; Vedin, H.; Vejen, F. Manual for operational correction of Nordic precipitation data. *DNMI Oslo* **1996**, *96*, 66.

41. Morin, S.; Lejeune, Y.; Lesaffre, B.; Panel, J.; Poncet, D.; David, P.; Sudul, M. An 18-yr long (1993–2011) snow and meteorological dataset from a mid-altitude mountain site (Col de Porte, France, 1325 m alt.) for driving and evaluating snowpack models. *Earth Syst. Sci. Data* **2012**, *4*, 13–21. [CrossRef]

water

MDPI

Article

An Integrated Approach for Site Selection of Snow Measurement Stations

Bahram Saghafian [1],*, Rahman Davtalab [2], Arezoo Rafieeinasab [3] and M. Reza Ghanbarpour [4]

1. Technical and Engineering Department, Science and Research Branch, Islamic Azad University, Tehran 1477893855, Iran
2. Department of Civil, Environmental and Construction Engineering, University of Central Florida, Orlando, FL 32818, USA; rdavtalab@saiengineers.com
3. Department of Civil Engineering, The University of Texas at Arlington, Box 19308, 416 Yates St. Suite 425, Arlington, TX 76019, USA; arezoo.rafieeinasab@gmail.com
4. Alberta Environment and Parks, Calgary, AB T2E 7L7, Canada; Reza.Ghanbarpour@gov.ab.ca
*   Correspondence: b.saghafian@gmail.com; Tel.: +98-912-327-7808

Academic Editor: Tommaso Moramarco
Received: 28 June 2016; Accepted: 7 November 2016; Published: 17 November 2016

Abstract: Snowmelt provides a reliable water resource for meeting domestic, agricultural, industrial and hydropower demands. Consequently, estimating the available snow water equivalent is essential for water resource management of snowy regions. Due to the spatiotemporal variability of the snowfall pattern in mountainous areas and difficult access to high altitudes areas, snow measurement is one of the most challenging hydro-meteorological data collection efforts. Development of an optimum snow measurement network is a complex task that requires integration of meteorological, hydrological, physiographical and economic studies. In this study, site selection of snow measurement stations is carried out through an integrated process using observed snow course data and analysis of historical snow cover images from National Oceanic Atmospheric Administration Advanced Very High Resolution Radiometer (NOAA-AVHRR) at both regional and local scales. Several important meteorological and hydrological factors, such as monthly and annual rainfall distribution, spatial distribution of average frequency of snow observation (FSO) for two periods of snow falling and melting season, as well as priority contribution of sub-basins to annual snowmelt runoff are considered for selecting optimum station network. The FSO maps representing accumulation of snowfall during falling months and snowpack persistence during melting months are prepared in the GIS based on NOAA-AVHRR historical snow cover images. Basins are partitioned into 250 m elevation intervals such that within each interval, establishment of new stations or relocation/removing of the existing stations were proposed. The decision is made on the basis of the combination of meteorological, hydrological and satellite information. Economic aspects and road access constraints are also considered in determining the station type. Eventually, for the study area encompassing a number of large basins in southwest of Iran, several new stations and relocation of some existing stations are proposed.

Keywords: snow measurement network; site selection; satellite images; Iran

1. Introduction

Water demand increase in urban, agriculture and industry sectors all over the world continues to intensify the pressure on water resources. In the southwestern of Iran, the Karkheh, Dez, Karun and Marun river basins provide over a quarter of surface water resources in the country and even supply water for the central arid part of Iran via transboundary water transfers [1].

Nowadays, hydrological and water resources modeling play a key role in water resources management [2]. On the other hand, development and calibration of hydrologic and water resources models rely on adequate and accurate hydro-climatic observation data [3]. Reliable data may be acquired through appropriate design, installation and operation of an optimum network of hydro-climatic stations. Designing a proper measurement network results in the representative location and optimum number of stations so that interpolation techniques can lead to acceptable estimation of spatially distributed factors with adequate accuracy [4,5].

Among the hydro-climatic variables, snow is one of the most important factors in the hydrology of mountainous areas. Snow, as a renewable water resource, remains on the ground for weeks or months. Since the lag time of river flow from snowmelt is more than that of rainfall, management of this reliable resource is relatively easier. The amount of snowfall and snowmelt depends on both geographic (elevation, slope and aspect, latitude, etc.) and synoptic factors. Storing and melting snow is a continuous process especially in southwest of Iran with low latitude and relatively ephemeral snow. Therefore estimating accumulated snowpack and its spatiotemporal pattern is complex. Furthermore, due to limited access to high altitudes of mountainous areas, snow measurement is considered as one of the most difficult data collections in water resources. The position of snow gauges and the type of snow measurement instruments should be selected in a way to represent the spatial and temporal variations of snow characteristics. Selecting representative locations for automated snow-pillow (ASP) stations or manual snow-course (MSC) stations will ensure access to the reliable data required for snow storage estimation.

Nowadays, satellite images and remote sensing (RS) techniques are widely used for the estimation of snow cover areas. Derived snow cover areas from satellite images accompanied with field observations can provide a unique tool for estimation of snow storage, river flow forecasting and flood control required in water resource management.

ASP measurement networks have been developed in many countries. The snow measurement network in western U.S. states or SNOTEL [6], provincial or local networks in Canada, such as ASP measurement network in British Columbia [7], and the ASP measurement network in Pakistan [8] are good examples of available operating networks.

Nonetheless, there are insufficient studies on site selection of snow measurement networks at regional scale and a few existing studies are limited to local scale, focusing mainly on a single station [6]. According to the World Meteorological Organization [9], the number of snow measurement stations or snow courses and their positions depend on topography, the type and aspect of vegetation, the objectives of data collection, and economic considerations. WMO recommended one snow measurement station for about 2000 to 3000 km^2 in low homogeneous region and one for about 5000 km^2 for more homogeneous and plain regions [9]. For developing a snow measurement network, it is recommended to establish a temporary and dense network that measures the snow parameters for four to six years continuously in order to detect snow cover characteristics, related to the physiographical parameters. Then, according to the required accuracy, operation constraints and cost minimization, the network density may be reduced [4].

In this study, areas will be prioritized according to the need for snow measurement stations using field observations that are supported by satellite snow cover images. Furthermore, through assessment of the existing snow measurement network in each of the altitudinal range, installation of new stations, removal/maintaining/relocation of the existing stations MSC stations, or upgrade to ASP stations is proposed. The paper is organized as follows. Case study, data acquisition, and methodology, including the meteorological, hydrological, and remote sensing criteria used are described in Section 2. Results are presented in Section 3. Section 4 provides conclusions and recommendations.

2. Data and Methods

2.1. Study Area

The Zagros Mountains are one of the major snowy regions of Iran, which include the four large river basins of Karkheh, Dez, Karun and Marun. The study area encompasses the Karkheh basin, upstream of Karkheh Dam, with approximately 43,000 km^2 in area and 5.2 km^3 mean annual runoff; Dez Basin, upstream of Dez Dam with 17,300 km^2 in area and 8.4 km^3 mean annual runoff; Karun basin, upstream of Gotvand Dam with 32,000 km^2 in area and 12.5 km^3 mean annual runoff; and two tributaries of the Marun basin, first upstream of Shohadaye Behbahan Dam and second upstream of Jarreh Dam with 6800 km^2 in area and 2.3 km^3 mean annual runoff, in total. The area lies between the geographical latitude of 30°15' N to 34°55' N and 46°05' E to 52°03' E longitude. The mentioned river basins play a major role in food production and economy of more than 4 million people [10,11]. Figure 1 shows the location of these major river basins.

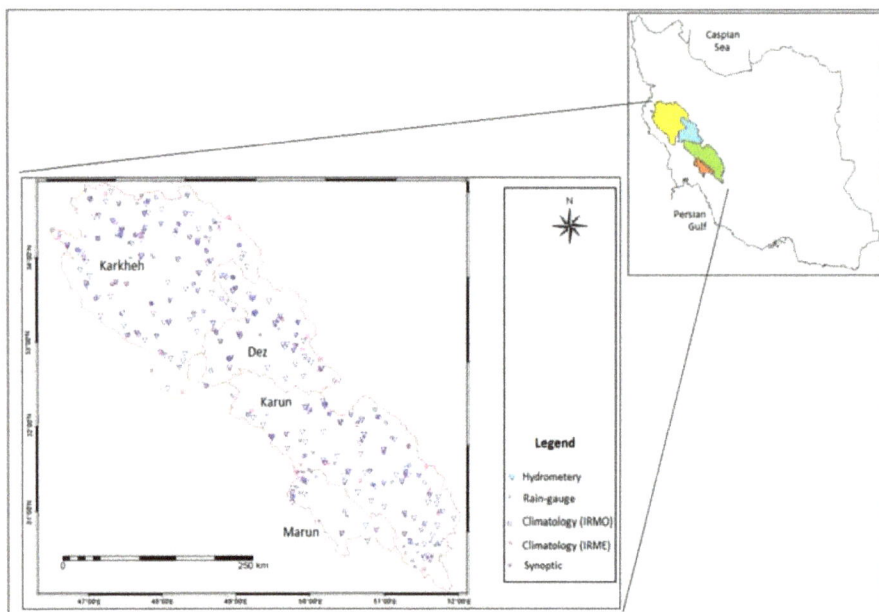

Figure 1. Location of the study basins and distribution of the meteorological and hydrometric stations.

2.2. Data Collection

The data used for this study comprise the monthly air temperature from 44 climatological stations (including the stations installed by Iran Ministry of Energy, IRME and Iran Meteorological Organization, IRMO), daily and monthly precipitation of 152 rain-gauge stations, daily temperature and 3-h precipitation depth and 6-h precipitation type (rainfall or snow) measured at 20 synoptic stations, snow water equivalent and snow depth at 45 available MSC stations accompanied by daily discharge at 40 hydrometric stations with long term data (Figure 1). For monthly temperature and rainfall, a 25-year data period is identified as a normal climate period (1974–1999). Missing data are reconstructed using linear correlation with the nearest neighboring station. The snow water equivalent may also be estimated through analysis of the amount and type of precipitation in the form of rainfall or snowfall, at synoptic stations (detail information is available in Saghafian and Davtalab, [12]). The only snow measurements available from MSC stations that are located in the Karun and Dez Basins. Data sampling in the mentioned stations is performed manually once a year. However, the sampling time varies between late February and late March. Therefore, the sampling period lasts about one month in any given year. Accordingly, the above-mentioned data are used for spatial analysis.

Quantity and quality of daily discharge data are assessed as well and the annual discharge series are obtained. The snow cover maps extracted from the National Oceanic Atmospheric Administration

Advanced Very High Resolution Radiometer (NOAA-AVHRR) series for the period of 1984 to 2003 have been used in this study. The maps are provided by Jamab [13] based on the methodology available in Porhemmat et al. [14]. The spatial resolution of the NOAA-AVHRR image is about 1.1 km, suitable for snow cover extraction over vast areas. There are two images for every 24 h, one of which corresponds to daytime. However, not all daytime images were suitable because of atmosphere noise and/or cloud cover. In all, 480 snow cover maps were available in Jamab's archive [13]. The snow cover maps are stored in binary format in GIS (1 denotes snow and 0 denotes non-snow pixel). The hypsometric maps, boundary of basins and sub-basins, digital elevation model (DEM), slope and aspect maps are derived from 1:250,000 topographic maps (Iran National Cartographic Center: http://www.ncc.org.ir) with 100 m contour interval. For instance, the contour map for Dez Basin is shown in Figure 2.

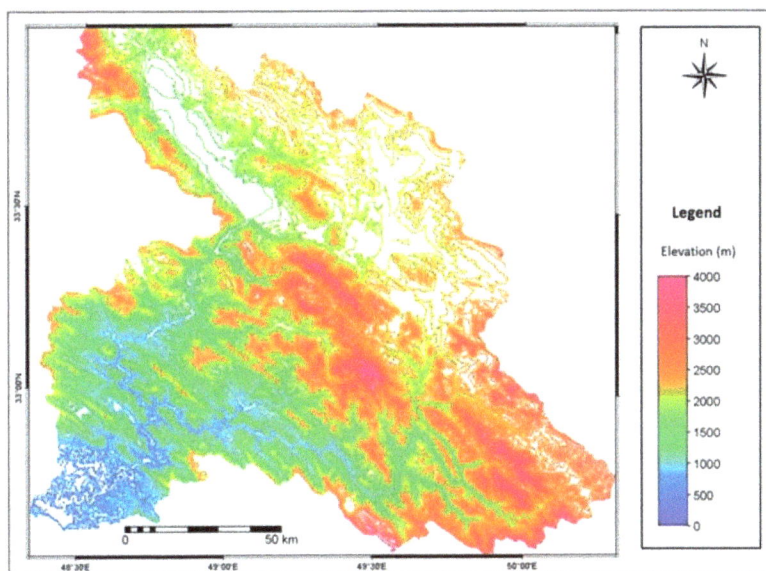

Figure 2. Elevation contour map of Dez Basin.

2.3. Methodology

The snow station site selection methodology is based on integration of meteorological, hydrological, and remote sensing studies. Meteorological analysis includes preparation of isothermal and isohyetal maps, determination of snowfall threshold temperature and precipitation estimation variance analysis [15]. Prioritization of the sub-basins has been done based on the contribution of snowmelt to the total annual runoff in each sub-basin. Analysis of remote sensing data includes the extraction of the FSO layers for two periods of snowfall accumulation and snowpack persistence. The details are given in the following sections of the paper.

2.3.1. Meteorological Criteria

Temperature is the most dominant meteorological factor in determination of precipitation type (snow and rain) and determines regions subject to snowmelt, snowfall, or both. Therefore, temperature and its spatial variation is a key factor in determining the proper location of snow measurement stations. In this study, the snowfall threshold temperature refers to the temperature under which the precipitation is assumed to be in the form of snow. Threshold temperature may vary from below 0 °C to 5 °C and may not be constant even in a single day [16]. Therefore, in many studies an average value is used as threshold temperature [2,17,18]. Daily precipitation data accompanied with the rain/snow observation at synoptic stations are used to estimate the threshold temperature [16].

In each station, through assigning the snowfall percentage in various temperature intervals, the temperature corresponding to 50% snowfall is determined as the threshold temperature. Based on the monthly and annual mean temperature for the study period at all stations, the spatial correlation relationships between temperature and elevation are found using DEMs. The areas below the threshold temperature are identified as potential snow storage areas using the isothermal maps. Hence, establishing snow gauges is considered in these areas.

In addition, the semi-variogram analysis [15] is conducted using observed snow water equivalent (SWE) data from snow measurement stations or MSC stations (average of SWE for the period of 1974 to 1999). The best variogram model is adopted to compute the spatial distribution of error variance. Estimation variance denoted by σ_e^2 demonstrates the variance of the error between observed and estimated values. In geostatistics, it can be computed from average values of semi-variograms using available supplementary functions [19]. In addition, for calculation of confidence levels, estimation variance is used in many other cases, such as designing an optimal measurement network, calculating the error reduction resulted from increasing the number of sampling, and assessing different sampling methods. Estimation variance can be defined with the following equation [20]:

$$\sigma_e^2 = 2 \sum_{i=1}^{n} \lambda_i \times \gamma_{oi} - \gamma_0 - \sum_{i=1}^{n} \sum_{j=1}^{n} \lambda_i \times \gamma_{ij} \tag{1}$$

where λ_i is weight of data at point i, γ_o is the variogram value for $h = 0$, (where h is the distance between paired points), γ_{oi} is the variogram value between the considered point and point i, γ_{ij} is the variogram value among i and j samples and σ_e^2 is the estimation variance at the ith point.

Further explanation about the error variance map is available in [19]. Error variance maps are used as the main reference in determining the new locations of a gauge or omitting redundant gauges. Therefore, snow measurement stations can be established where error variance is high. The reduction of spatial error variance corresponding to adding one station can also be determined.

2.3.2. Hydrological Criteria

In mountainous areas with a mixed snow-rain precipitation regime, the contribution of snowmelt to the annual runoff is an important factor. Because of the inherent uncertainty in the separation of flow components, a simple water balance model is used [21]. Precipitation over the melting season, annual runoff coefficient and snowmelt runoff determined for each hydrometric station are accounted in the water balance model, which can be expressed as follows:

$$R = (H_w + P) \times C \tag{2}$$

where R is the total runoff over the snowmelt season in millimeters, H_w is the depth of SWE at the beginning of the snowmelt season in millimeters, P is the precipitation over the snowmelt season (spring and summer) in millimeters and C is the average runoff coefficient that can be computed based on precipitation and discharge data. P is calculated based on monthly isohyetal maps. H_w represents the unknown of the model.

Having determined R at hydrometric stations, a regional model is developed to estimate R at ungauged locations. Basin characteristics such as area, average elevation, and the percentage of areas with elevation above 2000, 2500 and 3000 m are used as independent variables of the regional model [22]. As a result, a map of R values at sub-basin scale is produced such that sub-basins may be prioritized based on snowmelt runoff contribution. This map is used as one of the criteria in selection of snow measurement stations.

2.3.3. Remote Sensing Snow Observation Criteria

Time series of snow cover maps were available from Jamab [13]. These maps had been extracted from NOAA-AVHRR images for the period of 1984–2003. Detailed description of image geo-referencing, noise reduction, and snow cover extraction are available in Porhemmat et al. [14]. In brief, the methodology of snow detection by Porhemmat et al. [14] has two steps of split and merge technique on AVHRR satellite images, channels 2, 3 and 4. In the first step, two split-and-merge clustering/labeling procedures were used in order to separate clear land pixels from pixels contained by either snow or cloud as described by Simpson et al. [23]. In the second stage, the same procedure was used only on pixels containing cloud, snow or a mixture of both. At this stage, pixels containing snow were distinguished from those containing cloudy pixels. To improve and control this computerized procedure, the above methodology was combined with threshold techniques and filter analysis of AVHRR channels data, especially channel 2, 3, and 4. Due to the high contrast between snow covered and snow free surfaces, channel 1 facilitates the definition of snow pack boundaries. Also, channel 3 and channel 4 analyses were effectively used as the snow/cloud discrimination channels. Validation of the NOAA snow detection algorithm was performed against eye interpretation via on screen digitization of simultaneous NOAA and Landsat TM images.

The FSO derived from the mentioned snow cover maps is a key criterion for identifying the potential snowy areas. The following relationship for calculating FSO has been proposed by Saghafian and Davtalab [12]:

$$FSO(i) = \frac{\sum\limits_{l=1}^{T} X(i,l)}{T} \times 100 \qquad (3)$$

where *FSO(i)* is frequency of snow observations for the *i*-th pixel, T is the total number of snow cover maps over the study period and X is a binary number equal to one if it is snow and zero otherwise.

In this study FSO is calculated for two time periods of December–February and March–May. Since the study area receives snow from early December to late February [12], FSO during these months is considered as snowfall probability (FSO-SFP). Also, almost all snow storage of study area melts during March until end of May. Therefore, FSO during these months is considered as snowpack persistence probability (FSO-SPP). FSO-SPP is dependent on temperature, additional precipitation and topographic conditions of the land [24,25]. Snow may disappear quickly due to temperature increase and/or shallow depth of snow. Snowmelt and loss of snow storage occurs faster along southern aspects because of direct sunshine [24,25]. Since spring and summer are high water demand seasons in the study area, FSO-SPP or snowpack persistence is important for meeting the water demands.

2.3.4. Synthesis of Criteria in Site Selection

As previously mentioned, the study area is divided into 250 m elevation intervals. Establishment/relocation of new/existing snow measurement stations is analyzed within each interval. The area ratio of each elevation interval to the total basin area is considered as a weight index. The weighted average snowfall probability in each interval is also calculated, using FSO-SFP map. This criterion is used as one of the main factors in snow station site selection.

The areas with FSO-SFP higher than the weighted average snowfall probability are identified in each elevation interval that has no snow measurement station. If spatial distribution of the identified areas were like patchy snow cover, those small polygons were eliminated. Then the meteorological and hydrological criteria are taken into account so that the areas with no or minimal snowfall potential and/or sub-basins with low snowmelt runoff volume are eliminated in further analyses.

However, if a snow measurement station is already operational in an altitudinal range, the appropriateness of its present location is assessed based on the criteria such as FSO-SFP, FSO-SPP and field visit.

In addition, the spatial pattern of stations is investigated according to the estimation variance maps. Therefore, new stations are likely to be proposed in areas subject to high estimation variance. The FSO-SPP map is further used as a reference in very large areas of high variance to reduce the field visit area. Clearly, areas with FSO-SPP higher than the weighted average snowpack persistence probability are appropriate areas for establishing new stations, since they provide the snow storage data for simulating snowmelt runoff over dry seasons.

After determining suitable areas at regional or macro scale, topography, road access, and proximity with residential areas (to facilitate the operation and maintenance of stations) are the main factors for site selection of snow stations at local or micro scale. The final step for site selection was field study to determine the exact point for installing the stations based on WMO recommendations [26]. The following criteria has been considered during the field study for selecting the representative location for the snow course: leaving clear space within the radius of at least equal to the height of the nearest obstacle, absence of strong wind and snowdrift (as much as possible), safety of the station against avalanche, no significant changes in elevation in a radius of 50 m (if possible), no reverse slopes and hard surface conditions for ease of leveling, good drainage, homogeneous vegetation cover and land use in the vicinity, and having enough open space to avoid snow interception.

For sake of brevity, only Dez Basin results will be presented.

3. Results

3.1. Classification of Regions Based on Meteorological Criteria

Isothermal maps are developed using the digital elevation model (DEM) and the temperature gradient relationships. The average threshold temperature is 2.4 °C at most synoptic stations, which is assigned as the threshold for the entire region (more explanation is available in Saghafian et al. [16]). The area with average monthly temperature less than 2.4 °C are designated as the potential snowfall areas and areas with temperature below 0 °C are classified as the potential area for snow persistence. Therefore, the regions with temperature between 0 °C and 2.4 °C represent areas prone to coupled snowfall-snowmelt processes. Figure 3 illustrates the spatial distribution of potential snowfall, snowmelt, and snow persistence areas in February in Dez Basin.

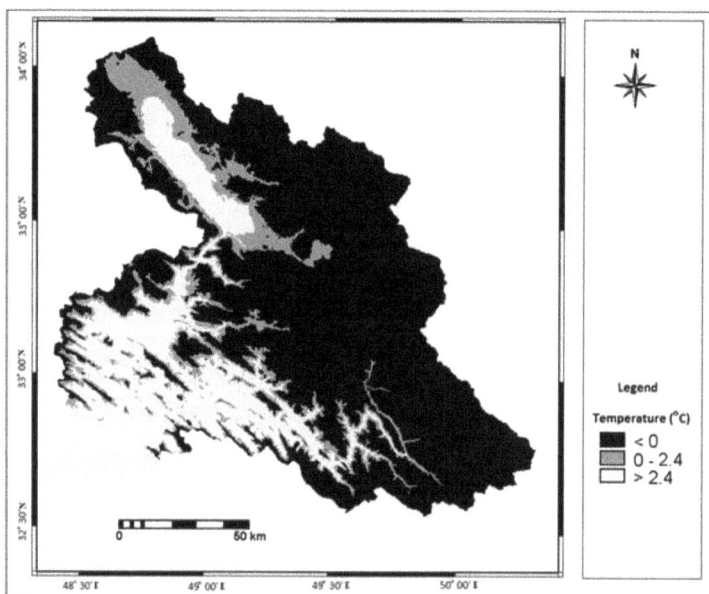

Figure 3. Potential areas for snow persistence ($T < 0$ °C), snowfall ($0 < T < 2.4$ °C) and snowmelt ($T > 2.4$ °C) in Dez Basin in February (for the period of 1974–1999).

Semi variogram analysis is carried out using the average of SWE data over the historical period, and the spatial distribution of estimation error variance is extracted using the best variogram model. Since the SWE data are available only for Dez and Karun basins, the performed estimation error variance analyses are limited to these basins. Figure 4 shows the error variance in the Dez Basin, areas with higher error are candidates for establishment of new stations.

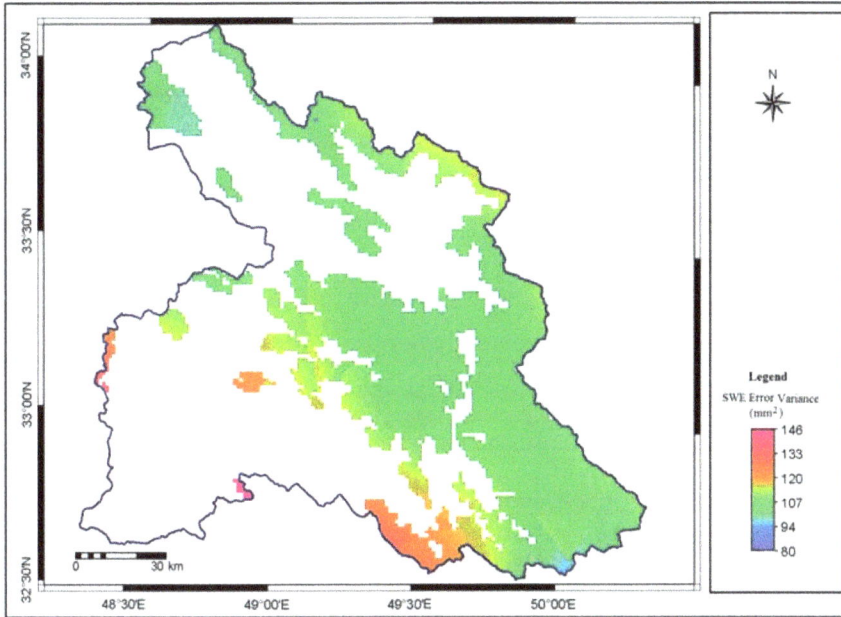

Figure 4. Error variance map of snow water equivalent for Dez Basin.

3.2. Hydrological Criteria

Snowmelt runoff volumes at 40 hydrometric stations are calculated based on the water balance model. Figure 5 shows the location of the selected stations in Dez Basin and all the hydrometric stations are shown in Figure 1. A regional regression model between snowmelt runoff volume and the percent of basin area higher than 2500 m is then built in order to provide estimates of snowmelt runoff at sub-basin outlets without hydrometric station (more information is available in Ghanbarpour et al. [22]). Snowmelt runoff volume and the corresponding priority in Dez Basin are given in Table 1 as an example.

Table 1. Snowmelt runoff volume and the corresponding priority in Dez Basin.

Sub-Basin	Snowmelt Runoff (mm)	Priority
Upstream of Tang Sezar	4.7	7
Upstream of Sepidan	12.4	6
Upstream of Sezar	44.7	2
Upstream of Tang Bakhtiari	42.5	3
Upstream of Zard Fahreh	84.2	1
Upstream of Daretakht	25.5	4
Upstream of Doroud	22.7	5

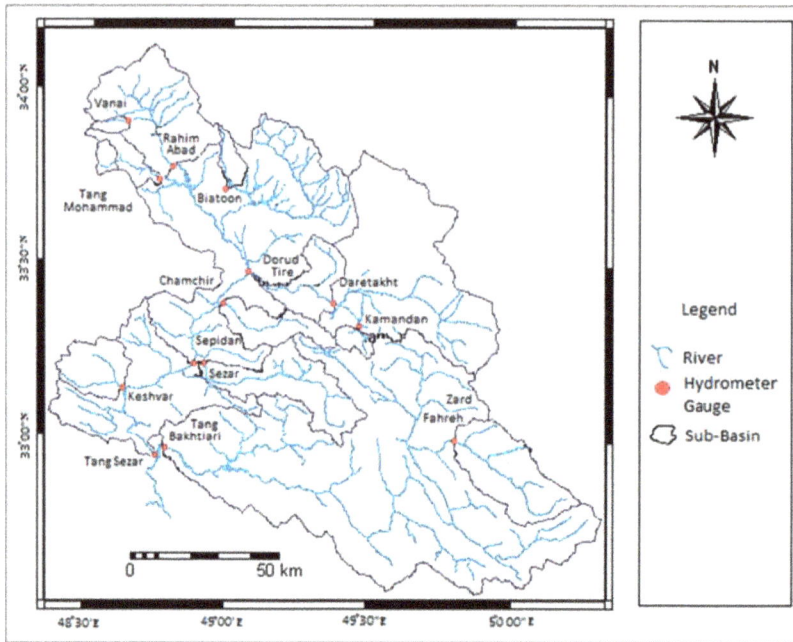

Figure 5. Selected hydrometric stations in Dez Basin.

3.3. Snowfall and Snowpack Persistence Probability Criteria

As previously indicated the FSO-SFP and FSO-SPP maps are derived based on Equation (3) over the snowfall and snowmelt seasons, respectively. These two layers represent the spatial distribution of the probability of the accumulation and persistence of snow that are key factors in site selection. Figures 6 and 7 show FSO-SFP and FSO-SPP maps in Dez Basin, respectively.

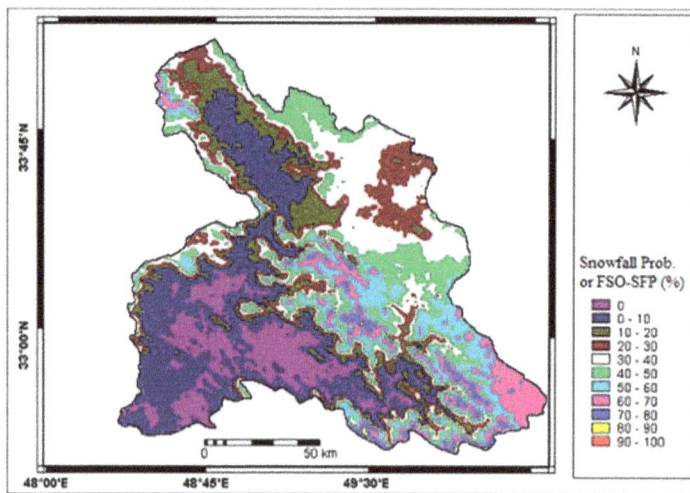

Figure 6. FSO-SFP Map in Dez Basin (for the period of 1984–2003).

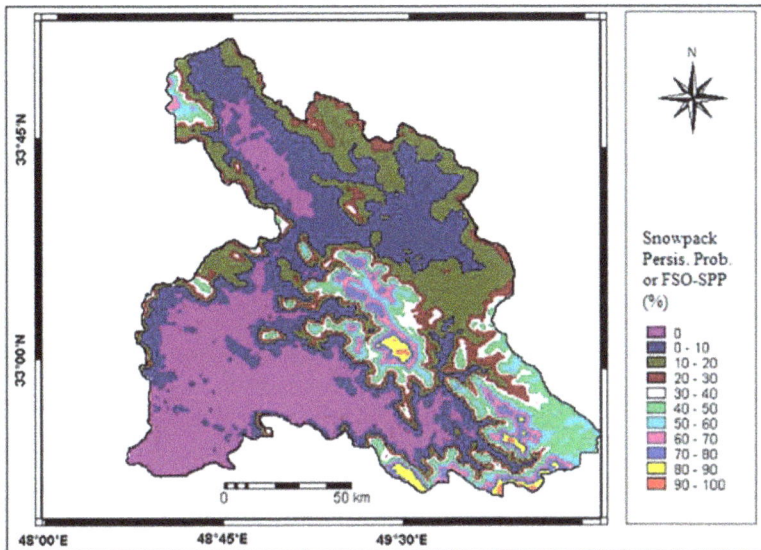

Figure 7. FSO-SPP Map in Dez Basin (for the period of 1984–2003).

3.4. Site Selection and Station Type

Site selection for snow stations is performed by integration of all available criteria including meteorological, hydrological and remotely-sensed variables. Site selection using the systematic and comprehensive approach described in this study will help decision makers to implement economically sound plans for optimizing a new set of stations or upgrading the existing network. This will be the case even where there is no historical snow data in the basin. In the study area, there is no historical snow data or MSC station available in Marun and Karkheh basins. Therefore, site selection of the snow stations is performed by integrating hydrological, meteorological and remotely-sensed variables. However, there are some operational MSC stations located in Karun and Dez Basins. Therefore, the refinement and optimizing of the existing network is performed based on the distribution of current stations and the error variance map, in addition to other variables mentioned above.

To make the proposed methodology more operationally practical, we choose to start the integration process from the altitudinal range of 250 m intervals and then generalize it to the basin scale. During the integrating phase, there is more emphasis on satellite images in terms of FSO-SFP and FSO-SPP criteria, due to the lack of homogeneous and long-term snow data. In the next step, meteorological and hydrological factors are also considered in order to compensate the impacts of snow data shortage. Since the site selection has been done in each individual elevation boundary, as an example, one of the altitudinal ranges of Dez Basin will be described in detail. This example shows step by step procedure that is followed in this research for all four major basins as follows.

The altitudinal range 2250–2500 m approximately covers 12% of the Dez Basin. The weighted average of FSO-SFP in this interval is about 40.8%. There are 11 MSC stations in this interval as listed in Table 2. According to this table, the distribution of stations over elevation is relatively good. Dalooni and Barfian stations have the highest and lowest snowfall probability or FSO-SFP in this altitudinal range, with 57% and 38%, respectively. In addition, Dalooni and Yazdgerd stations have the highest and lowest snowpack persistence probability or FSO-SPP, with 38% and 25%, respectively. Based on hydrologic criterion and road access, the ranking of each station is carried out for this altitudinal range. For example, in terms of road access, Ghaleh-Rostam and Yazdgerd stations are the most undesirable locations.

Table 2. Existing manual snow-course (MSC) stations in altitudinal range of 2250–2500 m in Dez Basin.

MSC Station	Elevation (m)	FSO-SFP (%)	FSO-SPP (%)
Vanaee	2250	42	30
Ghalavardeh2	2300	44	17
Vazmeh-dar	2300	48	32
Ghaleh-rostam	2300	32	10
Aziz-abad	2300	46	20
Barfian	2350	38	20
Yazdgerd	2500	44	25
Palang-dar2	2400	46	26
Dalooni	2500	57	38
Chahar-cheshmeh2	2400	46	18
Gardanehe-khakbad	2450	39	27

To make a final decision, the weighted average snowfall probability in this altitudinal range (rounded to 40%) is used as a criterion to determine the appropriate area for installing the stations. Thus, we set a threshold FSO-SFP of 40% to assess the present network. The selected zone for new stations is shown in Figure 8 along with the current station network. According to this figure, the existing network in this elevation has a desirable distribution, however, Ghaleh-rostam and Vazmeh-dar stations are close to each other and located at same elevation (2300 m). Therefore, one of the two stations should be removed. Vazmeh-dar station has more priority than Ghaleh-rostam station based on the FSO-SFP and FSO-SPP criteria, the ratio of winter precipitation to the annual precipitation, potential of snowfall, and road access. The two stations have similar contributions in snowmelt runoff according to the hydrological criteria. Therefore, Vazmeh-dar station is kept.

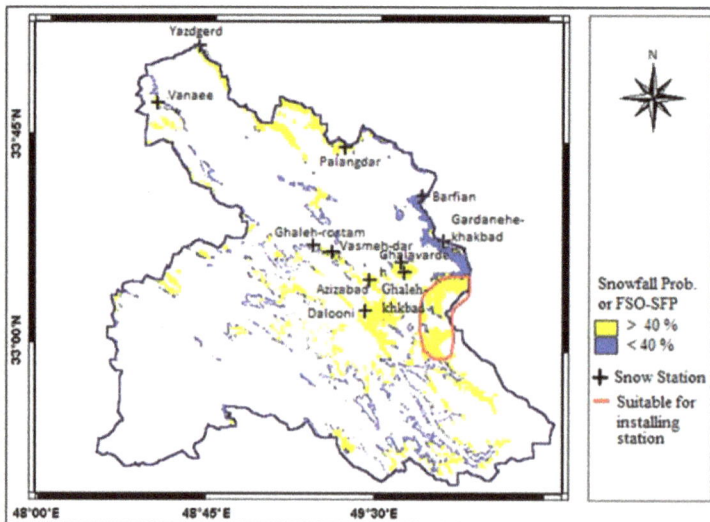

Figure 8. Selected zone for new snow stations and the existing station network in 2250–2500 m elevation interval in Dez Basin.

According to Figure 8 and the error variance map in Figure 4, spatial distribution of snow measurement stations in the 2250–2500 m elevation interval is generally appropriate while the error is relatively high in the west of the basin (shown by a thick polygon). Moreover, the average of snow persistence or FSO-SPP in this area exceeds 20% (Figure 7). By checking on topography and road access, one location is proposed for a new station in the mentioned area.

4. Conclusions

In this paper, a comprehensive approach for site selection of snow measurement stations in a mountainous area is proposed, based on an integration of meteorological, hydrological and remote sensing analysis. The approach is applied to a vast region encompassing four large river basins in the southwest of Iran. Snowfall and snowpack persistence probability maps are derived based on satellite images and used in a systematic site selection approach. Meteorological and hydrological variables such as basin average snowmelt and snowfall threshold temperature are included in the analysis to overcome lack of historical snow data availability.

The comprehensive site selection methodology for snow measurement stations not only ensures reliable snow data availability for the purpose of sustainable water resource management, but also provides an economically sound approach to optimize and maintain a cost-effective measurement network.

According to Moss [27], who emphasized economic considerations in network design, access (distance) to roads is considered as an indirect economic criterion for site selection of snow stations at local scale.

As a result of this study, six automatic snow-pillow stations (ASP) and four manual snow-course stations (MSC) in Karkheh basin; 17 ASP stations, 14 MSC stations and six control points in Dez Basin; 16 ASP stations, 10 MSC stations and 10 control points in Karun basin; and four ASP stations in Marun basin are proposed.

There is limited literature on design or evaluation of snow measurement networks in large spatial scale. The authors believe that the methodology described in this paper is a step forward for such studies, especially in data-poor regions where satellite information can play a significant role. However, there is room to improve on snow station network design in the future. One attractive alternative is to examine different scores for each input variable depending on their suitability, accuracy, and importance.

Acknowledgments: The authors acknowledge Jamab Consulting Engineering Co. for providing the snow cover map archive. Special thanks are also dedicated to the Khuzestan Water and Power Authority (KWPA) for partial support and providing companions in field trips.

Author Contributions: Bahram Saghafian and M. Reza Ghanbarpour proposed the original methodology for snow station site selection; Rahman Davtalab and Arezoo Rafieeinasab collected and analyzed the data and provided the results; Field trip were conducted by M. Reza Ghanbarpour and Rahman Davtalab; Bahram Saghafian, Arezoo Rafieeinasab, M. Reza Ghanbarpour and Rahman Davtalab jointly wrote the paper and addressed the comments.

Conflicts of Interest: The authors declare no conflict of interest. No one else was involved in the collection, analyses, or interpretation of data produced in this study; in the writing of the manuscript, nor in the decision to publish the results.

References

1. Iran Ministry of Energy (IRME). *Karoun, Dez, Karkheh and Maroun River Basins Water Comprehensive Study—Second Study*; Iran Ministry of Energy (IRME): Tehran, Iran, 2012.
2. Gyawali, R.; Watkins, D.W. Continuous hydrologic modeling of snow-affected watersheds in the Great Lakes Basin using HEC-HMS. *J. Hydrol. Eng.* **2013**, *18*, 29–39. [CrossRef]
3. Larson, L.W.; Peck, E.L. Accuracy of precipitation measurements for hydrologic modeling. *Water Resour. Res.* **1974**, *10*, 857–863. [CrossRef]
4. World Meteorological Organization (WMO). *Snow Cover Measurements and Areal Assessment of Precipitation and Soil Moisture*; Operational Hydrological Report No. 35; World Meteorological Organization (WMO): Geneva, Switzerland, 1992.
5. Quanta Consulting Engineering Co. *Development and Modernization Plan of Iran Climatological Station*; Iran Meteorological Organization Publication No. 1; Quanta: București, Romania, 1976.
6. Molotch, N.P.; Bales, R.C. Scaling snow observations from the point to the grid element: Implications for observation network design. *Water Resour. Res.* **2005**, *41*. [CrossRef]

7. Brown, R.D.; Walker, A.E.; Goodison, B. Seasonal snow cover monitoring in Canada. In Proceedings of the 57th Eastern Snow Conference, Syracuse, NY, USA, 17–19 May 2000.

8. Bell, W.W.; Parmley, L.J.; Walk, H.; Afzal, H. Inflow forecasting for Pakistan's major reservoirs. *Int. Water Power Dam Constr.* **1994**, *46*, 21–26.

9. World Meteorological Organization (WMO). *Guide to Hydrological Practices No. 168*; World Meteorological Organization (WMO): Geneva, Switzerland, 1994.

10. Davtalab, R.; Madani, K.; Massah, A.; Farajzadeh, M. Evaluating the Effects of Climate Change on Water Reliability in Iran's Karkheh River Basin. In Proceedings of the World Environmental and Water Resources Congress 2014, Portland, OR, USA, 1–5 June 2014; pp. 2127–2135. [CrossRef]

11. Ashraf Vaghefi, S.; Mousavi, S.J.; Abbaspour, K.C.; Srinivasan, R.; Yang, H. Analyses of the impact of climate change on water resources components, drought and wheat yield in semiarid regions: Karkheh River Basin in Iran. *Hydrol. Process* **2014**, *28*, 2018–2032. [CrossRef]

12. Saghafian, B.; Davtalab, R. Mapping snow characteristics based on snow observation probability. *Int. J. Climatol.* **2007**, *27*, 1277–1286. [CrossRef]

13. Jamab Consulting Engineering Co. *Karoun, Dez, Karkheh and Maroun River Basins Water Comprehensive Study—First Study*; Jamab Consulting Engineering Co.: Tehran, Iran, 1991.

14. Porhemmat, J.; Saghafian, B.; Sedghi, H. An algorithm to mapping snow, cloud and land in NOAA AVHRR data, formulation, verification and evaluation. In Proceedings of the Fourth International Iran & Russia Conference, Shahrekord, Iran, 8–10 September 2004.

15. Goovaerts, P. *Geostatistics for Natural Resources Evaluation*; Oxford University Press, Inc.: New York, NY, USA, 1997.

16. Saghafian, B.; Davtalab, R.; Kafayati, M. Comparison of methods for determining the snowfall threshold temperature and potential area affected by snowfall in the Karkheh, Dez, Karun and Marun river basins. *Iran Water Resour. Res.* **2016**, *10*, 31–39.

17. Fleming, M.; Neary, V. Continuous hydrologic modeling study with the hydrologic modeling system. *J. Hydrol. Eng.* **2004**, *9*, 175–183. [CrossRef]

18. Garcia, A.; Sainz, A.; Revilla, J.A.; Alvarez, C.; Juanes, J.A.; Puente, A. Surface water resources assessment in scarcely gauged basins in the north of Spain. *J. Hydrol.* **2008**, *356*, 312–326. [CrossRef]

19. Saghafian, B.; Bondarabadi, S. Validity of Regional Rainfall Spatial Distribution Methods in Mountainous Areas. *J. Hydrol. Eng.* **2008**, *13*, 531–540. [CrossRef]

20. Severino, E.; Alpuim, T. Spatiotemporal models in the estimation of area precipitation. *Environmetrics* **2005**, *16*, 773–802. [CrossRef]

21. Rango, A.; Martinec, J. Snow accumulation derived from modified depletion curves of snow coverage. In Proceedings of the Exeter Symposium on Hydrological Aspects of Alpine and High Mountain Areas, Exeter, UK, 19–30 July 1982; IAHS Publ. No. 138; IAHS Press: Wallingford, UK, 1982; pp. 83–90.

22. Ghanbarpour, M.R.; Saghafian, B.; Saravi, M.M.; Abbaspour, K. Evaluation of spatial and temporal variability of snow cover in a large mountainous basin in Iran. *Nord. Hydrol.* **2007**, *38*, 45–58. [CrossRef]

23. Simpson, J.J.; Stitt, J.R.; Sienko, M. Improved Estimates of the Areal Extent of Snow Cover from AVHRR Data. *J. Hydrol.* **1998**, *204*, 1–23. [CrossRef]

24. Martinec, J. The degree day factor for snowmelt runoff forecasting. In Proceedings of the IUGG General Assembly of Helsinki, IAHS Commission of Surface Waters, Helsinki, Finland, 25 July–6 August 1960; IAHS Publ. No. 51; IAHS Press: Wallingford, UK, 1960; pp. 468–477.

25. Akyurek, Z.; Sorman, A.U. Monitoring snow-covered areas using NOAA-AVHRR data in the eastern part of Turkey. *Hydrol. Sci. J.* **2002**, *47*, 243–252. [CrossRef]

26. World Meteorological Organization (WMO). *The Planning of Meteorological Station Networks*; Technical Note No. 111; World Meteorological Organization (WMO): Geneva, Switzerland, 1970.

27. Moss, M.E. *Concepts and Techniques in Hydrological Network Design*; Operational Hydrology Report No. 19; World Meteorological Organization: Geneva, Switzerland, 1982.

MDPI
St. Alban-Anlage 66
4052 Basel
Switzerland
Tel. +41 61 683 77 34
Fax +41 61 302 89 18
www.mdpi.com

Water Editorial Office
E-mail: water@mdpi.com
www.mdpi.com/journal/water

www.ingramcontent.com/pod-product-compliance
Lightning Source LLC
Chambersburg PA
CBHW051851210326
41597CB00033B/5855